Productive Safety Management

*I would like to dedicate this book to the countless numbers of families who have lost loved ones through workplace accidents. May we understand their pain and so be inspired to work with relentless passion to prevent workplace deaths and injuries in future.*

T.M.

# Productive Safety Management

## A strategic, multi-disciplinary management system for hazardous industries that ties safety and production together

Tania Mol

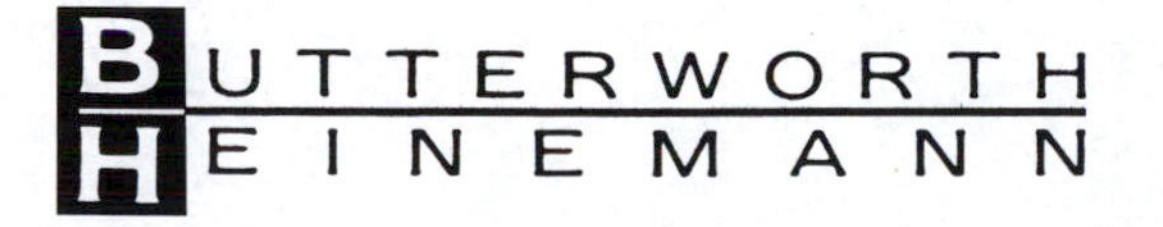


AMSTERDAM BOSTON HEIDELBERG LONDON NEW YORK OXFORD
PARIS SAN DIEGO SAN FRANCISCO SINGAPORE SYDNEY TOKYO

Butterworth-Heinemann
An imprint of Elsevier
Linacre House, Jordan Hill, Oxford OX2 8DP
200 Wheeler Road, Burlington MA 01803

First published 2003



**British Library Cataloguing in Publication Data**
A catalogue record for this book is available from the British Library

**Library of Congress Cataloguing in Publication Data**
A catalogue record for this book is available from the Library of Congress

ISBN 0 7506 5922 X

For information on all Butterworth-Heinemann publications visit our website at www.bh.com

Typeset by Replika Press Pvt Ltd, India
Printed and bound in Great Britain By Biddles Ltd *www.biddles.co.uk*

# Contents

# Foreword

**Productive Safety Management** is an exceptional book in the scope of its treatment of the subject, the new perspectives it provides on the nature of risks, and how it deals effectively with the involvement and interaction of people engaged in the control of risk.

The holistic theme of productive safety management is developed logically and systematically. Accident causation is discussed and the Entropy Model, a new and insightful perspective, is presented. This leads on to organizational decision making and the importance of external and internal strategic alignment for the enterprise. A critical factor identified is alignment of employee goals with those of the enterprise.

The functional operation of the enterprise is examined in four well-structured elements, and risk considerations are discussed in the context of these elements and their integration. The important functions of risk quantification and control strategy formulation are addressed competently. The human resources essentials for productive safety management are thoughtfully presented, through management commitment and leadership, training, competence, and engendering a positive safety culture, to quality assurance and behavioral audits.

The book concludes with the formulation of the productive safety management plan on a strategic basis, which includes developing risk management initiatives using the four-fold strategy presented in the first section. The strategies are appropriately complemented with performance measures.

The introduction of this book is timely in the light of the important and ongoing changes in legislation that regulates safety in both high-risk industries and industry in general. There has been a move away from detailed frameworks of standards to the more open ended 'duty of care' principles, which originally derived from the Robens Committee Report, *Safety and Health at Work*. These were subsequently embodied in the legislation in the UK. The Robens principles have been influential in Europe, Canada, Australia and South Africa.

Corporations and managers accustomed to operating under prescriptive laws, which changed only in detail and increasing regulatory complexity from the 19th to the late 20th centuries, have struggled to reach an adequate understanding of the full implications of the new obligations. The critical point is that the duty of care obligations are not static but become wider and more demanding with changing technology, increases in knowledge of hazards and higher community expectations. Moreover, senior corporate officers and line managers holding positions of statutory responsibility have growing concerns over the increased onus on them to direct, manage and control the corporation to discharge the legislated obligations.

The requirement to manage risks effectively is now an explicit obligation and some jurisdictions specify the framing of safety management systems to achieve these ends. Accordingly senior corporate and line management need to equip themselves with an adequate knowledge and capacity to carry out these duties and maintain productive performance in the enterprise. Productive Safety Management provides a substantial resource to widen the knowledge base, improve risk management and capacity and to more effectively involve and motivate the workforce in these processes.

Productive Safety Management will be of great interest to management and safety professionals in the USA, particularly in hazardous industries such as mining, which operates under highly prescriptive legislation enforced on a strict liability basis. A critique of this law and its enforcement was given by Laura E. Beverage in a paper at the MineSafe International 1996 Conference in Perth, Western Australia. The administration of the US Mine Safety and Health Act is legalistic and adversarial, and does not engender an ethos of continuous improvement, but rather a defensive attitude to avoidance of liability. In the USA, industry preference is to progress to the Robens style performance-based approach to risk management, and this approach is inherent in the principles of this book.

It is widely recognized that enacted law is by no means the only driver for improved safety performance by corporations. Cost imperatives (including civil action liability) are an increasingly important factor, and a range of further drivers includes corporate reputation, industry peer pressure and trade union sanctions. The overwhelming consideration is the rising tide of community expectations that reflect a better educated and informed workforce. In the past, traditional workforces in high-risk enterprises such as mining, tended to have a rigid acceptance of major hazards. As long as the potential for disaster was properly accounted for, a spectrum of serious injuries and occasional fatalities was an undesirable but inevitable outcome of working in these hazardous environments.

This outlook has been replaced by the expectation of a safe and healthy work environment in all phases of operations. Hazards and the associated risks are a challenge, but risks must be eliminated or compressed to an acceptable residual level in accordance with the principles of productive safety management. The importance of effective involvement of employees in the process is a key element, and involvement engenders their alignment with corporate goals and values.

One of the strengths of this book, over and above the values of the approach and principles, consists of the development and presentation of ideas in a format and style that will be comprehensible to everyone in the enterprise. This includes the general manager, line managers, supervisors and safety professionals, as well as general employees. It is eminently readable.

In its comprehensive treatment it offers new insights and makes effective use of models and logic diagrams to clarify these concepts. The reader is not confronted with difficult reasoning and complex algorithms. In addition, each chapter is thoroughly referenced.

Over the past 20 years, 17 of which were as State Mining Engineer heading the mining safety regulatory authority in Western Australia, I have made an extensive study of published safety material from many sources. This includes work undertaken nationally and internationally. I have also attended and presented papers at many conferences on the subject. From that perspective I believe that this book will make a substantial contribution to the body of required reading and study, and will assist individuals and organizations in the critical task of risk management.

It is not possible in a brief Foreword to touch on more than a few highlights in innovative concepts. Some examples include:

- the Entropy Model coupled with strategic alignment principles
- capacity building through effective training and the capacity reservoir of competence
- the application of the 'reasonableness test' by management in the introduction of systems and procedures.

The fundamentals for effective safety management are knowledge, capacity and motivation. **Productive Safety Management** will be an asset to any organization aspiring to continual improvement and excellence – whether mature and established or climbing the learning curve. I commend it without reservation.

J. M. Torlach

# Preface

*Productive Safety Management* described in this book is a multi-disciplinary, total management system that embraces occupational health and safety (OHS), human resource and environmental management, and engineering fields.

The book has wide application and has been written for managers and supervisors working in hazardous industries, OHS practitioners, university lecturers, graduate and post-graduate students, management professionals and engineers.
The aims of the book are to:

- provide a management system that allows production, safety and quality to be pursued concurrently in hazardous industries;
- improve the understanding of the nature of risk and its impact on company systems;
- improve risk management so that injuries are prevented and damage is minimized;
- describe, by way of practical example, the sources of risk in hazardous industries and how these can be managed;
- provide a method for ranking relative levels of risk so that managers are able to confidently allocate resources for risk reduction and control;
- enhance the understanding of organizational decision-making as it relates to the development of safety systems;
- explain that legal compliance and social responsibility are prerequisites for sustainable business;
- give managers the behavioral skills to implement the management system effectively;
- explain how to develop a strong productive safety culture;
- explain how to build workforce competencies and maximize the return on investment in training; and
- present a performance management system that provides structure, accountability and feedback using measures that balance productivity, safety, quality, financial and customer, compliance and social responsibility objectives.

The book is structured into four parts. Part 1 (Chapters 1 and 2) describes the tools that establish the framework for *Productive Safety Management*. These tools are called the entropy model and the strategic alignment channel. The entropy model illustrates the relationship between organizational systems and risk. It shows that the probability of an accident depends on (i) the level of inherent danger or residual risk that can not be completely eliminated by the firm, and (ii) the level of risk caused by the degradation of organizational systems (entropic risk).

The strategic alignment channel explains organizational decision-making on three levels. Firstly, companies are influenced by factors in the external environment, which in particular, require them to be legally compliant and socially responsible. In response, firms have to develop systems that are aligned with the general setting in which they operate. Secondly, they have to make decisions internally about the resources allocated to these systems. Thirdly, to achieve their goals companies need to gain workforce commitment and therefore, decisions have to be made about internal management systems. The channel explains that to build a sustainable business these three levels of decision-making need to be aligned.

In Part 2 (Chapters 3 to 7), the entropy model is applied to hazardous industries. The model shows that organizational systems are made up of processes, technology, the physical environment and human resources. These are collectively called system factors. Each system factor is subject to inherent danger and degradation. For example, heavy machinery poses an inherent threat to the safety of workers due to its size and mobility. In addition, it suffers from wear and tear and this deterioration poses further risks. Chapters 3 to 6 describe the sources of residual and entropic risks in each of these four system factors and how to manage them. Chapter 7 presents a method to determine relative levels of risk to improve the quality of risk management decisions and to make best use of the firm's limited resources.

Part 3 (Chapters 8 to 10) addresses the behavioral and cultural considerations involved in the implementation of *Productive Safety Management*. Managers and supervisors play a critical role in this process so before implementation can begin their commitment has to be gained. A management training overview is presented in Chapter 8 to explain how to develop supervisors' skills so that they become effective decision-makers and demonstrate appropriate leadership behaviors. Chapter 9 discusses the dissemination of the approach to the operational level and focuses on employee skill development. The issues discussed include induction, building workforce competencies, the optimization of employee involvement, and maximizing the return on investment in training using employee suggestion programs. Primarily, the process presented focuses on building the organization's capacity to continuously improve its performance and manage risk. Part 3 concludes with Chapter 10 which explains how to undertake a behavioral audit to determine whether managers have completed the steps necessary to develop a productive safety culture.

Part 4 (Chapters 11 and 12) discusses the development of the productive safety management plan containing strategies for the adaptation of firms' current OHS management systems to fit the entropy model, risk management interventions and cultural development programs. Chapter 11 presents the basic principles of strategic management, including the formulation of strategies, initiatives and action plans accompanied by measures and targets. The performance management system that measures and evaluates the implementation process is provided in Chapter 12. It contains four evaluation tools based on a scorecard approach which balance production, safety, quality and other achievement indicators. The tools measure corporate performance, the health of the organizational culture, business unit performance, and the achievement of goals set by managers and supervisors.

The author's wish is that this book will have provided the reader with a heightened level of risk awareness and a passion to contribute towards the elimination of fatalities and injuries in the workplace. It is hoped that the management of firms, particularly in hazardous industries, will appreciate that production, safety and quality can be pursued concurrently in a climate of co-operation and prosperity. The author also wishes to provide a caution that whilst this book presents a comprehensive approach to risk management, the extent to which this system, or any other system for that matter, is effective in preventing injuries and damage depends on the commitment of all employees, both managers and workers.

Tania Mol

# Acknowledgements

My greatest and most heartfelt thanks go to God for giving me a privileged life in which I have had the opportunity to write this book. God has provided a loving and secure environment in which I have been free to undertake this journey. Along the way I have had the enduring support of my husband Omer who has worked tirelessly for his family. He has been a great listener, an invaluable source of critique, and motivator/whip-cracker when the journey seemed a struggle. I thank God also for our wonderful children, Evrim, Tamer and Kenan. It is a great pleasure to see them developing into young adults ready for their personal destinies to unfold before them. I thank my husband and children from the bottom of my heart for the wonderful experience of sharing our lives together.

My gratitude also goes to Jim Torlach, who is an expert in occupational health and safety in the Australian mining and petroleum industries, for very kindly giving his time unconditionally to review the final manuscript. Above all else, I thank Jim for sharing the vision that organizations can be safe, productive places in which managers and employees work together in a positive atmosphere for a common purpose.

There are family, friends and colleagues that have enriched my life over many years who have been particularly supportive in the two years taken to write this book. I thank them for their encouragement and most of all, their senses of humor, which released me from the isolation of undertaking this project. In particular, it has been my privilege to have the friendship of Ray and Sandy Matthews. It was Ray's accident that initiated my desire to explore more effective ways of managing workplace risk. Sharing their experience has provided a very personal reminder of the impact that such events have on people's daily lives.

Tania Mol

# About the author

Tania Mol is a Director of Align Strategic Management Services Pty Ltd, a management consulting business in Perth, Western Australia. Her concern for employee health and safety stems from a long family association with the Australian mining industry, known to be a sector with a high level of workplace risk. Her father was an underground miner during the 1960s, her husband is a mining engineer, and many friends have been employed by this sector. She has known injured workers personally and this has provided the impetus to resolve the challenges faced by hazardous industries.

From a professional perspective, she also has concern for organizational achievement. In particular, with qualifications and experience in the human resource management and public health fields, she strongly advocates that firms should value the competencies and contributions of their employees. Tania believes that hazardous industries can pursue production and safety concurrently by managing risk effectively, adopting a strategic approach and developing a positive culture that allows workers to take ownership of safety and performance goals. Above all else, Tania shares a vision with those passionate about health and safety, that employees should be able to go to work, make a contribution towards the success of their company, and come home safe and well to family and loved ones.

# Introduction

On 4 January 1994, I received a phone call from a friend whose husband, Ray, had just been injured at work. I was living in the Pilbara region of Western Australia at the time and the majority of the people in the town of 4000 were employed by the mining industry. The news of a fatality or injury spread rapidly in this small, close-knit community. Ray was working in the mine's mobile plant workshop at the time of the accident. He was helping to maintain a bulldozer that had been raised on to a hydraulic platform for servicing. Ray was working at floor level near the rear of the dozer which was equipped with a heavy ripper. During maintenance, the support pins that secured the ripper to the dozer gave way and crashed to the floor towards him. Ray was lucky to survive that day. He managed to avoid being crushed, but when he raised his hands to protect himself the bones in his wrists were shattered by the impact.

This injury, along with two fatalities during 1994, raised the question, 'Why were these accidents happening?' The mining company promoted safety as a high priority, yet serious incidents continued to occur despite the firm's efforts to prevent them. Was the general understanding of how accidents happen and the nature of risk sufficiently understood to ensure effective risk management?

The health and safety of workers is and will continue to be a significant social and economic issue in Australia and other industrialized nations. Innumerable media reports provide reminders that organizational systems fail from time to time. As a result employees are injured and businesses incur damages and costs. Perhaps existing approaches to occupational health and safety management were lacking in some respect and a fresh perspective of risk management could make a difference.

At the time of Ray's accident I was studying management practice. The Bachelor of Business in Human Resource Management (HRM) covered occupational health and safety (OHS) and I used the opportunity to gain a better insight into hazard control. One of the challenges of the course was to analyze the causes of accidents. This involved evaluating the theoretical models used to describe the conditions that precede such events. At face value this appeared to be an academic exercise, however, closer examination revealed that these models have led to the current understanding of how accidents occur and also have provided the rationale for the development of OHS management systems. From the operational perspective, I questioned how useful these models were to supervisors responsible for safety in the workplace. Could they apply the models to explain the risks present in the factory or on the construction site? These theories did not show how risk is affected by the firm's activities and did not provide a strategic direction for managing risk. There may be benefit, therefore, in reconsidering the assumptions made about accidents, how they are caused and how they can be prevented.

One of the strongly held views in OHS circles is that all injuries are preventable. It is also acknowledged that accidents are unplanned events. In addition, there is an inherent risk in all activities from driving a car to swimming in the ocean. If all risk cannot be eliminated then is it possible to prevent all injuries? The ultimate conclusion has to be that workplace injuries are not acceptable and in order to prevent them the nature of risk needs to be understood better.

Why is it that risk cannot be completely eradicated? Human-beings try to manage such dangers by attempting to control the environment and how they interact with it. Running counter to these efforts is the tendency for systems to degrade. This concept was originally floated in the chaos theory[1] and has also since been applied mathematically to occupational accidents.[2] The risk associated with degradation is part of daily life, thus the constant need to organize, tidy up and maintain our possessions and surroundings.

In the workplace, systems also tend to degrade with time because organizations are subject to these natural laws. There are many examples of this, for example, plant and equipment suffer wear and tear which makes them less safe to operate. Sometimes shortcuts are taken during work processes that are more risky than following standardized routines in which the risks have been assessed. Employees become fatigued when they work long hours and this reduces their ability to remain vigilant. As the company's systems deteriorate, the probability of an incident must increase.

Could maintenance strategies be used to prevent the risk caused by degradation and would it eliminate this type of risk? Is degradation the only type of risk that a company has to control? It is accepted that some workplaces have an inherently high level of hazards, for example, the underground coal mining industry. Companies are prevented from eliminating risk by limited resources (time and money). In fact all businesses and the community as a whole are restricted in their risk minimization by these economic constraints. For example, how much would consumers be prepared to pay for a car that was 'guaranteed' to save them from injury in a collision at 100 km per hour? How much time and cost would a company be prepared to dedicate to the development of such a car? Limited resources cause society to accept a level of risk that is noncompressible in the short term. Economic factors and the conditions of natural systems are therefore the cause of inherent or residual risk. Society can, however, reduce this risk through technological and knowledge advances in the longer term, for instance, through the development of safer vehicles and better planning/design techniques. Consequently, maintenance strategies cannot completely eliminate risk, but can significantly counter the deterioration of organizational systems.

Although it is often treated as such, clearly risk is not a singular construct. There are in fact two categories of risk – the risk caused by degradation and the risk inherent in all systems. How could this be modeled in a way that would be useful to managers, supervisors, OHS practitioners and educators? The challenge was to illustrate these risks in terms of accident causation so that all parties having a stake in OHS management could apply it practically in the workplace. The purpose of this model would be to provide a better understanding of risk that would lead to more effective management and a reduction in incident rates. The model developed in this book shows this relationship between organizational systems and the probability of an accident according to the level of inherent risk and degradation. It is called the entropy model and is explained in Chapter 1.

'Entropy' is a scientific term meaning a measure of degradation or disorganization of the universe.[3] For the purpose of the model, this definition is refined to mean 'the degradation of a company's system factors' and is referred to as 'entropic risk'. These system factors are processes, technology, the physical environment and human resources. The danger inherent in all systems is 'residual risk'[4] which cannot be compressed in the short term. The entropy model shows that degradation of system factors leads to higher risk levels. This means, for example, that when shortcuts are taken, when technology is poorly maintained, when the physical environment becomes deteriorated, and when people become inattentive, the likelihood of an accident rises. The model explains the need to upkeep organizational systems to eliminate or control hazards to reduce these risks. It explains why companies that consider safety to be a priority are proactive in maintaining these systems.

The model also shows that entropy leads to poorer safety and performance outcomes. The degradation of system factors represents a reduction in safety and also introduces systemic inefficiencies. For example, poorly maintained equipment does not operate as well as properly maintained equipment. Fatigued employees are not as productive and safe as are alert employees. In expressing the relationship between rising risk and inefficiency, the model is a breakthrough in current perceptions of risk and its effect on business activities. It shows that safety and performance are compatible goals and therefore confronts the mindset found in some firms in hazardous industries, such as mining and construction, that production and safety conflict.

How would the model fit into current OHS practices? Could an OHS system be devised using the model or was it just a concept? Extensive work has gone into developing current OHS systems and practices and this has been a collective effort of many dedicated practitioners over decades. In devising a practical methodology for applying the model, I considered it important to preserve this work, to support it and to build on it. Any management system based on the entropy model would have to embrace current practices.

An analogy illustrates how it fits into these current systems. Some traditional approaches have resulted in OHS management systems that are like a brick wall. Each brick is a single unit that is part of the system, but not integrated with the next brick, for example, emergency procedures and incident investigations. It is a wall without mortar and at worst, results in lists of activities and paperwork. It is not always clear how one activity relates to another and consequently, safety manuals become compilations of papers and forms without a clear structure. There tends to be a checklist of components expected to be included in the system.

Does it mean that if items on the list are ticked off – if the firm carries out the activities and fills out the paperwork – that it is operating safely? That accidents continue to happen suggests that there must be a more effective way of managing risk. Conceptually speaking, the reason for this is the lack of 'mortar' to hold the practices together, which can only be remedied using a systems approach. The entropy model facilitates this. It forms the basis of a multidisciplinary management system embracing current OHS practices, environmental management, systems maintenance and the human resource management (HRM) practices used to develop safe and productive organizational cultures. This approach provides the 'mortar' between the bricks and merges risk management into a fully integrated structured system. To explain how it does this, the analogy of the brick wall is expanded.

If some traditional OHS management systems lack a bonding agent, what are the characteristics of the required mortar? Firstly, it is consistent. This means that the system needs to be applied to all business activities in the firm. It also has to be part of the company's safety culture. The culture is the set of shared customs, beliefs, practices, values and ideologies[5] held by the employees of the business. Secondly, mortar sticks, so therefore, the system needs commitment at all levels of the organizational hierarchy to make it effective. A prerequisite for sustained commitment is that the system must embrace and be driven by the shared goals and values of management and the workforce. These include the sanctity of human life, quality of life, every individual's right to be respected and ethical business behavior. Finally, because the mortar permeates all levels, it should be part of the 'bigger picture' which means looking at how the system drives the business' core activities. Effectively the OHS management system must be strategic and aligned with other systems.

It is in relation to this last point – making safety management strategic – that a further challenge was identified. I wondered how well the decisions that go into developing an OHS management system are understood. For example, why is it that some companies take a minimalist approach? They do as much as required to meet minimum legal compliance, whilst other firms inject extensive resources into safety management and still fall short of achieving zero accidents. This was a question about the underlying fundamentals of organizational decision-making. Specifically, why do companies undertake OHS management at all and what drives the standard of safety from a country, industry or firm perspective? It seems a self-evident question and the obvious answer relates to preventing accidents and meeting legal requirements. This does not, however, help managers to understand their eventual choices in the decision-making process, nor does it fully explain the rationale behind current OHS management practices and the legislative framework.

To improve this understanding the 'strategic alignment channel' was developed. The channel shows that organizations make decisions at three levels. Firstly, the company has to fit the external environment. To do this, business activities need to be legally compliant and socially responsible. In the least, this means that they must fulfill their duty of care to workers and the general public and comply with environmental standards. Secondly, each firm has to manage the internal environment to achieve results and this involves making decisions about resources. These are financial, physical and human capital – simply money, infrastructure/equipment and people. The final level of decision-making is how to use these resources effectively by sourcing the competencies and enlisting the commitment of employees. Consequently, judgements are made daily at the operational level about what work is done, how it is done and how organizational relationships function. It is these relationships and the method of work that affect the culture of the business.

There are a number of reasons for calling this decision-making model the 'strategic alignment channel'. Primarily, it is widely accepted in the OHS management discipline that safety should be a strategic part of the business' operations because it affects the 'bottom line'. The costs of getting it wrong can be enormous in terms of lost productivity, compensation, rehabilitation costs and litigation. The reason for including 'alignment' is that decisions that are synergistic and congruent achieve the best results. Channeling resources and efforts in a common, positive direction increases the return on those resources. Alignment also implies that all business activities from production to marketing should be integrated into the safety system and culture. An example is the case of a purchasing officer who sustained multiple injuries when the vehicle he was driving at 90 km per hour rolled over twice on the road to a mine site.[6] The driver had moved to the side of the road to avoid a pool of water when the vehicle's right side wheel hit a rut in the road causing it to slide and roll over. This incident highlighted the need to address safety issues in all areas of the firm, not just core business activities where hazards are most prevalent. It also indicates that all employees, not just those at the 'coal-face', need to be provided with an understanding of risk and given the necessary competencies to work safely.

A further reason for the emphasis on alignment is that effective decision systems, particularly in relation to risk management, require a two-way flow of information and ideas between the strategic and operational levels of the firm. The flow is therefore top-down and bottom-up. This means encouraging employee participation. In fact, in Australia, legislation has been enacted to support consultation between management and the workforce over safety issues.[7] The rationale for this interventionist approach is to encourage firms to provide forums where employees' goals and values and organizational goals and values can be aligned to achieve a common sense of purpose in relation to risk management. The strategic alignment channel explains how to achieve alignment at all three levels – between the firm and the external environment, within the firm in terms of its resourcing, and within management–workforce relations. The channel is discussed in Chapter 2.

The brick wall analogy forms a picture of the characteristics that make up a comprehensive OHS management system. Current practices, which are the bricks of the wall, need to be integrated into it. The structure requires a bonding material – the mortar – to make it strong. This is provided by a definitive risk management strategy from the entropy model, referred to as the 'four-fold strategy'. It involves, firstly, the control of entropic risk (degradation) through immediate corrective action and secondly, the development of maintenance systems for long-term prevention. Thirdly, residual risk is addressed by managing it in the short term and fourthly, developing strategies for compression in the longer term. Further, a comprehensive OHS management system requires effective decision-making processes to create alignment of OHS from the strategic to the operational level and an organizational culture that reinforces shared goals and values. This can be likened to rendering the brick wall. It makes the system stronger and more stable, because it involves the commitment of management and employees to higher standards of safety and systems quality, whilst pursuing production output.

The entropy model makes it easier to understand risk and how to manage it. The channel explains the reasoning behind the development of OHS management systems and how these are implemented. The concepts are illustrated in a manner that can be understood and applied by managers, supervisors, OHS practitioners and employees and readily communicated to future safety officers, engineers and other professionals by educators. These components are the foundations or educational tools of the total management system called 'productive safety management' presented in this book. It is a multidisciplinary approach that addresses the key management areas: productivity, safety, system factor quality, financial and customer, compliance and social responsibility. These management areas are called 'alignment indicators'. The firm has a strategic and congruent management system when these alignment indicators are balanced and, in particular, productivity, safety and quality are pursued concurrently.

It was mentioned earlier that this new management approach must embrace traditional methods of hazard control. What impact would productive safety management have on these OHS practices? Could these practices be, for instance, restructured to fit the entropy model? According to the model, the organization has four system factors. These are processes (work practices), technology (plant, equipment, tools and chemicals), the physical environment (locational and structural factors) and human resources (people). In the productive safety approach, current OHS practices are restructured according to these four system factors. The result is a greatly simplified framework for the administration of OHS. For example, when an incident is investigated, it is analyzed according to each system factor to find out which of them was degraded at the time of the accident. In addition, the investigator considers the contribution of residual risk to the event. Part 2 of this book – Chapters 3 to 6 – describes the allocation of these traditional practices to the four system factors of the entropy model. The necessary tools, for example, incident investigation forms and job safety analysis forms, have been included to assist practitioners to implement this approach. These chapters also describe in detail the sources of risk within each system factor. For example, design limitations are a source of technological residual risk. The final chapter of Part 2 – Chapter 7 – presents a method to quantify relative risk levels for different activities. It shows the firm how to determine the baseline levels of residual risk and susceptibility to entropic risk using a ranking system. Once these are known, managers can more confidently allocate resources using the four-fold strategy to reduce risks as far as practicable and to manage remaining risks effectively.

Thus far it has been explained that productive safety management involves a new perspective on risk management, a better understanding of OHS decision-making, and the adaptation of current OHS tools to fit the entropy model. These changes are, however, insufficient to create the behavioral modifications needed to reduce the residual and entropic risks within the human resources system factor, that is, the risks that workers introduce into the workplace. It also does not address the issue of management leadership that has a direct impact on the safety culture. For productive safety management to be a total management system it needs to be fully integrated, embracing both systems and behavioral change. The human dimension of safety needs to be considered and this involves building a 'productive safety culture' using effective HRM strategies, with particular attention paid to competency development and communication systems.

The implementation of these organizational changes requires planning. How do the entropy model and channel affect the planning process? The model directs the company to develop strategies that shift system factors – processes, technology, the physical environment and human resources – towards optimal safety, performance and quality by managing risk. The strategic alignment channel helps management select appropriate strategies by explaining the three levels of decision-making that occur in the firm and how these can be aligned. Appropriate interventions are, in the first instance, legally compliant and socially responsible leading to the alignment of the firm with the external environment. Internally, strategies need to effectively balance resources and reinforce the common goals and values held by management and the workforce. During this planning process, specific objectives are 'fleshed out' into initiatives and action plans and are accompanied by measurable targets. The channel also explains the need for feedback cycles to determine whether the company is on course to achieve its goals.

The product of this planning process is the 'productive safety management plan'. It includes systems and behavioral change strategies, the purpose of which are to make OHS management strategic and to deliver a high standard of safety, performance and system factor quality. It also ensures resources, practices and actions are aligned. Cyclic feedback systems are used to enhance results from period to period and this leads to progression along the 'organizational learning curve'. The plan not only provides a structure for strategy implementation, but also allows the firm to learn from the process and from the outcomes so that it is better able to effectively implement change in future planning cycles.

The concept of organizational learning is fundamental to productive safety management from the strategic perspective to the operational level. For this reason, initial implementation focuses on educating management and the workforce about the nature of risk and enhancing their knowledge, skills and abilities (KSAs), so they are better able to mitigate it. This is likened to building a 'capacity reservoir' – an investment in employee competencies and vigilance. The reservoir stores the organization's learning and risk management potential. Conceptually, once this basin is sufficiently full, workers have the capacity to understand and solve safety and productivity issues and manage risk more effectively. Their knowledge 'spills over' and results in suggestions that lead to improvements in safety, performance and quality. The name given to this overflow of potential is 'resourcefulness'.

In keeping with a fully integrated approach, the concept of resourcefulness helps managers develop OHS strategies. It shows them that greater improvements in organizational systems are achieved when resourcefulness is maximized and degradation is minimized. For example, in an intrinsically dangerous environment such as an underground mine, it is extremely important to control degradation otherwise the level of risk begins to escalate alarmingly because residual risks are already high. It is also important to have very competent employees capable of understanding the hazards involved and how these are best managed. In high-risk workplaces, therefore, entropy has to be prevented and employee participation strongly encouraged. In addition to structural interventions, risk management should therefore involve increasing employees' abilities to contribute, encouraging their participation and using their input or suggestions to reduce and manage risk. The 'productive safety formula' is an educational tool to communicate this idea. It shows, as a mathematical construct, that system factor performance and safety equals resourcefulness less entropy. The implication of the formula is that in dangerous operations, the firm needs to develop its employees' levels of competence and resourcefulness through training more so than in low-risk workplaces, to counter the hazards that workers face on a daily basis.

Training is an important organizational function that illustrates the interdependence of disciplines in the productive safety approach, in particular overlapping both OHS and HRM. Alignment is achieved when other functions within these disciplines reinforce each other. Accordingly, in this book, productive safety management extends into the HRM field with the design of general management tools to fit the system and to support the concurrent pursuit of production and safety. This makes it a holistic management approach. Strategies that will be discussed include identifying productive safety as a key result area (KRA) and sharing 'ownership' of it using position descriptions. Each employee's level of responsibility for safety and efficiency is thus defined within these documents, which are used as the basis of recruitment and selection, remuneration determination, skills analysis, dispute resolution and other management practices. The reservoir concept is also extended and integrated into the total training function that involves formal instruction and on-the-job experiences to build management commitment and leadership, develop appropriate supervision and decision-making, and enhance employees' competencies and risk awareness. The development of these HRM systems is covered in Part 3: Behavioral change.

In Part 4, it is explained how the strategies for systems and behavioral change can be included in a productive safety management plan. Within the plan, strategies are accompanied by initiatives and action timetables which detail how the approach is to be implemented. Using the measurement system, progress is monitored and achievement evaluated by identifying targets at the beginning of the planning cycle and comparing results against these targets.

What types of measures are used in the plan? The entropy model illustrates that the primary objective of the firm is to lift system factors towards optimal safety, performance and quality. The implication is that the condition of system factors has to be measured in addition to productivity and safety outcomes. How does this compare to current approaches to safety performance measurement? In many firms, financial incentives or bonuses are linked to organizational achievement, and in some cases, safety incentives are geared to lost time injuries frequency rates (LTIFRs). Does the measurement approach used in productive safety management support these practices?

The entropy model implies that the use of the LTIFR as the primary or sole measure of safety performance will give a very limited view of the firm's success in managing risk. A company may, for example, achieve its LITFR target within a given period with its internal systems considerably degraded. 'Luck' can therefore be a factor in determining the result. Reliance on this figure may cause the firm to overlook its level of degradation and risk exposure. According to the entropy model, a significant level of deterioration indicates that there is a high probability of accidents occurring in the forthcoming period, so therefore, system factor quality should be measured in addition to the LTIFR and other safety measures. The performance management system must therefore allow for multiple measures. In addition, for average-sized firms, the LTIFR numbers tend to be too small to provide meaningful information. In very large organizations or as an industry aggregate the figures can show valid trends. Accordingly, to address this limitation, the system described in this book uses the alignment indicators – a balanced mix of safety, production, quality and other measures – in a scorecard-style structure. The total result achieved by the firm for the period as expressed in this scorecard is geared to the payment of incentives, so employees are paid for balanced outcomes – outcomes that reflect sound performance in terms of output, safety and quality.

The tools used for performance management are contained in the 'achievement cycle', in Chapter 12 of Part 4. The scorecard approach is applied at three levels. At the corporate level, the 'business alignment scorecard' is supported by the 'cultural health scorecard'. The second and third tiers of performance management are the 'business unit alignment scorecard' and 'leader's scorecard', respectively. These tools facilitate goal setting that aligns organizational systems strategically

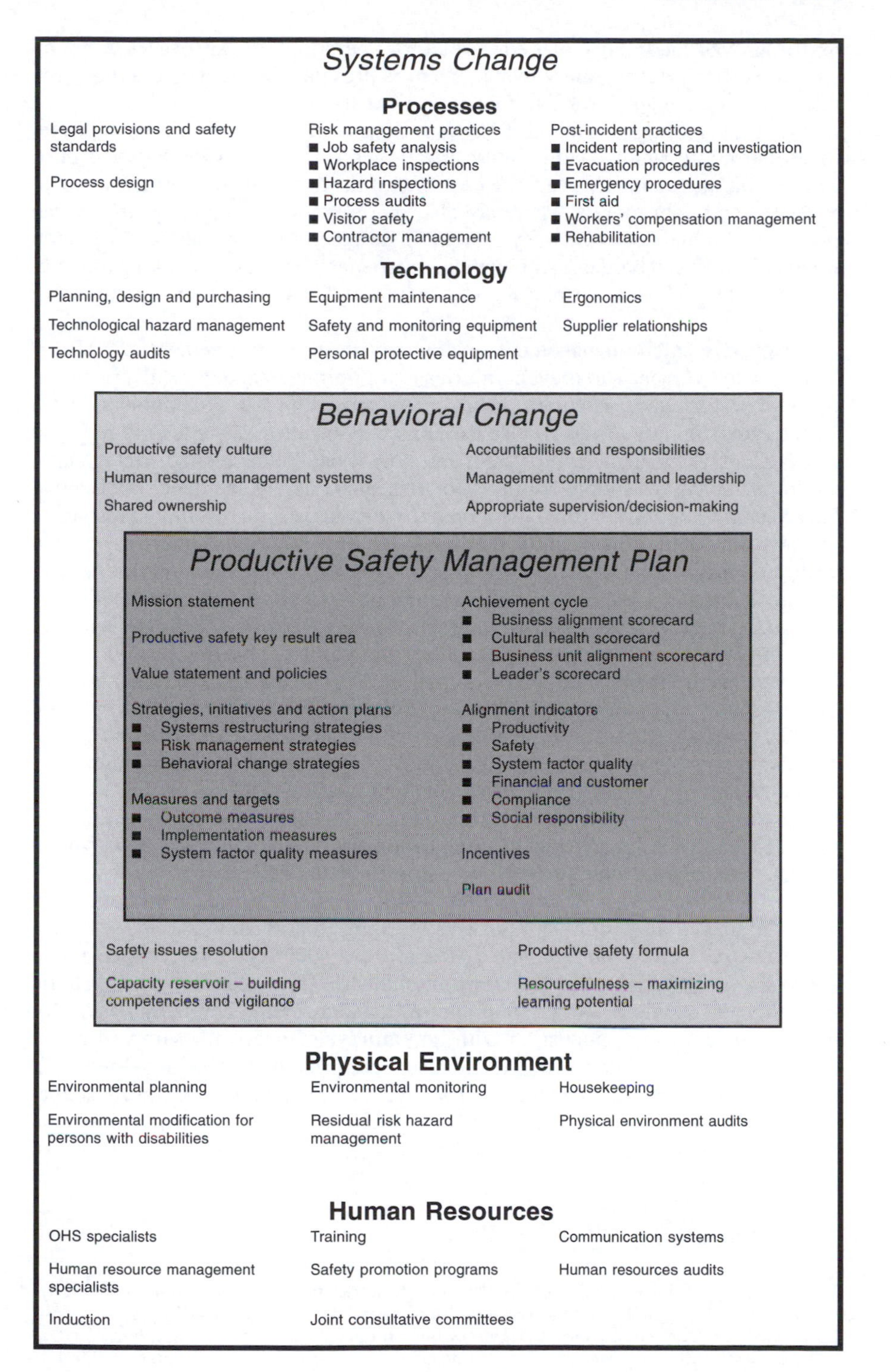


**Figure (i)** Overview of productive safety management

and improves the quality of system factors operationally. An overview of the structure of productive safety management is provided in Fig. (i), and it may be useful for the reader to refer to this from time to time.

When an employee is injured at work the losses can be immense. The employee, his or her family and the community suffer, just as they did when Ray was hurt. In addition, the company bears the cost of lost productivity, rehabilitation, legal expenses and poor morale. There are also secondary costs incurred through the public health and social security system, insurance premiums and other hidden financial and social burdens. Although there are many businesses working ardently to reduce these losses, despite their efforts, accidents and injuries continue to occur.

Productive safety management sets out to lift OHS management to a higher level of achievement – to prevent injuries and minimize damage whilst pursuing positive performance outcomes. It begins by providing key stakeholders with a better understanding of risk. It also explains why businesses are legally required to have an OHS management system and how it can be developed strategically. Comprehensive yet simple methods for risk analysis are included so effective hazard control and risk management plans that make best use of limited resources can be formulated. Concurrently, the strategies for behavioral changes associated with organizational cultures in which employees are both safe and productive, are presented.

Making a difference – delivering positive change in OHS management and reducing loss and suffering – has been the motivation for writing this book. The approach has been designed to be compatible with the extensive, valuable work undertaken by OHS professionals to date so that companies can use it to enhance existing systems no matter how basic or sophisticated these are. The underlying philosophy of this approach embraces the values collectively held by humanity. Fundamentally, this includes the belief that employees have the right to a workplace where they can be productive, contribute to the success of the organization, and return to their families safe and healthy. In addition are beliefs that reflect shared expectations of the value of work which are not only the extrinsic rewards of employment, such as remuneration and benefits, but also intrinsic rewards including recognition, social identity and positive working relationships.

Productive safety management is therefore a socio-economic management approach, designed primarily for hazardous industries, that takes into consideration both their mechanistic and organic nature. It balances the company's need for profit, efficiency and innovation with its employees' needs for safety, rewards and a positive workplace climate. It also balances short-term and long-term objectives by increasing the level of proactive management in risk control so that sustainable businesses are developed in which production, safety and system factor quality are pursued concurrently.

## References

1. The Society for Chaos Theory in Psychology and Life Sciences (2000) Available on website: www.emn.bris.ac.uk/research/nonlinear/faq-[2].html#Heading12
2. Hocking, B. and Thompson, C.J. (1992) Chaos theory of occupational accidents. *Journal of Occupational Health and Safety – Australia and New Zealand*, 8(2), Australia, pp. 99–108.
3. Fowler, F.G. and H.W. (eds) (1986) *The Oxford Handy Dictionary*. Chancellor Press, London.

4. Stellman, J.M. (ed.) (1998) *Encyclopedia of Occupational Health and Safety*, 4th edn. International Labor Office, Geneva.
5. Nankervis, A.R., Compton, R.L. and McCarthy, T.E. (1993) *Strategic Human Resource Management*. Nelson Australia Pty Ltd, South Melbourne, Australia.
6. Department of Mineral and Petroleum Resources Western Australia (2001) *Incident Report No. 290 – Light Vehicle Incident on 12/01/97*. Available on website: http://notesweb.dme.wa.gov.au/exis/fyinew.nsf/e7f8c6f8d521d0fec82563e70032a1c1/bfa379f9d95835874825643600dc6d8?OpenDocument
7. Mathews, J. (1993) *Health and Safety at Work – Australian Trade Union Safety Representatives Handbook*. Pluto Press Australia, New South Wales, Australia.

# PART I
# The productive safety management tools

The theoretical models that explain how accidents happen have been used to underpin current OHS management systems. These models have strongly emphasized the role of human error as a major contributing factor in safety deviations. They contain references to 'unsafe acts', 'mental condition of worker', 'physical condition of worker', 'perceptual skills' and other individual-centered terms. As a result, it has become easy to blame the worker when something goes wrong.

In this part, an innovative risk management model – the entropy model – is presented. It is one of two tools that provide the foundations of the productive safety approach. The model is a paradigm shift away from this human error view of accident causation. It shows that as a business' system factors – processes, technology, the physical environment and human resources – degrade, the probability of an accident rises. It also illustrates the impact of residual risk which is the inherent danger in all systems that can not be completely eliminated. In Chapter 1, some of the current models will be discussed to show how the stress on worker error has emerged. The entropy model will then be described. It takes a system's perspective to hazard control and 'buries' the concept of the 'careless worker', which is necessary for OHS modeling to embrace the fundamental fact that people do not want to be injured at work. Employees seldom expose themselves to high levels of risk willingly or prefer to act recklessly. Instead they tend to behave according to the demands of the organizational system and its culture.

Systemic weaknesses such as inadequate training, production pressures, excessively demanding tasks, high-risk environments, faulty equipment and long work hours contribute to accidents. These are, in large measure, not matters directly controlled by the worker. The entropy model provides a balanced perspective of these contributing variables and explains how risks associated with system factors can be managed effectively. The model provides the four-fold strategy for risk management that when applied reduces the probability of an accident and shifts the business' systems towards optimal safety, performance and quality.

The second tool used in productive safety management is the strategic alignment channel. It explains that there are three levels of alignment that need to be addressed through organizational decision-making. This challenges the current understanding of alignment as a concept that has, to date, been thought of as a single stream between the top and bottom levels of the firm. This means that theoretically strategies formulated by the executive team filter down to the shop floor. In practice, however, this can only be effective if workers unequivocally embrace organizational goals and values. In reality, employees' goals and values often conflict either overtly or covertly with those of the company creating a lack of fit or discontinuities internally.

The singular concept of alignment also does not consider that societal values have changed. The semi-authoritarian management styles of yesteryear are no longer acceptable or appropriate in today's workplace. The failure of some firms to adjust to this shift in values has led to the operational level being a conflict zone. Consequently, the supervisor has the difficult role of balancing the divergent objectives of management and subordinates. This becomes particularly evident when production and safety are adversarial at the 'coal-face'. An example of this operational conflict is when the supervisor pushes output targets because of pressure from the top and workers refuse to carry out the task because it compromises their safety. Such conflict does not need to be overt to create a problem. When disunity is subtle it can still create a climate of distrust and resistance that hinders the firm's ability to maximize its return on its investment in employees.

The channel is a useful tool, not only from an OHS perspective, but also in terms of total management systems, for a number of reasons. Firstly, it explains why firms are required to have a safety strategy and how the organization can align it and other business strategies with the external environment. This is referred to as 'external strategic alignment'. The channel also clarifies how internal decisions are made about the allocation of resources to the OHS management system and how to achieve consistency and congruence from the strategic to the operational level, referred to as 'internal strategic alignment'.

Finally, the channel is used to correct areas of discord through the development of common goals and values jointly held by management and employees. These are used as the basis for decision-making under productive safety management using a tool derived from the channel called the 'reasonableness test' which is explained later in the book. This test addresses the remaining issue of alignment – 'internal goal alignment'. Fundamentally, decisions that pass the test may be implemented, whilst those that do not are discarded. This leads to a consistently applied set of standards underpinned by shared values and beliefs so that all employees direct their effort towards common objectives.

Internal goal alignment is also concerned with maximizing employee participation and potential so that a 'productive safety culture' is developed. The channel, therefore, is a shift away from the singular concept of alignment. In relation to the development of effective OHS management systems it explains that external strategic alignment leads to legal compliance and social responsibility. Internal strategic alignment secures sufficient resources for OHS management. Finally, internal goal alignment ensures that safety systems and practices encourage employee participation, high levels of vigilance and behaviors that shift system factors towards optimal safety, performance and quality. The strategic alignment channel is discussed in detail in Chapter 2.

Chapter 1

# Risk – Can it be eliminated?

What is risk? The dictionary definition is 'chance of bad consequences; or; expose to chance of injury or loss'.[1] Risk has become a term with extensive implications in society and is used in reference to stock market volatility, public health and safety management, and to the potential for failures of organizational systems. It is linked with negative outcomes, such as regret, losses and damage. In OHS management, risk is a consequence of the presence of hazards. A hazard is a set of circumstances that may cause harmful consequences. The probability of it doing so, coupled with the severity of the harm, is the risk associated with it.[2] In this field, it is the likelihood of loss including injury to persons and/or damage to property or the environment. The term can be further described according to time horizons and severity, for example, imminent risk and serious risk. It is understood as a broad concept with no clear classifications; in other words, there have not been different categories of risk to date as such.

In the workplace, and in fact in the wider community, the presence of risk in itself is not considered to be a cause for concern. It is the degree of risk that matters. Similarly, the perception of 'safety' also centers on the level of the threat to health and wellbeing. What 'safe' means to most people is that the dangers associated with a particular activity are 'negligible' and to make something sufficiently safe means to reduce the risk to an 'acceptable' level.[2] The cutoff that determines admissible risk varies from person to person, so it is dependent on perceptions, tolerance of risk and on the circumstances.

Firms manage safety by developing strategies to control hazards and the associated risks. A simple and effective method is to 'spot the hazard, assess the risk, and make the changes'.[3] Tools have also been developed to quantify or prioritize risks such as the Risk Scorecard and to select appropriate interventions, for example, the Hierarchy of Controls. (These tools will be discussed in Chapter 3.) Once these changes have been made, however, has the risk been eliminated? Interventions can only reduce the threat to an 'acceptable' level, as explained later in this chapter. That accidents continue to occur despite the best efforts of companies to prevent them suggests that 'acceptable' risk is a gray zone. Legislation has been used to add clarity to this issue. The Workplace Health and Safety Act 1995 (Queensland), for example, defines the conditions under which workplace health and safety is ensured and therefore, the upper boundary of acceptable risk by stating that:

> 22. (1) Workplace health and safety is ensured when persons are free from –
> (a) death, injury or illness caused by any workplace, workplace activities or specified high risk plant; and
> (b) risk of death, injury or illness created by any workplace, workplace activities or specified high risk plant.[4]

The legislation relates risk to the potential consequences it has for human health and identifies the sources of risk as the work environment, processes undertaken and technologies that are known to be hazardous. Thus far, society has not sought to determine whether there are different categories of risk within this broad understanding or definition of risk that may determine what is the acceptable level.

## Accident causation models

Why is it that risk has not been differentiated further? The answer lies in the theories and models used to explain how accidents happen. These show how threats are translated into an injury or loss. An example is the domino theory developed in 1931. It suggests that one event leads to another, then to another and so on, culminating in an accident. The domino theory found that 88 per cent of accidents are caused by unsafe acts of people, 10 per cent by unsafe actions and 2 per cent by 'acts of God'.[5] Interestingly, the 'acts of God' concept alludes that there may be a level of risk that is not controllable and therefore, noncompressible.

The domino theory was a very simple model based on a singular concept of risk. Subsequent theories have been more sophisticated and comprehensive. The structure of accidents model, shown in Fig. 1.1, is an example. This identifies immediate causes and contributing causes of accidents. Immediate causes involve unsafe acts and unsafe conditions. Contributing causes include safety management performance and the mental and physical condition of the worker. The model acknowledges the importance of systems management, for example, provision of safety devices and interventions to correct hazards. This is, however, overshadowed by a strong emphasis on the operator as the primary instigator of accidents with the employee choosing to act unsafely, for example, by not wearing personal protective equipment or being in an unfit state physically and/or mentally.

Some other models, including the biased liability theory and accident proneness theory, stress worker ineptitude to an even greater extent. The former suggests that once a worker has been involved in an accident, the chances of the same worker becoming involved in further safety violations either increases or decreases compared to other workers.[5] The biased liability theory suggests that there is always a subset of workers who are more likely to be involved in accidents. These models take responsibility

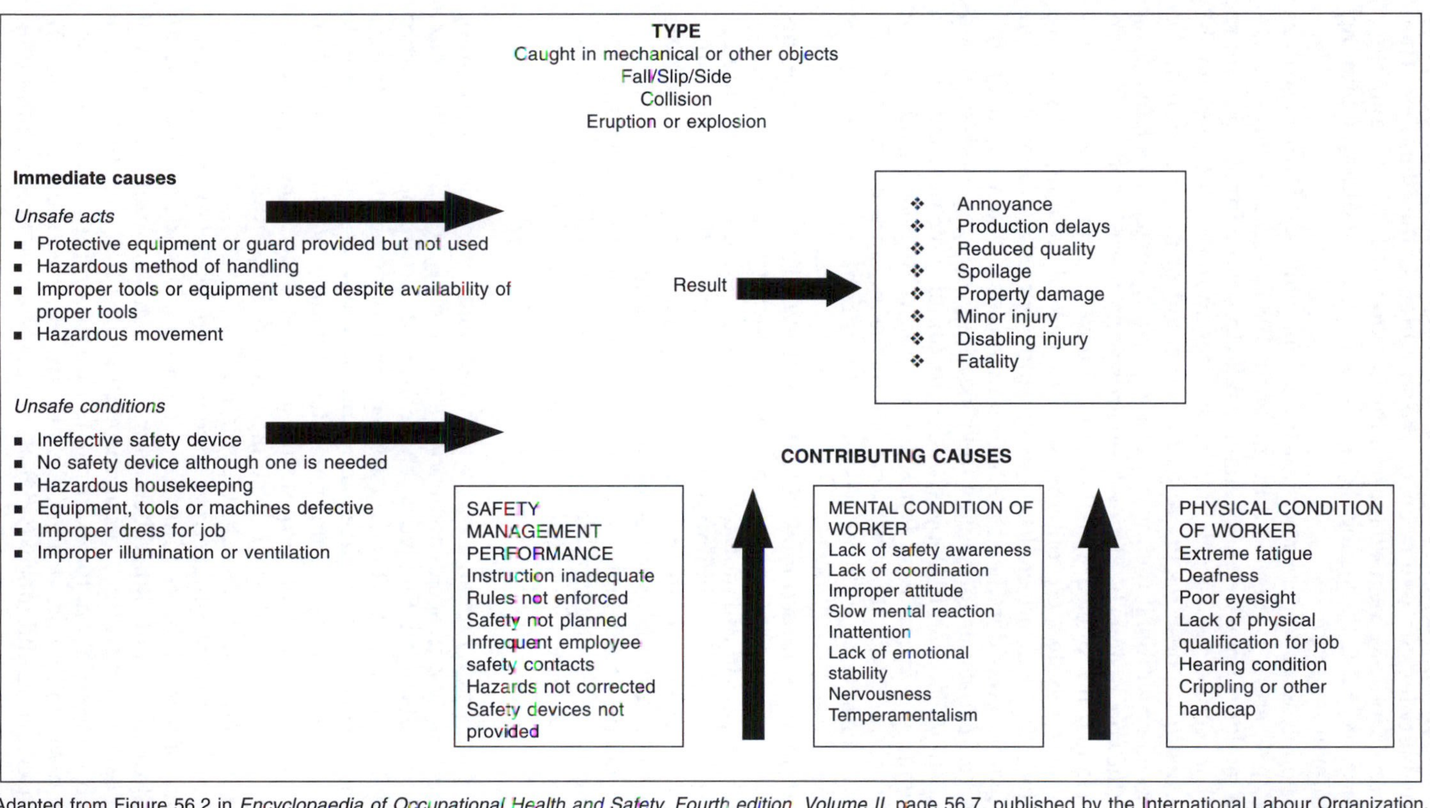


Adapted from Figure 56.2 in *Encyclopaedia of Occupational Health and Safety, Fourth edition, Volume II*, page 56.7, published by the International Labour Organization.


**Figure 1.1** Structure of accidents model

away from management to provide a safe work environment and place accountability for safety squarely on the shoulders of the employee. They do, however, indicate the need for training to reduce risk tolerance amongst the workforce. From these models, risk is taken as a 'condition' stemming from human behavior.

The notion of worker error became the primary focus of further accident modeling. A study undertaken in the early 1980s in Australia of work-related fatalities revealed that behavioral factors were involved in more than 90 per cent of fatal accidents.[5] This raises an important question. To what extent did these results reflect the method of investigation at the time? In other words, if the incident investigators focused on behaviors, would the conclusions be skewed towards worker error and away from system faults? It is unclear whether these 'human errors' were symptoms of underlying systemic problems, such as lack of training or working conditions that put affected operators under excessive demands. Did investigators consider the impact organizational systems had on the employees' capacities to manage the risks they were exposed to? Based on such models, a shift in perspective to underlying problems rather than symptoms is not easy when the emphasis of investigation is on 'unsafe acts' and 'unsafe conditions'. Logic has it that proportionally, 50 per cent of the assessor's attention is drawn to the behavior of the employees involved.

In addition, as argued more recently by Trevor Kletz, many accidents occur, not because they cannot be prevented, but because the organization did not learn, or did not retain the lessons, from past accidents. Organizations need to improve corporate memory to avoid repeating accidents.[6] Accordingly, there is a much stronger argument for considering the human dimension of accident causation at a macro or organizational level, rather than the individual worker level, in order to prevent incidents from recurring. Although it may be possible to retrace the steps of cause and effect back to the individual employee, either a manager or an operator, in the case of most incidents, it is not particularly helpful as a means of prevention.

This concept of 'human error' was broadened by the development of the human factors in accidents model, shown in Fig. 1.2. The interaction of individuals with the work environment, equipment and other contributing factors leads to adverse effects on work systems. These trigger a sequence of events ending in an accident. The model assumes for example that worker error leads to equipment design faults and limitations. Poor maintenance practices or further mistakes may exacerbate these faults, such that the combination of these factors with inappropriate operating procedure may result in a safety deviation.

The model indicates that errors have to be prevented to eradicate these negative outcomes. It identifies two main types of error. Firstly, errors occur through lack of skilled behavior, for example, unsafe operation of a chainsaw. Secondly, they occur through skilled behavior that leads to lapses of concentration, for example, failure to carefully evaluate where a branch will fall before using the chainsaw. The two types of errors are skill-based errors and mistakes.[5] Both are centered on human fallibilities in terms of competence or vigilance.

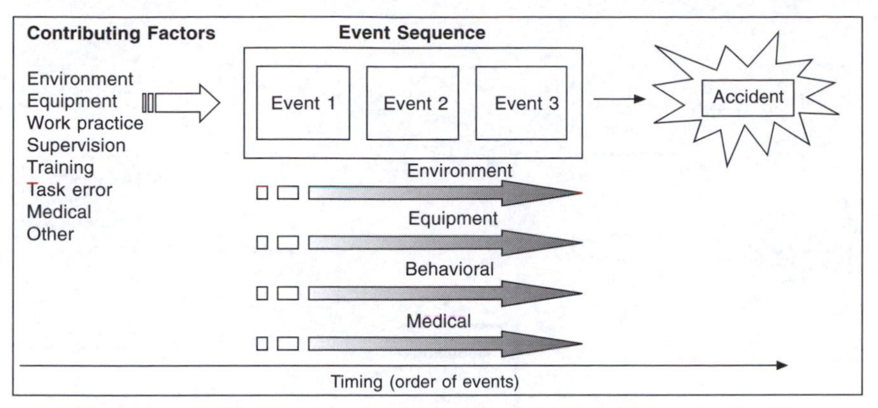


Adapted from Figure 56.3 in *Encyclopaedia of Occupational Health and Safety, Fourth edition, Volume II*, page 56.8, published by the International Labour Organization.
Copyright © 1998 International Labour Organization.

**Figure 1.2** Human factors in accidents model

The benefit of the model is that it indicates the need for training to enhance skills and safety consciousness because of the two categories of risk – the risk associated with poor skills and the risk associated with complacency. Again the focus is on the individual not on the system. The model attributes all system faults to human error and does not consider broader contributing factors, such as the natural tendency for equipment and the environment to degrade. It implies that factors become unsafe only through inappropriate behaviors. In addition, it does not account for the economic constraints that limit the inclusion of additional safety features at the technology or project design stage.

Despite these weaknesses the model does have some notable benefits. It helps managers appreciate the importance of competence in all areas of business operations. Sound skills mitigate the likelihood of these errors. The complete elimination of human error as a means of preventing accidents is, however, fanciful. From this point of view, the model does not fit with the practicalities of running an organization. For example, managers have to control risk based on limited and observable information and as a result, they may implement risk minimization strategies, which given this constrained knowledge, may seem entirely appropriate. The unforeseeable, however, sometimes leads to an accident. Under these circumstances, can the managers be accused of making an error or acting inappropriately? For this reason, OHS legislation requires employers to manage risks as far as 'practicable',[7] which takes into account the information and resource restrictions faced by the firm.

In the 1980s, a further group of accident causation theories was developed. These focused on individual perceptions of risk and motivational factors, an example of which is the risk homeostatic accident model, shown in Fig. 1.3. It applies particularly to road traffic. This model introduces the notion of a target level of risk – that people have a degree of risk that they accept, tolerate or choose.[5] The variations in the extent of caution people apply in their behaviors affects their health and safety.

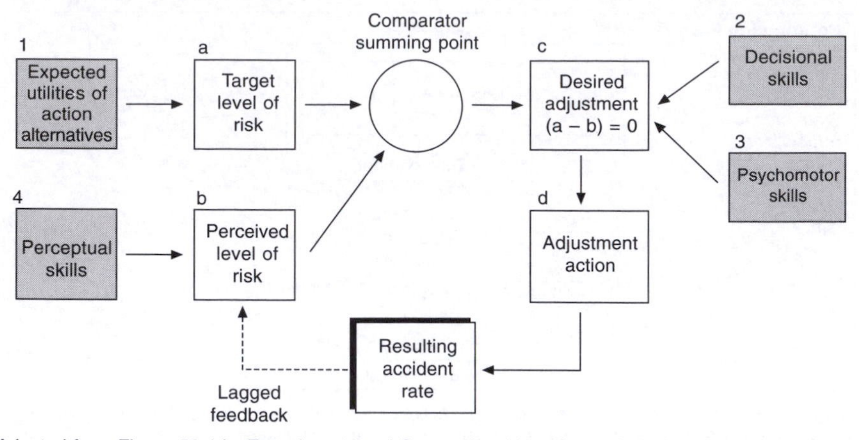


Adapted from Figure 56.4 in *Encyclopaedia of Occupational Health and Safety, Fourth edition, Volume II*, page 56.12, published by the International Labour Organization.


**Figure 1.3** Risk homeostatic model

Using this model, strategies are developed to reduce the level of risk people are willing to take.

The model indicates that time-averaged accident risk is independent of environmental factors and operator skills.[4] What this means is that regardless of other factors, such as driver experience and road conditions, the driver compares the current risk level against her acceptable level of risk. Decisions are made about the costs and benefits of taking the risk before any action is taken. For example, people tend to take additional risks when driving to an appointment if they are likely to be late.

According to this model, risk is a matter of individual perception. It highlights the need to look at the individual's motivation for taking the risk and this makes sense in the road traffic context. At work, however, this focus on the person is very limiting. Do employees choose to take additional risks or do factors in the workplace encourage risk-taking? In the organization, the worker has less control over contextual variables, such as deadlines and equipment condition, than the driver does in the traffic. The driver can choose to go 60 km per hour even if the rest of the traffic is going at 80 km per hour. The worker, on the other hand, cannot make decisions about production output and reduce the work rate in a team environment. Such matters are generally considered to be management prerogative.

Research based on this model showed that when something is perceived to be safer, people tend to compensate for this by behaving in a more risky manner.[5] This explains why the road toll in country areas of Western Australia is higher than in urban areas. The number of deaths and the per capita hospitalization rates are greater for residents in rural areas than in the city.[8] People have a tendency to increase their speed on long straight stretches of road where there is little traffic and attention is also reduced. This combination has led to horrific accident statistics. The model indicates the need to influence negative and positive behaviors using punishment or encouragement strategies, respectively. Accordingly, in Australia, road accident prevention strategies have included penalties for speeding and the provision of free coffee for drivers at various rest stops on major country roads. In an organizational context, the model suggests that strategies can be applied to reward safe acts and reprimand unsafe acts.

The risk homeostatic accident model indicates that individuals take risks but not how they cope with imminent danger. Various other theories illustrate the problem-solving process that the worker goes through in the face of these immediate threats. The Hale and Glendon model, shown in Fig. 1.4, explains how individuals control risk. It suggests that the worker acts following a comparison of the current situation against the desired situation.[5] The problem is firstly recognized, analyzed, prioritized and alternative solutions developed. These alternatives are compared to find a preferred solution to implement. The employee also evaluates the effectiveness of the final choice with this including planning for unforeseen circumstances and residual risks. The inclusion of 'residual risk' acknowledges that some degree of danger is present at all times in all situations.

This model reinforces the three levels of processing involved in making choices about risks. These are:

- automatic, largely unconscious responses to routine situations (skill-based behaviors)
- matching learned rules to a correct diagnosis of the prevailing situation (rule-based behaviors)
- conscious and time-consuming problem solving in novel situations (knowledge-based behaviors)[5]

These three analysis levels indicate that to equip the worker with the necessary competencies to process information in risky circumstances and to act appropriately, the firm needs to enhance employees' knowledge, skills and abilities (KSAs) and safety consciousness. It identifies a key strategic area of safety management which is training.

Further accident deviation theories followed the Hale and Glendon model. These have become more notable with the development of standards. Some suggest that the worker reaches a point of lack of control and this leads to loss of control and an accident. These events are, therefore, caused by deviations from set standards. For example, an employee may be working on an elevated platform that does not have guardrails. In this case, the employee has lack of control. The standard would be to have adequate protective barriers in place. When the worker loses his balance this leads to loss of control which results in an accident. Danger is considered to be ever present and is kept under control by numerous accident prevention measures, including the design of equipment, employing operators who are skilled, clear procedures and structured organization.[5] The concept is useful in that it promotes the monitoring of deviations from standard practice and conditions. According to this model, risk relates directly to the hazards associated with lack of control, but it is not clear how organizational systems as a whole contribute to this.

Not surprisingly, with the development of technology, accident investigation has become increasingly sophisticated. More recently, the

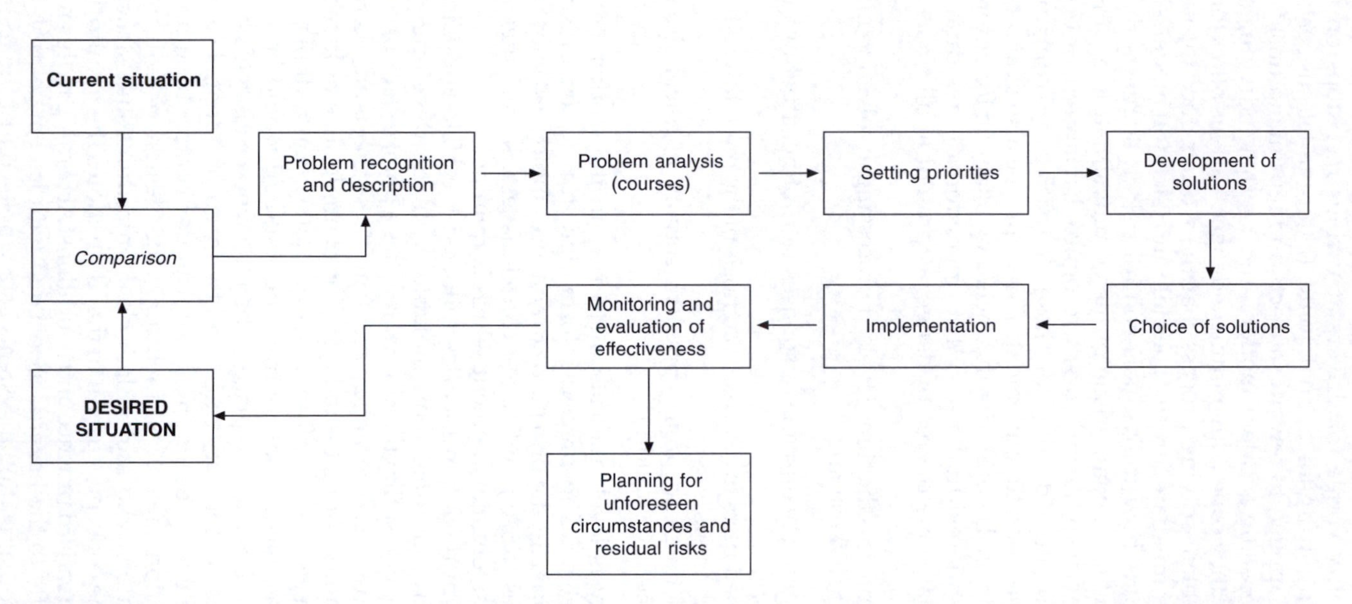


Adapted from figure 56.5 in *Encyclopedia of Occupational Health and Safety*, fourth edition, volume II, page 56.15, published by International Labour Organization. Copyright © 1998 International Labour Organization.

**Figure 1.4** Hale and Glendon model

Merseyside accident information model amongst others has been developed. It uses intelligent software to collect and analyze information on injury-related accidents,[5] then identifies appropriate corrective actions to prevent future recurrences.

The work undertaken by Reason in the 1990s has made a significant contribution to the understanding of organizational risks. His theory presents a strong argument that residual risks are not reducible by purely technological counter-measures.[9] The resident pathogens and risk management model identifies the various ways in which human beings contribute to accidents in high-risk technologies. These are through slips – where actions do not go as planned – and through mistakes which are deficiencies or failures of judgement. Slips and mistakes are both errors. There are also violations that are deviations from system rules and practices.

The model argues that the accumulation of human errors leads to active and latent failures. The former are felt almost immediately, whereas the latter only become apparent when they combine with other additive factors to breach the system's defenses. Reason has used the concept of technological systems having resident pathogens to explain major disasters such as Bhopal and Chernobyl. His model highlights the need to appreciate the presence of residual risk and develop strategies to contain it. It challenges managers to over-ride their assumptions that systems are 'safe' and to look for underlying variables, which combine inherent risks to the extent that causes systems to fail.

All of the models discussed have enhanced the understanding of accidents and how they occur. There has, to date, emerged a strong emphasis on the role of human error which has led to a sound case for training to develop employee competencies and safety consciousness. Many firms are investing in safety training to address these issues. A fundamental dilemma remains, however. This is the variability of perceptions of what is risky, what is safe and to what extent risks need to be reduced to make them 'acceptable'. In addition, the understanding of risk as a broad concept has diluted the significance of unforeseen and residual risks. In effect, this can create a delusion of safety. People tend to believe that once something has been done about a hazard, it is now 'safe' or 'safe enough'.[2]

A major weakness of the models is that they do not clearly illustrate how risk is affected by business activity. They do not provide managers and supervisors with a comprehensive strategic direction for risk reduction in the workplace. In addition, one of the perceptions that has resulted from these models is that all accidents are preventable, if human error is eliminated. Some current OHS literature perpetuates this emphasis on negligence, for example:

> Workplace accidents are caused by people. More accurately, they are caused by the things they do or do not do. Equipment and machinery will sometimes fail, and incidents may occur which cause accidents, but they are nearly always traceable to some degree of human error, negligence or ignorance.[10]

The focus on negligence is a very limited view of accident causation and is unconstructive when developing an OHS management system. It fails to acknowledge that the firm's efforts to manage business activities safely are countered by the natural degradation of systems resulting in opposing pressures that increase the level of risk. In addition, there are inherent dangers in all activities that cannot be totally controlled by the firm. The models identified these as 'acts of God' and 'residual risks'.

If the firm cannot eliminate all risk how can it prevent all accidents? The answer is that not all accidents are preventable. This is because of risk that is beyond human intervention. The implication is that strategies need to be devised to manage it and that workers need to be vigilant because of it. The majority of accidents and near misses can be avoided provided that the OHS management system addresses both natural degradation and these inherent threats. This is effectively a shift away from the broad singular concept of risk. The first step in developing such a system is to construct a model that managers can apply strategically and at the 'coal-face'. It must illustrate the relationship between the probability of an accident and organizational activities in the presence of these hazards.

## The entropy model

There are two categories of risks that are present in all natural systems including organizations. There is an inherent or residual risk that cannot be completely eliminated and the risk caused when systems degrade. The latter is referred to as 'entropic risk'. Entropy is a measure of degradation or disorganization of the universe.[1] For the purpose of the model and to apply the term specifically to the firm, it is defined as 'the degradation of a company's system factors'. These system factors are processes (work practices), technology (plant, equipment, tools and chemicals), the physical environment (locational and structural factors) and human resources (people). They cover all aspects of business activity regardless of the industry or the context. Every firm uses technology, human resources and the physical environment to generate a product or service. Processes determine the interaction of these factors.

The entropy model identifies these two types of risk and explains how they are affected by the condition of system factors. It is shown in Fig. 1.5 and begins by creating an organization within a void, independent of natural law in section (1).[11] This is an ideal context in which the firm always operates with perfect safety, performance and system factor quality. All factors are fully and effectively utilized with the accident rate and level of risk equal to zero.

In reality, however, the firm operates as a natural system. It is subject to universal laws that cause system factors to degrade with time, for example, technology deteriorates through wear and tear. The physical environment varies in humidity and temperature and this introduces suboptimal conditions for business activities. Infrastructure also deteriorates, for instance, corrosion of metal structures and erosion of road edges. Each time a process is carried out there may be deviations in practices that potentially introduce additional risks. Finally, human

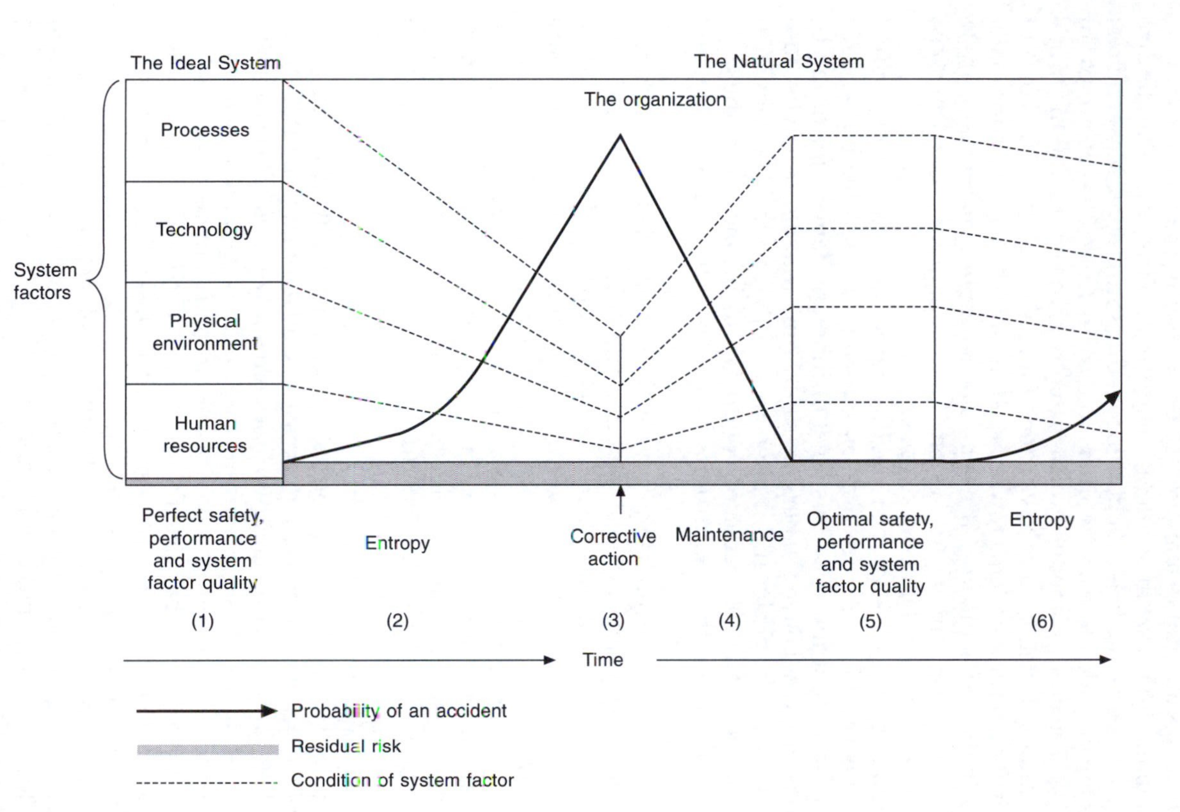


**Figure 1.5** The entropy model

resources experience degradation including fatigue, lack of vigilance or loss of physical capacity to cope with the demands of the task.

This entropy is shown in section (2) of the model as the rising line. In practice, it explains why in the workplace, and in other areas of life, human beings attempt to manage risk by controlling technology and the environment and how they interact with them. Firms also standardize work practices as much as practicable to reduce variations. Meal and rest breaks are provided to allow workers to recover from fatigue and loss of alertness. The entropy model indicates that running counter to the firm's interventions, system factors have a tendency to deteriorate. This concept – the drift of natural systems towards a state of disorder – was originally floated in the chaos theory.[12] It has also been applied mathematically to occupational accidents.[13] As system factors degrade with time, the probability of an accident rises exponentially.

In addition to the risk associated with degradation is a level of residual risk that is fixed in the short term, as shown by the shaded block. Each system factor contributes to this level of inherent danger, for example, technologies do not have 100 per cent safety built into the design because of resource, economic and know-how constraints. The physical environment is not 100 per cent conducive to business activities, for instance, poor weather conditions hinder building on construction sites.

In section (2) of the model that shows the entropic risk, system factors do not necessarily deteriorate at the same rate. It is possible to develop a number of versions of the entropy model to illustrate the variability of degradation. In Fig. 1.5, the rate for each system factor is similar and shown by the dashed lines, which correspond to the system factor. In practice, company's need to be aware of their susceptibility to variable degradation. An example can be given by comparing a new business venture, company A, with an established operation, company B. Company A has state-of-the-art infrastructure and technology, which have a reasonable lead-time before it begins to show signs of degeneration. Assuming that the company has implemented processes that were standardized at another operation's site, these are less likely to introduce entropic risks than untested processes. Consequently in this firm, technology, infrastructure and processes have low levels of residual risk and low tendencies towards entropic risk. If, however, the workforce is inexperienced in this new mechanized environment there is potential for suboptimal interaction between these employees and other system factors. This may result in deviations from standardized tasks, inefficient operation of technology and poor housekeeping.

When company A is compared with a well-established operation, company B, a contrast is provided. Firm B has older infrastructure and technology showing signs of wear and tear. The level of residual risk and the likelihood of entropic risk in these system factors are relatively higher than in the company A. Processes have continually been reviewed to minimize variations in practices and therefore, process-related entropy is low. The workforce is very experienced and stable, thereby maintaining this system factor at a high level of safety, performance and quality. The residual and entropic risks associated with the human resources system factor are low because of the high level of competence and vigilance of the workforce.

These examples indicate that each firm has a different set of circumstances that affect the condition of its system factors, which in turn, determines the levels of residual and entropic risks to which they are exposed. These businesses must therefore develop risk management strategies according to the nature and severity of these systemic hazards.

Referring back to Fig. 1.5, at point (3), the company recognizes that risk is rising. This may become apparent from a number of triggers, for example, a workplace inspection, a near miss or through hazard identification. The firm decides to take corrective action to address areas of degradation. Corrective interventions may occur at any level of decision-making in the company. An example of a strategic response is when an incident resulting from a vehicle brake failure leads to an immediate check of the brake systems of the entire fleet. At the operational level, an employee may notice that the oil warning light has come on in his vehicle and remedy the situation by adding oil to the engine and checking it more frequently.

The model shows that in the event of degradation of system factors, the first step is to take corrective action. Next it indicates that the firm has to prevent future recurrences of this risk. A program of maintenance begins, in section (4), to address these areas of entropy. Maintenance involves preventing deterioration in each system factor. Examples include job safety analysis (JSA) to reduce the hazards in processes, proactive maintenance of plant and equipment, introduction of sound housekeeping practices to maintain the physical environment, and training to reinforce safe work procedures and to upgrade employee competencies.

Effective maintenance lifts the organization to a state of optimal safety, performance and system factor quality, as shown in section (5). Here the firm has eliminated entropic risk. It cannot, however, eradicate all risk as the residual risk remains. As explained earlier, all system factors contribute to this residual risk to varying degrees, for example, dangerous work environments, such as underground mines, have a high level of inherent danger. This risk is attributable to characteristics such as rock instability, flooding and, in the case of coal mining, combustible materials and gases. Technology has inherent risk due to design limitations and characteristics including moving parts, for example, a circular saw has an unavoidable danger because it has exposed, sharp, moving components. Processes also have varying degrees of residual risk, depending on parameters such as complexity and the physical/mental demands placed on the worker. The more demanding the task, for example, climbing a firefighter's ladder versus climbing a stepladder, the more dangerous the process. Human resources introduce residual risks through their limited physical and cognitive abilities. Individuals do not have all the KSAs to deal with every possible situation that may arise in the workplace. (The characteristics of the residual risks inherent in each of the system factors will be explored further in Part 2.)

The model shows that optimal safety, performance and quality are stable for a time after maintenance. If, however, maintenance is not on-going, the effects of natural law cause the system to degrade again, in section (6). The company is potentially exposed to a cycle of improvement, maintenance, stability and then entropy again unless it proactively manages system factors on a continuous basis.

To prevent entropic risk the firm has to take corrective action and carry out on-going maintenance. Degradation can therefore be countered by a combination of reactive and proactive strategies. The firm cannot, however, nullify residual risk, thus it has to manage it. Residual risk can only be compressed in the longer term. This limitation is caused by the resource constraints that the company faces. Scarce resources, such as time and money, force the business to accept a level of inherent risk that is non-reducible in the short term. For example, it is cheaper to put a protective guard around a moving part of a machine than to replace the machine. This choice means that the firm has to accept a higher level of residual risk in the short term until it is prepared to expend financial resources to significantly modify or to replace the machine. In the meantime, this remaining risk has to be managed to prevent injuries or damage. The entropy model provides a clear direction for risk management. It proposes a four-fold strategy to address entropic and residual risks that involves:

(1) Taking immediate corrective action to eliminate entropic risk;
(2) Establishing maintenance strategies to prevent future entropic risk;
(3) Managing residual risk in the short term; and
(4) Compressing residual risk in the longer term.

In Chapters 3 to 6, the sources of entropic and residual risks in each of the four system factors will be identified. Chapter 7 will explain how the level of risk can be quantified in relative terms to allow effective risk management strategies to be developed and prioritized.

The entropy model illustrates the relationship between entropic risk and system factors. This relationship can be further described using a mathematical equation derived from the model. Each system factor is given a numerical value that represents its condition. The best result is 0 and the worst result is 10, corresponding with nil risk and extreme risk, respectively. The probability of an accident due to entropic risk is expressed as:

Entropic risk $= p \times t \times p_e \times h$

where: $p$ = processes risk score; $t$ = technology risk score; $p_e$ = physical environment risk score; and $h$ = human resources risk score. This equation explains the importance of minimizing risk in all four system factors. For example, if three of the four factors have low scores, but one factor is high, the probability of an accident remains fairly high because of the multiplication effect. Maintenance strategies are required for each of the four system factors. The best score of 0 is only possible in the ideal system where there is nil exposure to natural forces of degradation. The earlier comparison of a newly developed enterprise and an established firm can be used to illustrate how the equation assists managers to understand the company's vulnerability to these risk factors.

In the new enterprise, company A, entropy was low in the technology, physical environment and process system factors (though these had not been tested in this operational environment) and high for human resources. A representative calculation would be:

$$\begin{aligned}\text{Entropic risk} &= p \times t \times p_e \times h \\ &= 5 \times 2 \times 2 \times 8 \\ &= 160.\end{aligned}$$

For the established operation, company B, with deteriorating technology and physical environment, with standardized processes and experienced employees, the calculation would be:

$$\begin{aligned}\text{Entropic risk} &= p \times t \times p_e \times h \\ &= 3 \times 6 \times 5 \times 2 \\ &= 180.\end{aligned}$$

The formula shows that the two firms have similar levels of exposure to entropic risk, though company B's is somewhat higher because of the age of the operation. This established company is heavily reliant on its workforce to manage its vulnerability whereas, the new firm relies on its state-of-the-art technology and environment. A risk management priority in company A, is training and development of employees, whereas in company B, it is environmental and technological maintenance. The formula and the entropy model illustrate the importance of preventing degradation to minimize the risk that increases the probability of incidents.

What are the benefits of the entropy model? Firstly, it represents risk in relation to organizational systems and secondly, it can also be applied to non-work situations as the four system factors cover all areas of human activity. This includes, for example, public safety and risk management. In relation to road safety, the model explains the need for effective traffic management processes, such as rules relating to speed, use of emergency lanes and other mechanisms that control the process and make it safer. It explains the importance of road design and maintenance, vehicle design and maintenance, and driver training. Strategies to counter driver fatigue and factors, such as alcohol, which reduce driver alertness and competence, are also readily justified as means of preventing human resources entropy.

The model allows the firm to take a strategic perspective on risk management by providing the four-fold strategy. It explains that different tactics are required for entropic and residual risks. In addition, the model does not simply suggest that the presence of one or more variables will lead to injury or damage or conversely that no incidents mean that the system is safe. In effect, an organization may achieve satisfactory safety results, specifically, a low lost time injuries frequency rate (LTIFR), in a given period, with badly degraded system factors. The LTIFR and other outcome-driven safety measures are not sufficient indicators of the extent to which risk is being contained. The primary determinant of this is the condition of system factors.

An additional insight provided by the model is that it explains how the dangers involved in high-risk activities can be managed by monitoring the condition of these system factors. For example, skydivers use a number of criteria to prepare for a dive. They ensure that the weather is suitable which is an assessment of the condition of the physical environment. They practice the activity to reinforce a standardized process that minimizes entropic risk and manages residual risks. The equipment is checked a number of times before the jump. Finally, they maintain a high level of vigilance during the process. Even though the activity has a very high residual risk and potential for entropic risk, strategies can be implemented to mitigate it effectively. Likewise, the model indicates that firms need to manage high residual risks by reinforcing the need for constant vigilance and implementing criteria that defines when it is acceptable to carry out the activity and when it is not. It also indicates that the long-term solution is to compress residual risk as far as practicable rather than imposing high demands on the worker.

The entropy model does not attempt to 'guess' potential consequences. The reason for avoiding these assumptions is that the range of results of an incident can be extreme. Even the most dangerous conditions can lead to variable outcomes from no injuries to multiple fatalities. For example, there have been numerous traffic accidents where drivers and passengers have walked away from car crashes relatively unscathed, whilst similar circumstances have resulted in deaths. In an absolute sense, the repercussions of risk cannot be predicted, however, as will be explained in Chapter 7, the energy that determines the likely outcome can be estimated more accurately.

The model indicates that companies need to focus on the level of risk more so than trying to anticipate the consequences. High levels of risk automatically imply that the potential for undesirable events is serious. In practice, this means that in circumstances of high environmental, technological and process risks, workers need to maintain an acute level of competence and vigilance. The model also indicates that this is only possible for short periods of time because the probability of an incident increases dramatically when system factors become rapidly deteriorated, as shown by the formula given earlier. After some time, therefore, the demands from other system factors will exceed the individual's capacity to cope. When these demands are excessive it sends the human resources system factor into a sharp state of entropy. Consequently, when an employee is required to work in difficult conditions on a demanding task with equipment that is hard to operate, after some time, the excessive strain on the worker is extremely likely to result in an incident. An example is the case that involved an underground mine worker who had a fatal accident, when he fell 40 meters into an open stope (cavity), whilst removing a ladder in a stope rise.[14] (The accident occurred in a longhole stope design underground mine. A stope is an underground area where ore is mined which results in a void or cavity being created as the ore is removed in a series of steps. A stope rise is a vertical or inclined shaft from a lower to an upper level.[15])

The method of mining required working from a three-deck stage suspended in the cavity. The ladder was part of a continuous emergency ladderway installed on the footwall (wall or rock under the vein or orebody[15]) of the rise and had to be removed progressively as the stope was advanced upwards. (This mining method involves the blasting of overhead materials to extend the cavity upwards.) The deceased's colleague was on the middle deck of the stage preparing a sling to anchor the ladder. The deceased went down the stage access-ladder below the middle deck to unbolt the lowest section of the rise footwall ladder. This connection was not accessible from the stage because of the excessive gap between the bottom deck and the footwall. In the process of removing the ladder the deceased fell (together with a section of the ladder) into the open stope and sustained fatal injuries.

The contributing causes of this accident included an unsafe method of work, a hazardous physical environment involving dangerous heights and limited illumination, a poorly designed ladder construction, and the company had failed to ensure that the use of safety equipment was mandatory. The process of hauling up the entire ladder was physically demanding and this exceeded the worker's capacity to operate safely. The equation derived from the entropy model indicates that this firm should not have relied on its employee to mitigate the risks associated with the other system factors.

The remedy for this case and for all organizational risks is to implement both proactive and reactive preventative strategies. This means that risk management involves using resources to correct and maintain systems – to prevent entropy. Resources are also needed to develop high levels of safety consciousness to prepare workers to respond effectively in the event that residual risks translate into imminent threats. In the longer term, financial capital is required to purchase safer equipment and to modify the physical environment to reduce its inherent dangers. That the entropy model identifies two types of risk and provides a structured four-pronged approach to hazard management is a shift away from risk as a broad concept. This is the most significant difference between the entropy model and the other models described earlier.

This explanation of accident causation also shows that as the probability of an accident rises, both safety and performance suffer. It therefore challenges the perception held in some firms, particularly in hazardous industries, that production and safety are incompatible goals. In reality, when the impact of deterioration is considered in operational terms this relationship is evident. For example, when the workplace is untidy it becomes increasingly hazardous and also hinders workers from performing efficiently. In addition, deviations from standardized practices can introduce risks and reduce productivity. Likewise, when the components of a processing plant are aged and degraded, it is unlikely that the rate of output can be maintained at a high level. The application of the model to organizational management, therefore, allows the concurrent pursuit of production and safety through the maintenance of system factor quality and the implementation of the four-fold strategy to manage risks.

There are further distinctions to be made. The entropy model identifies degradation of system factors and the presence of residual risk as the cause of incidents. This is a very different perspective from the 'unsafe acts' versus 'unsafe conditions' approach. The shift focuses OHS management on systems quality, including management practices, rather than simply employee behavior and general work conditions. Consequently, there is much less emphasis on human error as the cause of accidents. This is an important step forward in both a practical and ideological sense as it accommodates workplace realities that have not received sufficient attention in previous models. The first is that employees do not want to be injured at work and secondly, they rarely choose to or knowingly act in an unsafe manner of their own volition. 'Unsafe acts' are usually symptoms of systemic problems such as insufficient skill-based training, inadequate knowledge-based training of the risks involved, work pressures, excessive physical or psychological demands associated with the task or conditions, or an unhealthy organizational culture. Finally, companies do not want employees to be injured. The emphasis on human error hinders management from exploring fully the underlying variables that lead to incidents that are often attributable to the quality of organizational systems. The model also provides an avenue for greater emphasis on the goals and values shared by management and employees in relation to health and safety.

The role of human error in accident causation is by no means excluded from the entropy model. It acknowledges that all system factors, including human resources, contain a degree of residual risk. This stems from employees' limited capacity in terms of KSAs and physical vulnerabilities. No individual is capable of having complete competencies to anticipate and deal with every activity or to solve every potential safety problem that may arise in the workplace. These limitations do not, however, act as an excuse for safety violations, which are intentional deviations from safe practices. Workers should be held accountable under such circumstances for outcomes to the extent that they have been provided with the competencies and risk training to perform safely and efficiently. The level of risk must therefore be within their capacity to avoid or to manage it.

The presence of inherent risk means that the firm experiences imperfections in all its system factors. For example, a state-of-the-art machine will still have systemic weaknesses because of the limitations of current technical knowledge and also because of economic constraints in product development and manufacturing. The machine may be maintained and operated to the highest possible standards, yet despite this, it still has a residual risk, that given a unique set of circumstances could contribute to an incident, as argued by Reason's theory.

Should human error be considered to be an additive factor in these situations? How far back along the manufacturing chain should culpability be investigated? The entropy model implied that where all reasonable and practicable measures have been taken to mitigate the risks associated with system factors – that is, the firm has sought to achieve optimal safety, performance and system factor quality – the pursuit of someone to blame is not a constructive exercise. This is because these situations occur in a context of residual risk partly caused by constrained technological knowledge, skills and resources. What is required instead is that measures are taken to effectively manage the risks associated with technology, the environment and the process. In addition, employees need to be provided with training to enhance competencies and safety consciousness. This is firstly to limit their tendency to be risky or to be 'risk-takers'. Secondly, training is required to prevent their behavior from having a catalytic effect on the degradation of other system factors; in other words, to give them the skills to interact with technology and the environment safely – to be effective operators. Finally, training in the nature and severity of risk allows them to manage noncompressible inherent risks by being vigilant.

In practice, the entropy model can be applied to identify the sources of entropic and residual risks in each of the four system factors. The four-fold strategy can then be used to develop remedies to correct or manage these risks. In the case of the longhole stope miner who was removing the access ladder, clearly there was a high level of residual risk associated with the physical environment, technology and the process. In addition, the worker did not adequately appreciate the risks involved. Entropic risk was mostly attributable to the lack of a standardized safe work practice and loss of physical control by the worker. To correct these, the firm should have ensured that the wearing of a safety harness was mandatory and that the task did not exceed the worker's physical capacity. Under uncertain and high-risk conditions, a job safety analysis (refer to glossary of terms) should have been undertaken before the task was attempted. Nevertheless, reliance on worker competencies can not be considered an appropriate on-going strategy. A long-term approach requires systems maintenance that, in this instance, could have included standardization of the process, the upkeep/modification of the equipment and work area, and the eventual installation of a safer ladderway system. In the short-term, the residual risk could have been managed by educating employees about the inherent danger associated with heights and falling objects. The application of the entropy model allows companies to more readily identify the types of risks associated with business activities that involve the interaction of these four system factors. This will be explained in detail in Part 2 of this book.

The entropy model is a significant contribution to the understanding of the nature of risk and its impact on organizational systems. Readers should bear in mind, however, that possible misinterpretations of the model might be made in the following areas. Firstly, the model does not explain the mental processes that the worker goes through that cause risk to be translated into an actual injury. It does not describe the worker's behavior that contributes to an accident. This is because the model focuses on the system and not on the employee. To gain a full appreciation of the parameters that lead to injuries therefore, OHS practitioners and managers need to be aware of factors like the variability in individual perceptions of the level of risk in a given situation. This needs to be rectified through training and on-the-job coaching. These different perceptions will also be present amongst management team members who determine the resources allocated to risk management strategies. To address this matter, in Chapter 7, a method is presented that allows managers to evaluate relative risk levels. Using the method, managers are able to more confidently direct resources to areas of higher risk. It provides a shared framework for the determination of relative risk levels and therefore, reduces these variabilities in individual perceptions.

Secondly, the reader should be aware that the entropy model presented in Fig. 1.5 does not explicitly illustrate the benefits of continuous maintenance and monitoring because it depicts maintenance as occurring within a block of time. To overcome this, another version of the model may be developed to show that entropic risk can be held in a low and steady state using a proactive approach to the maintenance of system factors. The benefits of continuous upkeep are implied however, because the model shows that once maintenance stops, systems again go into decline.

An additional matter implied by the model is that the levels of residual and entropic risks should be analyzed or measured. It is not possible to include all the variables that determine these levels in the model without making it highly complicated. As also explained later in Chapter 7, it is not feasible to ascribe an absolute value to the level of risk. It can only be quantified in relative terms; that is, that one particular hazard is more dangerous than another. Using the method, the firm can 'score' its current risk levels and compare its 'scores' from period to period, for example from year to year. Although the level of risk cannot be measured absolutely as such, the company can evaluate whether it is reducing residual risks and counteracting degradation more effectively. In addition, in Chapters 3 to 6 the auditing of system factors is discussed. Audits provide managers with information about current hazards and their control. These results can also be compared from period to period to determine whether risk management strategies are improving the quality of system factors.

Finally, readers should bear in mind that the concept of residual risk may cause some organizations to be less than diligent in their efforts to manage risk because incidents may be prematurely or conveniently attributed to inherent danger rather than systemic weaknesses. This may be particularly the case in workplaces such as underground mines where the level of residual risk is high. For instance, a fatality caused by the collapse of the mine roof may be causally linked to natural weaknesses in the rocks rather than lack of structural support and environmental monitoring. Concurrently, however, because the entropy model raises awareness of the significance of residual risk and its impact on safety and performance, there is more reason for firms to be diligent in the management of such risks.

## Summary

In this chapter, some of the current accident causation models have been presented. Most of these focus on individual risk-taking behavior, perceptions, judgement and competence. The emphasis is on how people get it wrong rather than on how people can get it right. It is therefore, not surprising that some OHS professionals have felt disempowered. These models tell them that their ability to create change and reduce the incidence of injury is constrained by individual behavior beyond their control. Clearly, there is a need to shift away from the individual to a systems approach to correct this. The entropy model provides this shift. In addition, to date, limited attention has been paid to the goals and values jointly held by management and the workforce in relation to safety and performance that determine whether a healthy organizational culture develops. This requires a new perspective on OHS management involving a strategic approach that integrates OHS into the firm's total management system.

In Chapter 2, the process of developing a strategic OHS approach will be discussed using the strategic alignment channel. The channel is the second tool underpinning productive safety management. It explains the external drivers that require firms to develop these systems in the first place. The major impact of these external forces is that businesses are required to be legally compliant and socially responsible. The channel also explains how decisions are made internally about the resources allocated to OHS and other management systems. Finally, it highlights the importance of employee participation to manage risk effectively and to attain high levels of organizational achievement in terms of safety, production output and quality. Chapter 2 will be followed by Part 2, in which the risks associated with the four system factors will be explored.

## References


1. Fowler, F.G. and Fowler, H.W. (eds) (1986) *The Oxford Handy Dictionary*. Chancellor Press, London.
2. British Medical Association. (1987) *Living with Risk*. John Wiley and Sons, Chichester, quotation from p. 56.15.
3. Department of Consumer and Employment Protection, Government of Western Australia (2002), ThinkSafe Campaign. Available on website: www.safetyline.wa.gov.au/PageBin/edcgen10065.htm
4. *Workplace Health and Safety Act 1995* (*Queensland*). Queensland Government Printers, Brisbane, Australia.
5. Stellman, J.M. (ed.) (1998) *Encyclopedia of Occupational Health and Safety*, 4th edn. International Labor Office, Geneva.
6. Kletz, T. (1993) *Lessons from Disaster – How Organizations have no Memory and Accidents Recur*. Gulf Publishing Company, Houston, Texas.
7. *Occupational Safety and Health Act 1984 (Western Australia)*. Western Australian Government Printers, Perth, Australia. Also available on website: www.safetyline.wa.gov.au/sub.htm#2
8. Injury Control Council of Western Australia website: www.iccwa.org.au/road.html
9. Reason, J. (1991) Resident pathogens and risk management. *Safety in Australia* 9, 3, pp. 8–15.
10. Robert-Phelps, G. (1999) *Safety for Managers – A Gower Health and Safety Workbook*. Gower Publishing Limited, Hampshire, quotation from p. 29.
11. Mol, T. (2002) An accident theory that ties safety and productivity together. *Occupational Hazards*, October, Penton Media Inc, USA, pp. 89–96.
12. The Society for Chaos Theory in Psychology and Life Sciences (2000) website. www.enm.bris.ac.uk/research/nonlinear/faq-[2].html#Heading12
13. Hocking, B. and Thompson, C.J. (1992) Chaos theory of occupational accidents. *Journal of Occupational Health and Safety, Australia and New Zealand*, 8, 2, pp. 99–108.
14. Department of Mineral and Petroleum Resources Western Australian (2001) *Significant Incident No.* 58. Available on website: http://notesweb.mpr.wa.gov.au/exis/SIR.NSF/6b390eea5649d21c48256097004aacb7/3cf3739659e88d19482561ea000cdc26?OpenDocument
15. Thrush, P.W. (ed.) (1968) *A Dictionary of Mining, Mineral, and Related Terms*, US Department of Interior, US Government Printing Office, Washington D.C, USA.

Chapter 2

# Organizational decision-making and alignment of management systems

How are strategic decisions about OHS management made? What factors influence a firm's choice of strategy and the resources allocated to safety management? How does the concurrent delegation of OHS and production responsibilities affect the role of the supervisor? All of these issues relate to organizational decision-making. Although the accident causation models and entropy model presented in Chapter 1 assist managers to address risks, they do not provide an understanding of the forces that influence how OHS decisions are made. These forces are important because they affect the development of the OHS legal framework within the broader community and also internal systems within the firm. This lack of understanding has led to difficulties in integrating OHS into companies' total management systems, to the extent that OHS is in many businesses peripheral to core management. The accident causation models also fail to explain external factors such as the development of legislative controls and how these affect firm behavior, and in particular, how they constrain the company's eventual choice of strategies and practices.

To develop an integrated OHS system, therefore, requires an understanding of two issues. The first is how accidents are related to organizational activity as illustrated using the entropy model. The second is why firms are required to have an OHS management system in the first place and how this system with resultant strategies, operational decisions and behaviors can be aligned to achieve optimal safety, performance and quality outcomes. The strategic alignment channel has been developed to explain these forces and processes. It will be described in this chapter.

As explained earlier, productive safety management is a total management system involving the integration of various disciplines. Integration, for the purpose of this book, means creating overlaps in management systems using common elements. The entropy model provides the framework for these elements by showing that the quality of system factors determines the level of risk and that this risk affects both safety and performance. The management of risk requires a multidisciplinary approach thus providing a focal point for decision-making and strategy development. Integration occurs when various management disciplines are drawn into an 'umbrella' management system that has a shared set of decision-making and performance criteria.

In the organization, the cumulative impact of decisions made at all hierarchical levels determines the firm's overall performance in terms of safety, output and economic achievement. These decisions are made strategically and operationally. At the strategic level, managers decide organizational priorities, for example, they determine budget allocations including those for capital projects, such as the replacement of old technology. This in turn has an impact at the operational level, for instance, work practices have to be developed to fit new technologies, potential risks have to be identified and decisions have to be made regarding how to manage these risks. On the shop floor, employees make decisions about how they carry out the work and how they interact with this technology and its physical environment.

How can the company ensure that decisions that lead to positive safety, performance and quality outcomes are made at both the strategic and operational levels? How can it also guarantee that decisions at various levels are compatible – that they are aligned? The firm firstly has to understand the business environment that constrains its choice of decisions. This environment contains a number of forces that influence company behavior and the determination of options. The business does not have full freedom to choose from an infinite number of strategies and action plans. Not only is it restricted by external forces, it is also limited by forces resulting from the internal dynamics of the company. The channel illustrates these forces and is shown in Fig. 2.1.

## External strategic alignment

The channel shows that the firm operates within a 'bigger picture'. It has both an external and internal environment. The former influences the organization's behavior by providing opportunities and by imposing constraints. These limits define the scope available for the development of OHS and other management systems. The forces shown specifically fit the business context of developed countries such as the United States and Australia. However, it can be adapted to reflect the external environment of any nation. The broad categories of forces shown have general application in most contexts, for example, political forces are relevant whether the firm operates in Australia, Malaysia or China.[1] The way in which the firm's behavior is affected will differ depending on the nature of the political system and the other factors in these countries.

External forces can be divided into five categories: economic; social, cultural, demographic and environmental; political, governmental and legal; technological; and competitive.[2] Economic forces include macroeconomic issues, such as interest rates, inflation, unemployment and currency values, and also microeconomics issues, such as demand for products and supply of inputs. These economic forces affect decisions about OHS management, for example, interest rates have an impact on the cost of technology projects. If the firm borrows to purchase new equipment, the interest rate will increase the 'true' cost of the project over its life. On the other hand, if the firm uses its available cash to purchase the technology, there is an opportunity cost as this money cannot

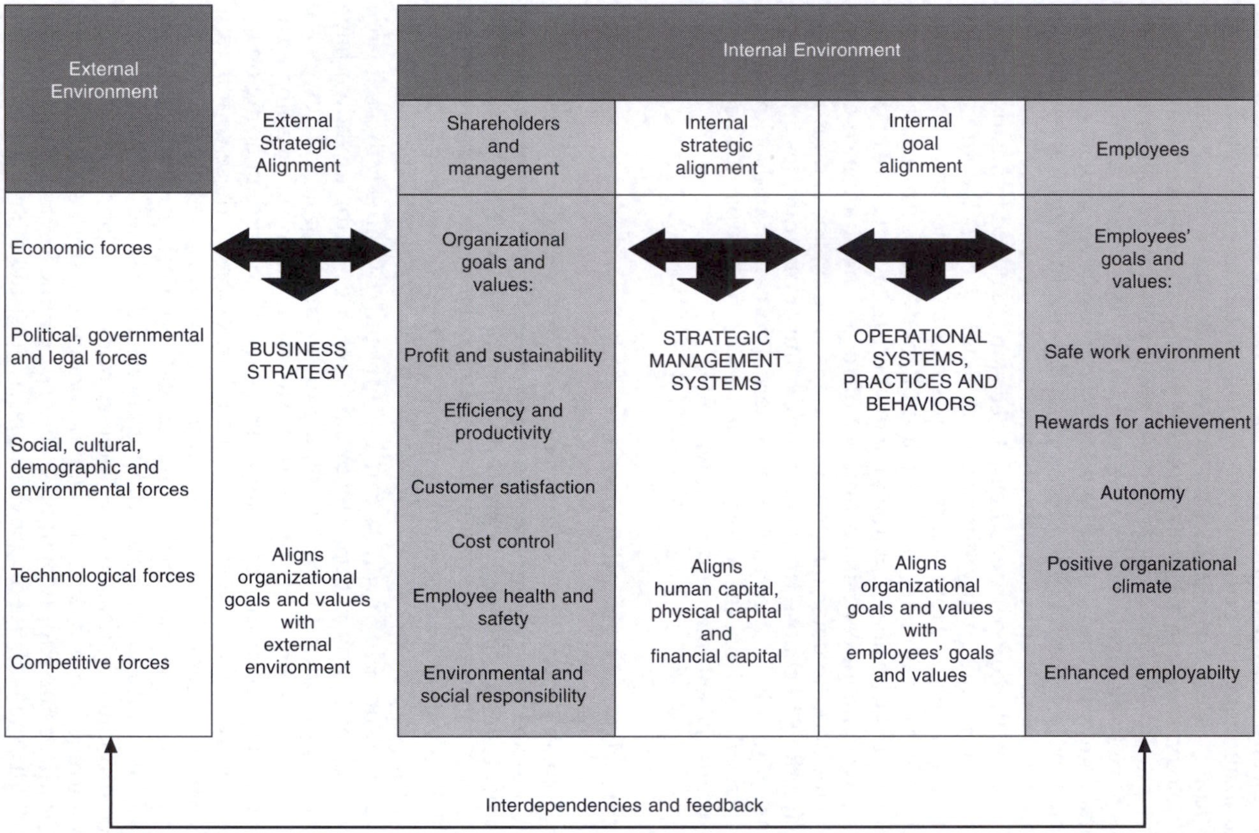

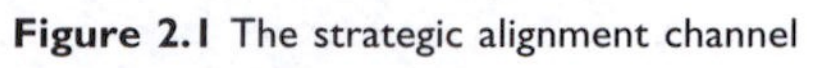


**Figure 2.1** The strategic alignment channel

be used for other purposes once it is committed, for instance, it cannot be used for safety training. The cost of a capital works decision, such as the replacement of old technologies, is also often accompanied by secondary costs stemming from the change in risk it creates. For example, a state-of-the-art, large-scale equipment may have a lower level of residual risk than an old technology. In the short-term, however, the new plant may increase susceptibility to entropic risk because of its impact on other systems. As a result, work practices will require modification, the physical environment will need to be adapted and employees will require pre-operational training. The introduction of new technology not only affects risk levels, but also requires additional expenditures to manage these risks. The firm, therefore, has to consider the full financial impact of its resourcing decisions given current economic variables.

These forces also have a major bearing on how the firm balances its capital resources – physical capital (infrastructure/equipment), financial capital (money) and human capital (people). For example, if wage rates are rising the firm may choose to purchase larger-scale equipment which allows higher levels of output with fewer operators. The firm, thus, may substitute physical capital for human capital. On the other hand, in countries where labor is relatively cheap, the firm may elect to be labor-intensive and use less technology. These economic decisions have a secondary impact on the level of risk within the firm.

Political, governmental and legal forces lead to the development of safety legislation and environmental protection laws, primarily driven by community expectations and the social dynamics that reflect them. These are important variables affecting OHS management and require the firm to be legally compliant. These forces also define the scope of OHS decisions. Legislation ranges from broad statements that describe employers', employees' and manufacturers' accountabilities and responsibilities in relation to safety and health, to specific enforceable operating standards.[3] Legislation has and will continue to be the most significant factor affecting OHS management. In Australia, companies are legally required to develop strategies that ensure that their operations meet the all-encompassing 'duty of care'.[4] (Specific provisions of current legislation will be discussed in Chapters 3 to 6.)

Legislation also affects the relationship between the firm and its workers. Industrial relations laws define the issues that may be negotiated between the parties. These also have implications for safety, for example, awards (a type of industrial relations agreement) define hours of work, rest periods, allowances for special duties which involve additional hazards, and other matters which have an impact on the length of and degree of exposure to risk.

The third category shown in the channel – social, cultural, demographic and environmental forces – are intertwined with employee and community values. Some of these values underpin current OHS management systems, for example, in developed countries the sanctity of human life, quality of life, and the rights of individuals to be treated respectfully, are strong community values. Companies have to be conscious of these values and also of emerging trends because they affect how the community evaluates whether a firm's operations are ethical. In relation to OHS, practices that result in repercussions for the environment and/or the health and welfare of employees or the general public can create a social backlash, as well as reduce the desirability of the firm as an employer.

The channel implies that it is insufficient for firms operating in or primarily based in industrialized, highly regulated countries to satisfy only minimum legal requirements. Companies need to consider all the forces that operate in the external environment, including social factors. A number of firms have felt the impact of these community values when they have sought to be compliant based on the legislation of the host nation. The environmental impact of the Ok Tedi mining operation is an example. Ok Tedi Mining Limited, previously operators of the mine at Tabubil in the remote western province of Papua New Guinea, admitted that the discharge of mine tailings and erosion from waste dumps had caused extensive flooding, killed fish in the river systems and harmed vegetation in surrounding areas.[5] The traditional landowners have since instigated court proceedings against the company for alleged damages to their homelands,[6] their lifestyle and subsequently their wellbeing. The legal system, therefore, also provides for the application of social values through the remedies available under common law.

Further, in some cases, community opinion is ahead of legal process. There may be neither applicable legislation nor an existing precedent under common law, so there is no clear upper boundary that defines the firm's responsibility for health and safety. This means that firms have a choice whether to be proactive or reactive in relation to strategy development. In other words, companies can decide to self-regulate above defined legal requirements or at the other end of the continuum, to risk a minimalist approach. Some firms also adapt to forecast changes in legislation. An example of this occurred prior to the introduction of tobacco smoke-free enclosed public places in Western Australia. In the time before and whilst the legislation was being reviewed by state parliament, a number of eating-houses made the operational changes proposed by the draft regulations. Some provided separate dining facilities for smokers and non-smokers, whilst others became smoke-free. These enterprises responded proactively to proposed changes that impact the health and safety of employees and customers. The Health (Smoking in Enclosed Public Places) Regulations 1999 have since come into force[7] and those businesses that were not proactive have had to change their operational practices in response to the legislation. Their choice of strategy has been reactive.

Many value-driven changes that occur in society have legal ramifications. Some, however, do not change the legislative framework, yet still affect a company's choice of strategy. An example is the rising levels of education amongst Australian workers. Employees are becoming more aware of their legal rights and also expect greater participation and self-efficacy at work.[3] This means that firms operating in this context have to develop their OHS systems using a management style that is appropriate to the current labor climate. During the 1940s and 1950s the workforce valued the 'protestant work ethic' which involved conservatism and company loyalty. Employers were able to apply authoritarian management at this time. The workforce since the mid-1980s, has valued flexibility, job satisfaction and loyalty to relationships.[3] Authoritarian management styles now lead to poor motivation, low morale and high turnover amongst today's workforce. Values have changed and consequently, social, cultural and demographic forces have affected management philosophy and, therefore, the way in which OHS systems can be implemented in the firm. New management strategies are therefore required to fit the current labor context and this simultaneously affects the management of safety.

Technological forces also have a very significant effect on organizational decision-making in this modern era. The firm can choose which combination of technology and labor it wishes to use to produce its output.[8] For example, in a car assembly plant, the firm may operate robotics or hire human resources or use a combination of these to manufacture its products. What impact does this have on OHS management? Firstly, robots can carry out dangerous or repetitive work. Employees need no longer be exposed to the hazards associated with these processes. In this way, technology can help to minimize risk. On the other hand though, technology can lead to increased hazards as it has allowed large-scale operations to be introduced in a number of industries, for example, chemical manufacturing plants and off-shore oil drilling. The severity of consequences for the health and safety of employees and the general public, in the event of a major incident, is much greater than for small-scale operations particularly, where toxic or explosive materials are involved. (This is an issue explored in greater detail in Chapter 4: Technology.)

Technological improvements also allow better monitoring of hazards. This includes, for example, seismic evaluation of slope stability in mining operations and detection of toxic fumes in chemical plants. The entropy model indicates that technologies have an inherent residual risk that has both short-term and long-term consequences. It may not always be possible to anticipate the long-term impact of technology on worker health. An example includes the recent debate over the use of mobile phones at work, which has resulted in the Australian Government putting out a guideline which states that:

> A great many scientific studies worldwide have not been able to demonstrate any harm caused to humans through using hand-held mobile phones. At the same time, no study has been able to totally eliminate the possibility of harm from the radiowaves put out by mobile phones during use. In the absence of concrete proof that no risk exists, this guidance note offers practical advice on reducing exposure.[9]

The extent of residual risk associated with plant and equipment is therefore not predictable until the consequences have been fully researched.

The final external factor shown by the strategic alignment channel is competitive forces. These involve firm rivalry and can shift companies towards either end – minimalist or self-regulatory – of the OHS management continuum. The minimalist approach is particularly evident in cases where firms compete for contracts based on price. As a result, safety is seen as a cost. For instance, an OHS specialist advised that he had conducted audits at remote mine sites in the eastern states of Australia where the contractor site operated at safety standards well below those of the company that owned the lease. The firm had a vastly different attitude towards safety for contractor sites than its owner-operated sites. This could be attributed to cost minimization and that the company's liability for safety was not as direct as if it had been an owner–operator. By law however, as the principal, it was not immune from prosecution in the event of injury or damage.

In the external environment the forces shown in the channel may have counteracting effects on organizational decision-making. Contrary to these price-driven competitive pressures, are legal remedies of two types. There are legislated penalties for specific breaches and there are potential costs of compensation resulting from injury or damage to a third party as a result of lack of due diligence. Companies are required to carefully select and manage contractors to avoid any such liability under the tort of negligence. This means that the owner-company may be held responsible for any physical damage or financial loss resulting from the inappropriate action or from the inaction of its contractors. Under the tort of negligence, the plaintiff (the party which has incurred damage or injury) has to establish that the company had a duty of care, that there was a breach of that duty, and that damage was caused by that breach.[10] The second issue covered by this tort is professional conduct. In this case, the law establishes firstly, whether there is a duty of care, and secondly, the standard of care required of the professional contractor or consultant under the circumstances.

A court of law uses tests to assess cases of negligence. The first of these is the 'reasonably foreseeable' test that considers whether 'a reasonable man in all the circumstances of the case, would have foreseen the likelihood of injury'.[11] The second assessment is the 'proximity test'. This requires that there is some relational closeness between the negligent act and the detrimental consequences caused by the act. For example, a member of an injured employee's family would be less likely to receive compensation for depression brought on 2 years after the employee's accident, than if the depression commences soon after the injury. There has to be a tangible link in the cause and effect relationship for the accused to be found guilty of negligence.

Effectively, legal measures discourage competitive pressures from being entirely cost-driven because of the potential consequences of noncompliance. Competitive pressures can also, however, encourage firms to be more safety focused. When this happens, it tends to be market-driven with the consumer demanding that the product or service meet or exceed minimum standards. In this case, these pressures force companies to develop more thorough safety strategies. Purchaser expectations can be a major driver for the development of effective OHS management systems, particularly amongst contractors. A simple example is the policy of some mining contractors to prefer a noise attenuated piece of equipment, all other things being equal, to reduce the risk to hearing loss amongst operators.

The channel identifies the major forces operating in the external environment that affect organizational decision-making. These forces constrain the firm's choice of strategy in the development of OHS and other management systems. The discussions above illustrated that sometimes these forces run counter to each other; for example, legal forces deter competitive forces from leading to cost-minimization strategies. The nature of these forces varies geographically, so management has to consider them within the context that the firm operates. This explains why the transition for domestic westernized business into the global marketplace, particularly where firms establish overseas operations, is a challenge[1] and why firms value local knowledge. It also explains why it is even more difficult for businesses operating in unregulated environments to establish operations in highly legislated countries, such as Australia.

Consequently, firms have to align their goals and values with the host nation's systems and social structure to achieve external strategic alignment. This alignment is formalized using business strategies usually contained in a business plan or using a number of related systems, for example, an OHS management plan and an environmental management plan. The two-way arrow between the external and internal environments, shown in the channel, indicates that the firm is both constrained by and accountable to the external environment. In relation to OHS, this accountability is defined in terms of legal compliance and social responsibility.

## Internal alignment

Once the company has achieved external strategic alignment it has to manage its internal environment. There are two key groups of stakeholders involved – shareholders and the management who represent them, and employees. Each group has a different set of goals that it hopes to achieve. Shareholders and management establish organizational goals that focus on outcomes including profit and sustainability, efficiency and productivity, customer satisfaction, and cost control. As discussed earlier, the external environment also imposes constraints that require the firm to set employee safety plus environmental and social responsibility, as organizational objectives.

Workers also have goals relating to their jobs as shown on the right-hand side of the channel in the internal environment. These include having a safe place of employment, a positive organizational climate and rewards for their contributions. In addition, with rising education levels, Australian workers want greater autonomy;[12] that is, more involvement in decision-making within their workplace. It has also been shown that workers seek to enhance their employability. This can be described as a 'boundaryless career' which allows individuals to move across boundaries of different employers and also within the current employer organization.[13] There is thus an increasing emphasis on skills, networks, aptitude and knowledge in both internal and external labor markets. Concurrently, there is less stress on security, status and salary as measures of success.[14] Employers, therefore, face pressures to provide employees with opportunities to develop a broader range of competencies than was the case in the past.

The channel identifies the main goals currently driving businesses in western countries and the labor markets in these countries. These goals are underpinned by a set of social ideals or values, some of which are common to both employees and management, particularly in relation to

OHS. The underlying values of these safety objectives include the sanctity of human life, quality of life, and the individual's right to be respected. These values are reinforced by the legal system and are part of the cultures of western nations.

Company cultures also depend on values. Each firm has its own set of beliefs to which it ascribes a meaning that is shared and understood by its employees. It is not possible to generalize about company values because these are to a large extent determined by the nature of the business, for instance, some organizations are service-oriented such as the public sector, whilst others are profit-oriented or private sector-driven. Companies in the same industry may have different values as described in their mission statements and enacted as they go about their business. This can be seen, for example, in the different industrial relations approaches taken by Australian mining companies with some favoring collective bargaining (negotiating work agreements with management and a group of employees) and others preferring individual contracts. Australian domestic operations are, however, subcultures of broader Australian society and consequently, social values that are integrated into Australian culture, apply as much to the firm as they do to the greater community. This same rationale will apply to the cultures of companies operating in other countries. A prerequisite for the firm to achieve internal alignment is, therefore, to define what its goals and values are.

Given that the two groups in the channel – shareholders/management and employees – have divergent goals, particularly concerning profit maximization attained through the minimization of labor costs, how can the company achieve internal alignment? The first step is to clarify the meaning of 'alignment'. This concept has been discussed in management literature to date as if it were like a two-headed arrow stretching between the upper strategic level to the lower operational level of the firm. This is shown in Fig. 2.2 in the left-hand section. Firstly companies establish their organizational goals. From these, strategic management decisions are made, which are enacted by middle and lower management layers. In this way, the company's goals filter down to the operational level. On the shop floor, these strategies provide the framework for the development of procedures and practices. Theoretically, as a result of this flow-down effect, strategies are reinforced by operational decision-making.

OHS management can be used to illustrate this singular concept of alignment. The management team identifies strategies, such as planned maintenance to improve the firm's safety record. This, of course, is also fundamental to efficiency and reliability. Procedures are implemented and, for example, could include daily checks of the condition of tires, oil levels and other routine matters to maintain a mining company's truck fleet. The supervisor decides when tires should be changed and other maintenance carried out. This singular concept of alignment assumes that employees will accept organizational goals as their own. It infers that there will be no conflict over operational issues. Corporate goals supposedly become aligned from top to bottom and employees embrace organizational goals as their own.

What happens though when there is a dispute, for instance, when the operator considers the tires of his truck to be unsafe and the supervisor

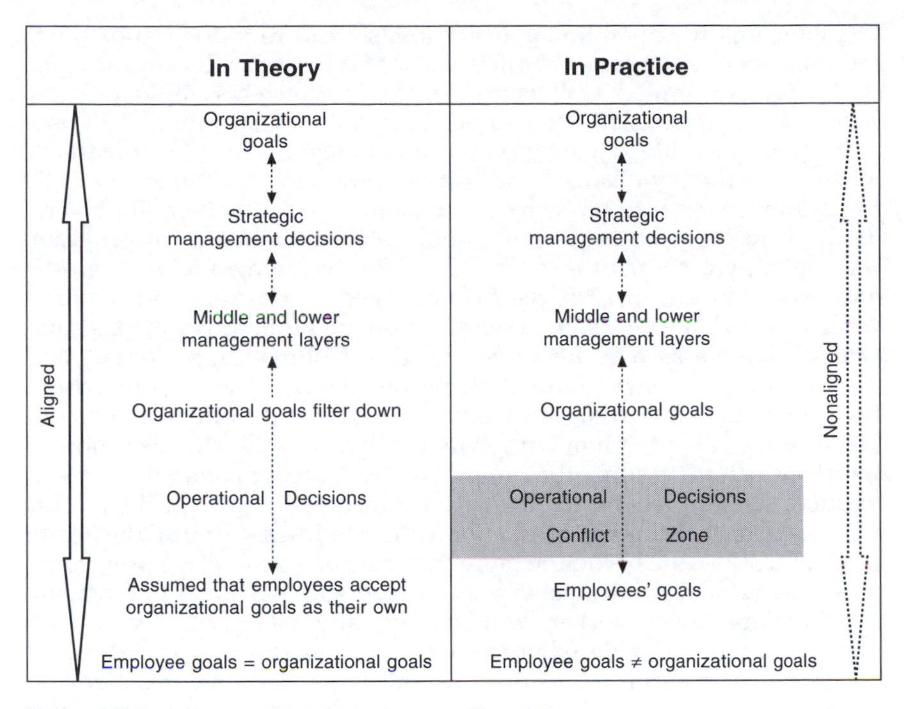


**Figure 2.2** Single stream alignment in theory and in practice

assesses the risk as 'acceptable'? The overseer may then apply legitimate authority to enforce the decision and justify it on the ground that additional mileage gained from continued operation reduces costs and achieves higher production output for his shift. As a consequence, where the employee stands his ground overt conflict results, or where the employee acquiesces, the conflict becomes covert. In either case, the worker has not accepted the firm's goals as his own and alignment fails to be achieved.

In practice, firms often neglect to address the divergent goals of management and workers. This causes blocks that lead to conflict, as shown in the right-hand side of the figure. Incompatible objectives cause breakdowns in trust and communication, so in effect, alignment is seldom achieved. As an example, at an open-cut mining operation, workers requested that all light vehicles used for maintenance carry a walkie-talkie in addition to the two-way radio whilst in the mine pit. The employees believed that this would make communication easier and operations safer, given that the two-way was usually busy with truck movement instructions. Despite the company's 'safety first' policy, the request was rejected on the basis of cost. It also highlights a comment made by a number of OHS consultants. Their experience has been that where the support of top management and employees had been gained for a safety management program, implementation generally stumbled at the lower management decision-making level. This is the level where production and safety become discordant objectives as a result of the current approach to operational decision-making and OHS management.

Conflicts at the operational level can also lead to industrial disputes and legal action. The case of Collins and AMWU versus Rexam Australia Pty Ltd is an example.[15] Collins was a diabetic, which his employer knew prior to his employment. He had been hired on a 7.00 a.m. to 3.30 p.m. shift and he was able to manage his treatment regime around these times. When the company changed the shift in 1992/1993 to 9.00 a.m. to 5.30 p.m., Collins was asked to work this roster along with other employees. He declined to do so because it would not suit his treatment program. The company accepted this at the time after receiving a letter from the employee's doctor. In 1995, the firm changed its position and required Collins to work the new shift system. Following numerous warnings and correspondence as a result of his refusal, Collins' employment was terminated. The court found that the business had acted unfairly in dismissing him and had contravened S170DE (1) of the Industrial Relations Act 1988 and awarded him with damages accordingly. The case showed that firms cannot assume that employees will accept company goals or practices where their needs or goals are incompatible with them. This will increasingly become an issue for firms to address because the health needs of workers are becoming more specific. For example, the prevalence of asthma-related conditions in westernized societies is on the increase and therefore, fewer workers will be physically able to cope with work environments containing what are now considered to be 'low' doses of allergens and chemicals. Firms will therefore have to take into consideration individual differences when managing health and safety to minimize workplace conflict over these issues.

Clearly, if alignment is to be achieved the issue of goal compatibility needs to be addressed. For this reason, and also because effective OHS management requires employee participation, the channel identifies internal alignment as two separate issues. These are internal strategic alignment and internal goal alignment. The former relates to the allocation of human (people), financial (money) and physical (infrastructure/equipment) capital resources, whilst the latter is concerned with identifying common goals and values jointly held by managers and employees, and using these as the basis for decision-making at all levels.

## Internal strategic alignment

As explained earlier, companies are constrained by limited resources. If a decision is made to employ more human capital, then financial capital has to be spent. Using an analogy, it is like having three buckets half-full of water. If water is poured from the financial bucket into the physical capital bucket and the human capital bucket there is no impact on the total volume of water available. Raising revenue by extending borrowings or equity are the only means of increasing resources. These options too are finite, particularly in the short term.

How do companies determine their levels of use for each resource? Once the company has evaluated the external environment and developed its business strategies, it decides what balance of resources it needs to achieve these strategies. It uses financial capital to acquire human and physical capital. This human capital is the collective knowledge, skills and abilities (KSAs) of the workforce.[16] The company determines how it will use people to achieve its business strategy and the key issues are how many people are required and what capabilities they should possess. In practice, it is not always possible for the firm to obtain all the competencies it requires. This depends on what is available in the labor market. In addition, the firm has to compete with other companies for high quality human resources. If there is a shortfall between the capabilities of the workforce and those required by the firm to operate efficiently and safely, it has to address this gap through training and development. To be justifiable, training expenditure must translate into competencies that result in better productivity, safety or quality outcomes. In other words, the financial capital expended must raise the level of human capital to maintain equilibrium in capital resource value.

Companies also have to make decisions about the physical capital that they use. For example, if the firm decides to purchase a new technology it has to spend financial resources. It will only do so if it believes that the new equipment will, likewise, contribute to the company's productivity, safety and quality. Secondary benefits, such as reduced maintenance costs, are additional system factor quality considerations that have to be taken into account. The firm's capital resources are shown in Fig. 2.3.

There is a flow-on effect between the external and internal environments of the firm. The channel indicates that the business has to achieve legal

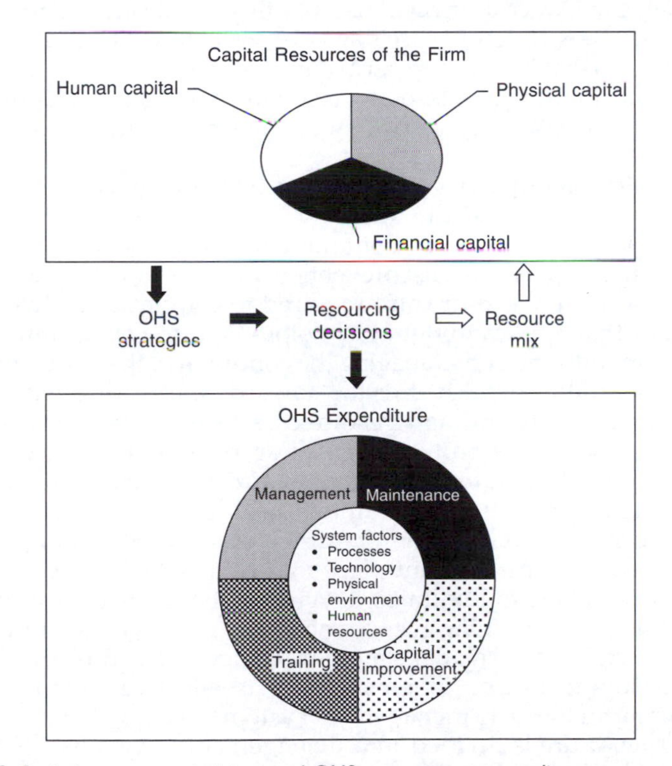


**Figure 2.3** Capital resources, decisions and OHS management expenditure

compliance and social responsibility through its OHS strategies. Forces in the external environment drive this requirement. Once these strategies are in place, choices have to be made concerning the level of resources required to implement them. These resourcing decisions, as shown in Fig. 2.3, affect the resource mix or the balance of financial, physical and human capital. The entropy model indicates that to manage risk effectively and to shift the firm towards optimal performance and safety, resourcing should focus on the quality of the four system factors. The areas requiring expenditure in safety management, as shown in the center of the OHS expenditure wheel, are processes, technology, the physical environment and human resources.

To minimize risk, each system factor needs to be maintained at an acceptable level of quality and this maintenance determines the types of expenditure incurred by the firm. To improve the efficiency and safety of processes, for instance, requires reviews such as job safety analysis (JSA) and audits, which are management activities. Once processes have been improved, employees have to be trained to implement the changes. Training is therefore, a secondary cost of process-related risk management.

There are also expenses related to technology. These resourcing decisions are of two types – maintenance and capital improvement expenditures. The former is a significant component of OHS costs required on a continuous basis to prevent degradation. Capital outlay is part of a longer-term strategy to reduce the firm's susceptibility to entropy and to compress residual risk. New technology has a lead-time before it begins to degrade and this means that in the short-term entropic risk is reduced. In addition, technological improvements often result in plant and equipment that have lower levels of inherent risk and concurrently greater capacity for output.

The physical environment also has these two types of OHS expenditure. Firstly, infrastructure and the operations site have to be maintained to minimize hazards. Examples of environmental maintenance include sound housekeeping, road repairs, rust prevention and removal of waste products. Secondly, capital projects are also required to compress residual risks. A further cost that applies to both technology and the environment is the expense of monitoring and managing the condition of these system factors.

OHS expenditure is also directed towards addressing entropic and residual risks in the human resources system factor. The primary expenditure to prevent and manage these risks is training which is a direct cost aimed at shifting human resources towards optimal safety, performance and quality. OHS training focuses on three main areas, the first of which is the interaction of employees with technology and the physical environment. This interface involves work processes and the aim of instruction is to minimize deviations that increase entropic risk. The second type of training focuses on hazard awareness and therefore encourages vigilance to manage inherent dangers. The final training area is the development of management skills to effectively plan, organize, lead and control the OHS management system. Overall, therefore, safety-driven expenditure is divided into management, maintenance, training and capital improvement.

The channel explains that external forces influence OHS decisions, including how the system is resourced. Within this context, the firm can, however, exercise choices about the extent to which it responds to such forces and therefore, the strategic decisions taken also reflect resource availability and how much risk managers are willing to accept. For example, if companies choose the minimalist approach, then there will be fewer resources allocated to risk management. In addition, firms make choices about the time horizons on which they operate, for instance, businesses that have the resources available, but choose not to utilize them to improve safety operate on short time horizons, are quick profit-takers and are highly risk-tolerant. On the other hand, businesses that are committed to safety and are proactive, assign resources to balance them against an 'acceptable' level of risk. Their focus is on sustainability and practicable risk management.

Conceptually, the channel in Fig. 2.1 shows that to make effective, strategically aligned decisions requires a sound understanding of the external environment, strategy development techniques and resource availability. The channel illustrates that these issues are the prerogative of management by the proximity of internal strategic alignment to these key decision-makers. They are placed nearest the external environment where they can access the information required to make tactical choices. The two-way arrow indicates that there is a continuous feedback process that allows for adjustment of the business strategy in response to changes in the external environment and also in the internal environment. Whilst continually scanning for information, the executive team develops its strategic management systems. When human, financial and physical capital are aligned through the development of such systems that concurrently pursue production, safety and system factor quality, the firm achieves internal strategic alignment.

## Internal goal alignment

Once the firm has achieved the strategic levels of alignment how can it encourage employees to accept or buy into its OHS and other business strategies? The decision whether or not to 'buy in' depends on motivational factors, specifically, the match between employee expectations and the rewards offered by the company. The channel identifies these motivators as 'employees' goals and values'. These show that workers have expectancies beyond those defined in the employment contract. They look for benefits additional to base salary and conditions. The channel indicates that workers, like shareholders and management, derive their goals from factors in the external environment, as shown by the black arrow at the bottom of the diagram, thereby illustrating the interdependencies between the internal labor group and society as a whole. In addition, there is a feedback cycle continuously in operation between the macro- and micro-social environment of the firm that results in employees refining their goals and values according to those of the broader society. Concurrently, feedback from within the organization affects these community mores.

As shown in the channel, employees' goals include a safe work environment, rewards for their contribution to the firm, greater autonomy, a positive organizational climate and enhanced employability. In reality, the employer is only obliged under the law to provide a safe work environment and negotiated conditions of employment. There are no legal obligations to fulfill the other expectations that employees hold.

Why would an organization consider expending limited resources to satisfy these additional employee goals? The first reason is to maximize the productivity of its workers and a second is to retain them. Retention is important because people cost money to recruit and train. Companies invest in training to achieve the high competency levels required for organizational efficiency and effectiveness. To maintain a high rate of return on this investment it needs to keep its human resources engaged. In addition, the firm has to compete for skilled employees in the labor market and cannot be certain of replacing a full contingent of those competencies lost when individuals leave the company.

The extent to which an employee's goals are met affects whether the worker chooses to stay or to depart from the firm. This decision is based on what the current job offers and what alternatives are available in the labor market. Employees look to the external environment for information to make this comparison. This is part of the interdependencies and feedback cycle between the external environment and employees shown by the channel. Figure 2.4 illustrates this process whereby the employee makes this 'loyalty' decision.

The left-hand panel of the diagram shows that the labor market, like all other markets, operates under the laws of supply and demand. These are not legislated laws, but natural laws that reflect collective human behavior, for example, when there is an under-supply of information technology professionals, the price they can demand for their services tends to go up relative to other professions. Supply and demand provides the basis for relative job value and is a key environmental factor affecting the 'loyalty' decision. Other issues in the external environment are economic conditions, industrial relations awards and agreements, and trade union influence.

Economic conditions, in addition to affecting the opportunities available external to the workplace, take into consideration the individual's ability to value-add his competencies through training, education and experience. These push the price of labor upwards. In the labor market there are also compensating differentials, that is, variations in the levels of remuneration received for differences in working conditions. Compensating differentials are also linked to levels of risk.

> Compensating differentials are required when the job is clearly very dangerous. Those who take great risks are usually rewarded with high wages. High-beam workers on skyscrapers and bridges command premium wages.[17]

This explains the differences in remuneration from industry to industry, for example, the Australian mining sector paid average weekly earnings of $1400 to male workers in 2000, whilst the retail sector paid approximately $650.[17] Figure 2.4 also indicates that industrial relations mechanisms such

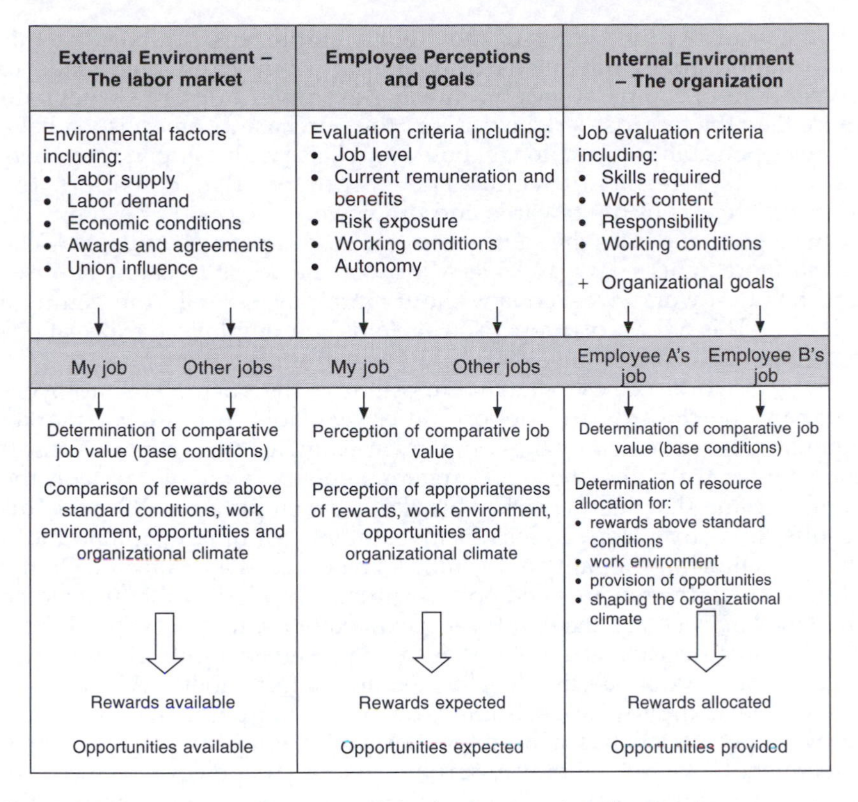


**Figure 2.4** The external environment and the development of employee goals

as awards and agreements have an impact on relative wages, as does the degree of union influence within an industry or a company. Employees compare the overall rewards and opportunities available within their current positions with those available in the labor market.

Workers also consider conditions within the internal environment, shown on the right-hand side of the diagram. Firms use job evaluation criteria, including skill requirements, work content and environmental factors to value jobs. Comparative job value is further influenced by organizational goals. For example, companies sometimes link sales targets to achievement at the operational level using commission geared to team-based or individual-based sales volume. Such performance-based incentives also include bonuses and share ownership. Using cost/benefit analysis, companies determine whether to provide additional rewards, make improvements to the work environment, introduce developmental opportunities and implement strategies to build a positive organizational climate. These factors determine the actual rewards and opportunities available to the employee. The diagram shows this as 'rewards allocated' and 'opportunities provided' by the current employer.

The worker, therefore, has two sources of information on which to develop expectations regarding rewards and opportunities. These are their options available in the external environment and those provided by the firm. At the center of the figure, employees compare all this information about other jobs against their own job, which leads to perceptions of comparative job value. For example, 'I deserve to get paid more than the records clerk because, as the purchasing supervisor, I have more responsibility. I work longer hours and liaise with company suppliers as well as internal officers.' The employee also considers the appropriateness of the rewards and other work factors. For example, 'I know a guy who does the same job as me in company X and gets $5000 a year more in his salary package. Mind you, he doesn't get any bonuses and he often works weekends without extra pay. Overall, I'm getting a fair deal.' This process of comparison explains how employees' expectations develop.

Why is workforce expectancy important to the business? Employees compare the rewards and opportunities available in the labor market against those provided by the company and against their expectations. If the worker feels that there are greater benefits available outside the company and that he has marketable skills, the employee will leave. This results in costly turnover, loss of investment in human capital, and a likely decline in the quality of the human resources system factor. On the other hand, if the rewards and opportunities provided by the firm match the labor market and the employee's expectations, the worker will stay. The 'loyalty' decision is therefore based on perceived equity and fairness.

The workforce's collective 'loyalty' decisions affect stability. When people leave, the company loses human resource competencies and local knowledge, which has a number of implications for organizational effectiveness. Firstly, employee turnover is very expensive. The cost of losses on training represents a significant overhead. For example, the Australian mining industry, in 1999, spent $A82.73 per employee on health and safety alone in comparison to all other industries combined which spent $A6.82.[18] For medium- and high-risk industries in particular, the cost of turnover can be even greater. The entropy model shows that the quality of human resources has a major bearing on systems performance and safety and that there is a direct relationship between employee competencies and the levels of residual and entropic risk. Workforce competency levels also affect the ability of the business to take corrective action when systems degrade. The firm therefore needs people with the capacity to identify risks, to assess these risks and to develop appropriate remedial action. In addition, high labor turnover has been identified as a major barrier to the implementation of effective OHS management systems.[19] A fully integrated OHS approach, therefore, must include strategies to address turnover as a risk variable. In addition, determinants of turnover, such as poor job satisfaction resulting from insufficiently addressed employee expectations, also need to be considered within the firm's total strategic approach, and in particular, its effect on productivity.

Employee retention is an important issue affecting the firm's ability to maintain its capital resources and to achieve the goals shown in the 'shareholders and management' column of the channel. This means that strategies have to be developed to influence workforce expectancy and the 'loyalty' decision. The channel indicates how this is done. It shows that employee and organizational goals should both be considered in the development of operational systems, practices and behaviors. These seemingly divergent goals can be aligned and made compatible because both groups are interdependent in the employment relationship. Neither can attain its goals without the input and support of the other. With this in mind, there is ample scope for the identification of common goals and values – those shared by management and employees – and for operational systems to be driven by them.

Central to the development of effective management systems particularly OHS systems are mutually held beliefs and reference points. Shared goals define what management and employees want to achieve – the task component of the work. Values identify how they want to achieve them – the relationship component of work. (The concept of work having both a task and a relationship component stems from sociotechnical theory.[20]) When the common ground is defined, employees are encouraged to buy into the OHS management system. This integrated approach leads to internal goal alignment, which is a separate alignment issue from the resourcing decisions that were shown to be a management prerogative. Internal goal alignment is concerned with employee participation and maximizing the potential of and return on investment in human capital through the development of effective operational systems, practices and behaviors.

What difference do common goals and values have in the workplace? Referring back to Fig. 2.2, it was shown that the singular concept of alignment based on the assumption that employees accept organizational goals as their own can result in conflict at the 'coalface'. The channel indicates that at this level, using a shared point of reference as the basis for decision-making minimizes disagreement by breaking down this conflict zone. In addition, positive workplace relationships are developed that, in turn, affect the organizational climate and employee satisfaction.[21] A supportive workplace atmosphere provides managers and employees with intrinsic (nonmonetary) rewards that meet expectancies. (In Chapter 8, specific management techniques that allow the application of common goals and values in organizational decision-making will be provided.)

The channel is an educational tool that, with the entropy model, underpins the productive safety approach. It explains why firms are required to implement OHS systems and that alignment is not a single-stream concept. It has three components – external strategic alignment, internal strategic alignment and internal goal alignment. This is a new perspective on organizational decision-making that has application to all areas of management and is particularly relevant to OHS where employee quality, participation and commitment are critical to the firm's safety performance.

The lack of internal goal alignment has had some significant implications for the management of OHS to date. In particular, it has had the impact of creating role conflict in positions in which supervisors are concurrently responsible for production output and safety. Firms have historically delegated these responsibilities to shop floor supervisors. This has caused decision-making at this level to be difficult and 'no win' when something goes wrong. For example, if the team leader chooses to continue operations when there is a safety issue and an incident results, she is the first point of accountability. On the other hand, if operations are ceased to rectify the safety issue and production output is not achieved for the shift, the team leader may also receive criticism from the hierarchy. The failure to have common goals and values driving production and safety as compatible outcomes means that there are no criteria against which to test the appropriateness of operational decisions. In addition, the problem with alignment being understood as a single-stream concept results in the assumption that employees will accept organizational goals and values as their own. At the 'coalface', supervisors have to deal with the reality that this is not always the case, particularly when safety issues arise. The onus of finding resolutions to these disagreements rests with the supervisory level. Further, when measurement systems are skewed towards short-term output, supervisors are put in a compromising situation that results in the tendency for production pressures to over-ride risk management.

An OHS training specialist summed up the consequences of this role conflict perfectly. He said that the implementation of safety management systems generally falls apart in the middle. The implication is that the supervisory level is usually the weakest link in the chain, which is by no means an indictment on the abilities of supervisors. It reflects instead a major weakness in the implementation of OHS systems. Traditional management has given accountability for safety to supervisors and also put them under pressure to achieve short-range output results. In addition, some companies have not focused on common goals or built effective strategies that encourage employees to take greater ownership of safety. Systems therefore, have not been appropriately developed to support the role of the supervisor in safety management and criteria have not been established to determine appropriate operational decision-making.

Productive safety management addresses these issues as a total management system. It not only provides the traditional foundations of OHS management, but also tackles those areas that tend to derail such systems. These include unclear roles and responsibilities, conflicting objectives, lack of strategic direction, insufficient planning, inappropriate measurement methods, and solely production-centered, authoritarian cultures.

The strategic alignment channel is an effective tool to communicate how firms can achieve congruence with the external environment and within internal systems. The reader should, however, be aware that the channel might be misinterpreted in a number of ways. As stated earlier, the external environment contains opposing forces. There are those that require firms to be legally compliant and socially responsible, whilst financial markets concurrently drive them to be cost minimizers. The desire to maximize shareholder return in the short term can lead to fewer resources being allocated for hazard reduction and control. The channel is not intended to imply that firms can satisfy all stakeholders concurrently. In reality, decisions are usually a compromise of these divergent demands. In addition, decisions have to balance short-term gains against long-term objectives. The channel is concerned with helping managers find this balance so that sustainable businesses are developed. Ultimately, however, these decisions are management's prerogative and it is their responsibility to attain this balance using corporate accountability as the determinant of firm behavior.

In the last two decades in particular, financial markets have had a very strong influence on organizational decision-making. Although the monetary costs of noncompliance are well understood, there remains some under-appreciation of the costs of failure to act with social responsibility. Further, the benefits of social responsibility have attracted little attention. As a result there is currently limited incentive for firms to embrace this behavior as a business ideology. Recent corporate collapses, the rise in litigation and scrutiny of financial reporting methods may stimulate a transition towards greater corporate accountability.

The reader should also beware of skeptics who would argue against the practicality of strategic management. In the explanation given, it was acknowledged that limited resources are a significant constraint on firms. For businesses that are marginal in their financial performance, the scope of decision-making may be restrictive and short-term focused. These companies have limited flexibility and may consider a strategic approach to be a 'luxury' they cannot afford. When taking a 'calculated' risk, they cannot however, forfeit their immediate or long-term responsibility for risk management and need to consider the repercussions of the risks that they accept and the potential consequences of failures.

To assist firms to overcome cultures that are solely production-centered, the channel establishes a framework for appropriate behavior. The cynic may attempt to denigrate the value of this framework using examples of practices that currently contradict it. For example, the channel implies that employees should give equal priority to their goals and not compromise their values. In reality however, workers in hazardous industries sometimes accept additional monetary compensation for the dangers they face. In this way, it could be argued that workers, to some degree, allow existing hazards to continue in their current state. As a result, these financial incentives create an imbalance preventing production and safety from being compatible objectives. In Chapter 8 this issue is addressed using the 'reasonableness test'. The basis of this test is a value statement. This is a statement of beliefs developed by management that acts as a testament of the company's commitment to the concurrent pursuit of safety, production and quality goals. As explained in Chapter 8, the value statement is used to test the appropriateness of both strategic and operational decisions to ensure that they are aligned with this company ethos. The value statement acts to filter out those decisions and strategies that prevent the development of a sustainable business based on balanced output and safety objectives.

Finally, the following clarification is made in order to shed light on the meaning of the terms 'alignment' and 'integration' in relation to the reference to OHS management in formal quality management system standards, such as ISO 9000. This standard states that it does not prevent an organization from developing integration of like management system subjects.[22] The Australian standard AS/NZSISO9000:2000, which is linked to ISO 9000, mentions the following two points regarding the compatibility of such systems:

> 0.4 Compatibility with other management systems
> This International Standard does not include requirements specific to other management systems, such as those particular to environmental management, occupational health and safety management, financial management or risk management. However, this International Standard enables an organization to align or integrate its own quality management system with related management system requirements . . .; and
> 2.11 Quality management systems and other management system focuses
> The various parts of an organization's management system might be integrated, together with the quality management system, into a single management system using common elements . . .[23]

From the wording provided above it is unclear whether 'align' and 'integrate' are interchangeable terms or whether they have two different meanings and represent two different approaches in these standards.

Productive safety management and the channel, in particular, help to clarify these matters. Firstly, the management system described in this book involves the integration of various branches of management such as OHS, HRM, environmental management, quality control and plant maintenance under one umbrella. As explained earlier, this integration is provided by the entropy model and results in a shared set of decision-making and performance criteria. (These criteria are captured in the performance management system presented in Chapter 12.)

Under productive safety management, integration and alignment are not interchangeable terms. Alignment as shown by the channel, is concerned with, firstly, creating a fit between the firm and the external environment, and secondly, the allocation of resources to strategic management systems. Thirdly, it requires operational systems, practices and behaviors that reinforce the compatibility of organizational goals and values with employees' goals and values. Overall, alignment depends on the congruence of organizational decisions from the strategic to the operational level. In addition for alignment to be achieved, internal systems must not contain contradictions. For example, if a company has a 'safety first' policy yet, at the same time, rewards the workforce using solely production-based incentives, their internal systems are incongruent. On the one hand, the firm expresses safety as its highest priority and on the other hand, production is most important.

Integration is a holistic term that involves creating overlaps in management systems that were previously specific to a particular professional field, using common elements. As explained later in this book, this is achieved using a value statement against which to test all organizational decisions, deriving strategies from a central key result area that applies to multiple disciplines, and using a performance management system to tie strategies and measures together. Under this integrated umbrella management system, alignment focuses on the compatibility of decisions, practices and behaviors associated with the management of the company. The channel is therefore an expression of the firm's accountability to key stakeholders including the community, shareholders and employees. From the productive safety management perspective, it would have been better for the Australian standard above to indicate that the international standard enables an organization to integrate and align (as opposed to 'or') its quality management system with related management system requirements.

## Summary

The entropy model and the strategic alignment channel are equally important tools in the productive safety management system. The model provides a strategic direction for risk management and explains the compatibility and interdependence of production, safety and system factor quality. The channel explains the separation of 'alignment' into three levels which is an important step forward in organizational strategy development. It explains the need for legal compliance and social responsibility in the formulation and implementation of OHS management systems. It also identifies resourcing decisions as management prerogative. The channel shows that firms have to balance scarce capital resources and this relates to the concepts of 'practicable' and 'acceptable risk'. Finally, it conveys that operational systems, practices and behaviors need to be based on common goals and values to minimize conflicts on the shop floor, to maximize employee participation and to achieve greater ownership of organizational outcomes at all hierarchical levels.

The channel is the driver of behavioral change. It complements the adaptation strategies identified by the entropy model. These were restructuring of traditional OHS practices to fit the model and risk management based on the four-fold strategy, collectively referred to as systems change. In Part 2 of this book, systems change will be discussed with each of the system factors analyzed to determine the sources of residual and entropic risk associated with them. In the final chapter of Part 2, a method to quantify these risks, in relative terms, will be presented and later used to develop risk management strategies that make optimum use of limited company resources.

Part 3 contains the behavioral change strategies needed to build a productive safety culture. In this part, it will be shown how HRM practices can be used to clarify accountabilities, develop competencies, and improve the quality of organizational decision-making to greatly reduce conflict that results from divergent objectives. The final part of this book – Part 4 – will present the productive safety management plan. It amalgamates these restructuring, behavioral and risk management strategies and develops measures that determine whether production, safety, system factor quality and other indicators of alignment have been achieved.

## References

1. Erwee, R. and Mol, T. (1998) International HRM issues in China, Malaysia and Australia: A contemporary comparison. In *Human Resource Management – Contemporary Challenges and Future Directions*. K.W. Parry and D. Smith (eds) USQ Press, Queensland, Australia.

2. David, F.R. (1995) *Strategic Management*, 5th edn. Prentice Hall, New Jersey, USA.
3. Smith, D. (1998) An occupational health and safety perspective. In *Human Resource Management – Contemporary Challenges and Future Directions*. K.W. Parry and D. Smith, (eds) USQ Press, Queensland, Australia.
4. *Occupational Safety and Health Act 1984 (Western Australia)*. Western Australian Government Printers, Perth, Australia. Available on website: http://www.austlii.edu.au/au/legis/wa/consol_act/osaha1984273/
5. Australian Broadcasting Corporation (2000) *BHP makes noise about Ok Tedi closure*. Available on website: www.abc.net.au/news/2000/05/07
6. Australian Broadcasting Corporation (2001) *Villagers sue BHP over pollution*. Available on website: www.abc.net.au/news/2000/04/12
7. Worksafe Western Australia (2001) *Guide to Western Australian Smoking in the Workplace Legislation*. Available on website: http://www.safetyline.wa.gov.au/pagebin/pg007688.htm
8. McTaggart, D., Findlay, C. and Parkin, M. (1992) *Economics*. Addison-Wesley Publishing Company, Sydney, Australia.
9. Worksafe Western Australia (2001) *Guide to Mobile Phone Use in the Workplace*. Available on website: http://www.safetyline.wa.gov.au/pagebin/pg007563.htm
10. *Contract Management* (1994) Management Education Program from the Association of Professional Engineers, Scientists and Managers, Australia.
11. Vermeesch, R.B. and Lindgren, K.E. (1992) *Business Law of Australia*, 7th edn. Butterworths, Sydney, Australia.
12. Marchant, T. (1998) The changing nature of work. In *Human Resource Management – Contemporary Challenges and Future Directions*. (K.W. Parry and D. Smith eds) USQ Press, Queensland, Australia.
13. Inkson, K. (1997) Organization structure and the transformation of careers. In *Advancement in Organizational Behavior*. (T. Clark ed.) Ashgate Press, Louth, UK, pp. 165–85.
14. McColl-Kennedy, J.R. and Dann, S.J. (2000) Success: what do women and men really think it means? *Asia Pacific Journal of Human Resources*, 38, 3, pp. 29–45.
15. Industrial Relations Commission of Australia, *Collins and AMWU v Rexam Australia Pty Ltd (960511)* DECISION NO:511/96. Available on www.austlii.edu.au/special/industrial/
16. Case, K.E. and Fair, R.C. (1989) *Principles of Economics*. Prentice Hall, New Jersey, p. 410.
17. Australian Bureau of Statistics, *Survey of Employee Earning and Hours, Australia* Catalogue No. 6305.0 released 11 December 2000. Available on www.abs.gov.au/ausstats/
18. Butterly, N. (2000) Mechanised mining creates new safety hazards for industry. *Gold Australia* 16 October, p. 1–2.
19. National Occupational Health and Safety Commission (2001) *Occupational Health and Safety Management Systems – A Review of their Effectiveness in Securing Healthy and Safe Workplaces*. Released April 2001. Available on website: http://www.nohsc.gov.au/Pdf/OHSSolutions/ohsms_review.pdf
20. Shonk, J. (1992) Teams. *Team-Based Organisations – Developing a Successful Team Environment*. Business One Irwin, Homewood, Illinois, pp. 1–12.
21. Alpander, G.G. (1991) Developing managers ability to empower employees. *The Journal of Management Development*, 10, 3, 13–24.
22. International Organization for Standardization (1999) *ISO 9001:2000 Quality Management Systems – Requirements*. International Organization for Standardization, Switzerland.
23. Standards Association of Australia (2000) AS/NZSISO 9000: 2000 Quality Management Systems. Standards Association of Australia, New South Wales, Australia.

# PART 2
# Systems change

Productive safety management involves two types of change – systems change and behavioral change. The former involves restructuring current OHS practices so that they fit the entropy model and the application of methods to quantify risk so that priorities for risk management can be established. Behavioral change is needed to create an internal business environment which allows these systems changes to translate into improved organizational safety, performance and quality through the competencies, vigilance and resourcefulness of employees.

Why is it necessary to restructure current practices? In Chapter 1 it was explained that existing accident causation models have been used to develop OHS management systems to date. A number of weaknesses were identified in these, including a lack of practical application to the organization. In addition, they were skewed towards human error as the cause of incidents and as a result, current practices tend to focus on the individual and not on the system. The restructure is, therefore, needed to shift prevailing approaches to a systems perspective. In the *Introduction* the analogy about the brick wall was presented to illustrate these weaknesses. It explained that traditional components of OHS systems tend to be like a wall built without mortar. Each brick is not clearly connected to the next. The restructure based on the entropy model allows a more orderly approach to these practices and establishes clearly where each brick fits into the system.

According to the model, the firm has four system factors that are applicable to all activities of the company. These are processes, technology, the physical environment and human resources. Restructuring involves the allocation of practices to the system factor that provides the best fit. For example, workplace inspections fall under processes and equipment maintenance is part of technology. This change strategy also requires the modification of tools, such as incident investigation questionnaires and JSA formats, which are provided in the relevant sections of Part 2.

Chapters 3 to 6 discuss each system factor in turn and identify the OHS practices that apply to them. An overall perspective of this restructuring process was provided in Fig. (i) in the *Introduction* and a summary of the specific systems change elements of productive safety management is shown in Fig. (ii). There has been extensive work undertaken by OHS professionals to develop current practices, and as a result, this book does not set out to 'reinvent the wheel'. It presents instead a better-formatted, systems-based approach to these management components. Productive safety is designed to improve the systems that most firms already have in place and provide greater understanding of the nature of risk and OHS decision-making using the entropy model and the channel, respectively. The restructuring and risk management strategies described complement the behavioral modification initiatives that will be discussed in Part 3.

In addition, Chapters 3 to 6 detail the sources of entropic and residual risk associated with each system factor. An in-depth exploration of risk variables is

## Systems Change

### Processes

Legal provisions and safety standards

Process design

Risk management practices
- Job safety analysis
- Workplace inspections
- Hazard inspections
- Process audits
- Visitor safety
- Contractor management

Post-incident practices
- Incident reporting and investigation
- Evacuation procedures
- Emergency procedures
- First aid
- Workers' compensation management
- Rehabilitation

### Technology

Planning, design and purchasing

Technological hazard management

Technology audits

Equipment maintenance

Safety and monitoring equipment

Personal protective equipment

Ergonomics

Supplier relationships

### Physical Environment

Environmental planning

Environmental modification for persons with disabilities

Environmental monitoring

Residual risk hazard management

Housekeeping

Physical environment audits

### Human Resources

OHS specialists

Human resource management specialists

Induction

Training

Safety promotion programs

Joint consultative committees

Communication systems

Human resources audits

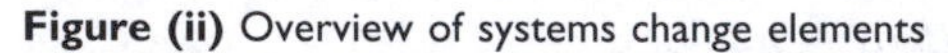

**Figure (ii)** Overview of systems change elements

undertaken supported by practical examples. Managers and workers are better equipped to manage hazards when they have a sound knowledge of the specific risks within work procedures, equipment, the work environment, people and how these interact. Incident reports are used to illustrate how these risks have translated into actual events in the workplace.

The final chapter of Part 2 – Chapter 7 – presents a method to quantify the entropic and residual risks within each system factor. The resultant relative values allow the hazards associated with various work activities to be prioritized. As a result, managers are more able to confidently allocate limited company resources to mitigate these risks. In addition, this risk quantification method permits the firm to set a baseline level of risk above which it considers risks to be unacceptable. The overall objective is to reduce this baseline over time. In Part 4, the baseline measures together with audit results and other safety measures, such as the lost time injury frequency rate (LTIFR), are included in the productive safety management plan.

Chapter 3

# Processes

Business activities involve the interaction of human resources, technology and the physical environment through processes. Processes therefore provide the interface (a point of contact or interaction) between other system factors. They can be divided into three categories which are:

(1) Process designs or work practices;
(2) Risk management practices; and
(3) Post-incident practices.

The first processes are those activities required to get the work done. These are developed prior to the firm becoming operational or when new tasks are added to the system. The second involves risk identification and monitoring activities such as job safety analysis (JSA), workplace inspections, hazard inspections and audits. Post-incident practices are the responses that follow events that are deviations from safe systems, for example, incident investigations and emergency procedures. They also include activities to rectify or manage injury such as the rehabilitation of affected workers.

When designing work practices a number of variables have to be considered. These are firstly, the levels of entropic and residual risk associated with current system factors, and secondly, the potential for degradation caused by the introduction of change. Prior to the implementation of new work practices, therefore, assessments are required of:

(1) The condition and characteristics of equipment that will be used to carry out the procedure;
(2) The state and characteristics of the environment where the work will be undertaken;
(3) The current competencies and attributes of individuals who will do the work;
(4) Any residual risks within these system factors; and
(5) The potential implications resulting from the interaction of these system factors.

New or modified work practices have a negative effect on safety and performance in the short term because employees require time to adjust to change and to learn new habits. In addition, some of the consequences of change may not have been anticipated causing unexpected rises in entropic risk. For example, if a company introduces a vehicle wash-down procedure in an area where the water run-off creates an additional hazard in an adjoining workplace then a new source of entropy is introduced. The lack of fit between the process and one or more other system factors accelerates degradation.

In addition, all processes have some level of human input. The greater this involvement, the greater the worker's vulnerability to the hazards associated with the interface between system factors, and the higher the likelihood of an incident involving the worker. Exposure according to the entropy model, and this is where it differs from the current understanding of exposure as a time-related variable, refers to the severity of the risks more so than simply the length of time of contact. The reason for this is that the model indicates that the probability of an incident relates to the absolute levels of residual and entropic risk at a given point in time. A worker, when exposed to hazards, may be injured in the first or last five minutes or at any time in the shift. The level of danger is treated as occurring in static time when giving a relative score to the risk using the method in Chapter 7. The worker's susceptibility to injury is therefore related to the level of residual risk plus the variability in the entropic risk in other system factors. The individual worker also has risks specifically related to current competencies. Areas in which the employee has a lack of knowledge, skills or abilities (KSAs) are a source of residual risk, while fluctuations in capacity are a worker-specific entropic risk. Worker exposure is therefore the sum of a complex set of variables that may be expressed as the relationship:

Worker exposure at a given cross-section of time
= Level of residual risk in system factors
+ Levels of entropic risk in system factors
+ Level of worker-specific competencies
+ Level of worker-specific degradation

When this rationale is applied, there is a number of implications. For instance, if the process involves exposing the worker to a chemical that causes respiratory problems, his physical capability will degrade according to individual tolerance and susceptibility to such hazards. It is, therefore, not assumed that prescribed chemical exposure rates are safe because they are dependent on individual reactions. The worker is thus subjected to risks as a result of his interaction with the physical environment, technology and other human resources while carrying out a set of procedures. Therefore before process risks can be managed three issues need to be considered. The first is to identify the hazards specific to each system factor and the interactions involved when the process is undertaken. The second is to evaluate how the hazards are affected by human parameters such as competence level, the ability to manage risk and attitudinal factors. Finally, the demands that these hazards place on the worker have to be understood before work practices can be designed to minimize these risks. According to OHS legislation, the onus is on the firm to take all reasonable measures to manage process risks. Hazard assessment at the process design stage is therefore mandatory.

## Legal provisions and safety standards

The strategic alignment channel indicates that companies are required to be legally compliant and socially responsible by forces in the external environment. Management must therefore have a sound knowledge of the legislative provisions that apply to their operations. In each company there should be a resource system which identifies the acts and regulations to which it must comply. The resource should also include relevant codes of practice and standards. Codes are documents describing how the workplace may comply with legislation and are generally negotiated between employers and unions. They are not legally binding but may be used as evidence in court proceedings.[1] There are two groups of standards – those supported by legislation which are legally binding and those which are intended as guidelines. It is important to make these resources readily accessible to employees, and firms should encourage workers to increase their levels of safety knowledge and consciousness.

The post-1980s approach to building a legal framework for OHS in Australia has been to develop 'enabling' legislation. From country to country, the 'style' of legislation may vary from 'enabling' to 'prescriptive'. The Australian approach has involved establishing all-encompassing provisions that define employers', employees' and manufacturers' duties of care. The relevant sections cover employers' responsibilities in relation to the development of work practices, for example, the Occupational Safety and Health Act 1984 (Western Australia) states that:

> 19. (1) An employer shall, so far as is practicable, provide and maintain a working environment in which his employees are not exposed to hazards and in particular, but without limiting the generality of the foregoing, an employer shall –
>
> (a) provide and maintain workplaces, plant, and systems of work such that, so far as practicable, his employees are not exposed to hazards;[2]

In this legislation, processes are referred to as 'systems of work'. Regulatory authorities have the power to bring companies before the courts. The courts then determine whether a firm has breached its duty of care by assessing if it undertook 'practicable' measures to prevent the incident. The Act defines this term as:

> 'practicable' means reasonably practicable having regard, where the context permits, to
>
> (a) the severity of any potential injury or harm to health that may be involved, and the degree of risk of it occurring;
>
> (b) the state of knowledge about
>
> (i) the injury or harm to health referred to in paragraph (a);
>
> (ii) the risk of that injury or harm to health occurring; and
>
> (iii) means of removing or mitigating the risk or

mitigating the potential injury or harm to health; and

(c) the availability, suitability, and cost of the means referred to in paragraph (b) (iii).[2]

Where the regulatory authority finds a breach, prosecution action may be initiated. The penalties imposed through this process are separate from claims of negligence by a third party against the company. In these circumstances a court of law will test whether the firm has been negligent using the 'reasonably foreseeable' and 'proximity' tests.

This legislation is performance-based because it allows and encourages firms actively to develop their own workplace-specific safety standards. Firms committed to self-regulation and to the achievement of higher levels of safety and social responsibility than required by external forces exhibit proactive strategies such as planned maintenance, inspections and system improvements. The extent of this self-regulation, which involves performing at a level above the minimum norms of legislation implying that little intervention is required from the regulator to define firm behavior, depends on the decision-making processes that occur within the business.

Across the business sector there are varying degrees of compliance and self-regulation. In Fig. 3.1, this range or continuum has been narrowed down to four levels which illustrate companies' choices in relation to the development of work practices. Some firms are noncompliant which means that they are highly tolerant or ignorant of risk and take a reactive approach on a needs basis. Safety is rhetorically referred to as an organizational

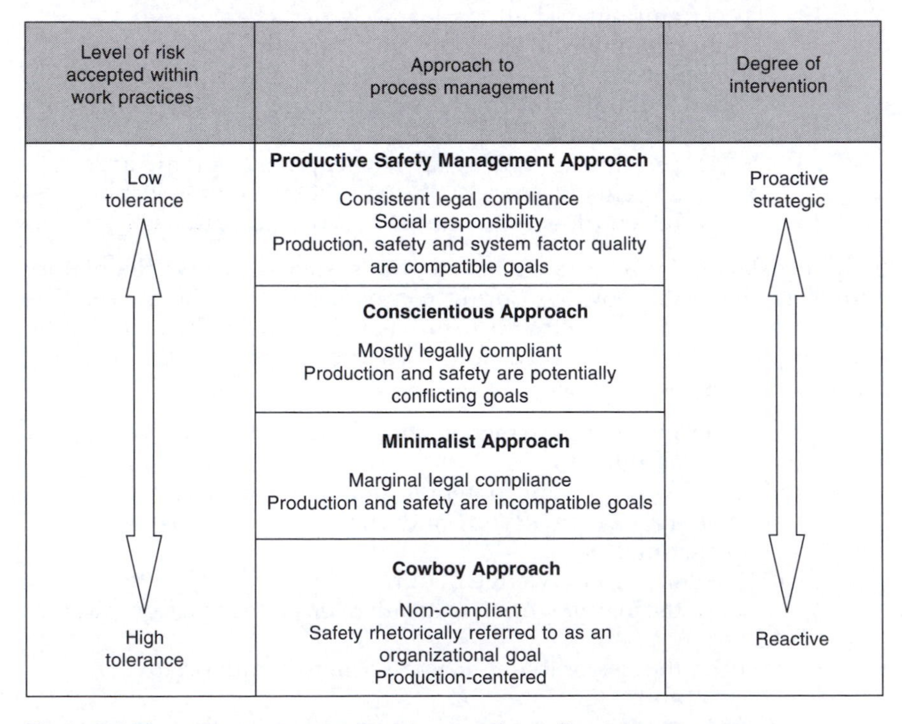


**Figure 3.1** The process management continuum and compliance outcomes

goal but the firm is fundamentally production-centered and focuses solely on short-term profit without considering long-term strategies. This is referred to as the 'cowboy approach'. The next category is the 'minimalists' who generally achieve legal compliance, however, at the 'coalface', production and safety are incompatible goals because the commitment to safety is only superficial. It takes second place to output goals.

From the regulatory authority's perspective this has implications at three levels in terms of how the regulations are used. The lowest level involves use of enforcement strategies to raise the performance of noncompliant firms. At the intermediate level are those firms that are generally abiding, however, through lack of knowledge or resources may slip below requirements, in which case the regulatory authority may serve as both watchdog and educator; thus the strong drive to disseminate information by organizations such as WorkSafe Western Australia. The upper level of regulatory application is for those firms that are safety conscious to apply the statutes as the benchmark of compliance above which they consistently perform. This allows the regulator to focus on information provision and support.

In Fig. 3.1 this 'conscientious approach' demonstrates low tolerance of risk and greater application of proactive strategies. The firm achieves a high level of legal compliance. Production and safety are potentially conflicting goals because OHS operates as a separate division of management and is not integrated into the total management system. At the top of the continuum is the 'productive safety management approach'. It is characterized by consistent legal compliance and social responsibility, commitment to self-regulation, and a total management system that reinforces production, safety and system factor quality as compatible objectives. OHS management is proactive, integrated and strategic.

Both the productive safety management and conscientious approaches involve self-regulation with the former being a holistic integration of OHS, maintenance, environmental and HRM systems. These firms set workplace-specific standards using the 'control loop',[3] shown in Fig. 3.2, which is the traditional mechanism for this activity. The first step is to set the standard. The results are then measured and compared to the standard so that if there is a discrepancy, corrective action is taken. The control loop is a very effective tool because it involves evaluation and modification of processes so that they are continuously improved. It is compatible with the entropy model as it helps to eliminate entropic risk and manage residual risk by defining the conditions under which it is acceptable to carry out work practices.

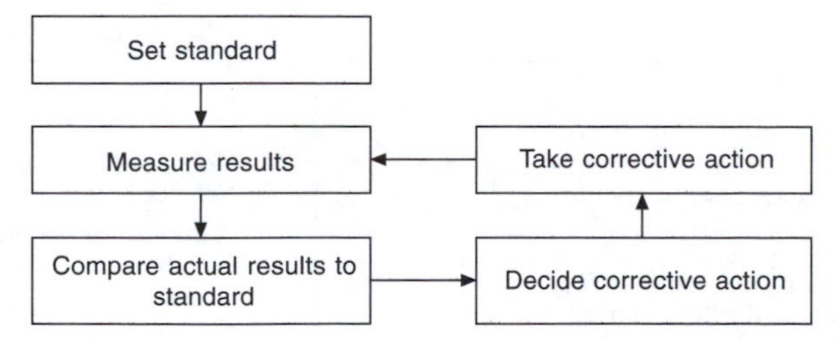


**Figure 3.2** The control loop

## Work practices

The entropy model indicates that all system factors, including processes, are subject to degradation and have an inherent residual risk, so therefore, it is important to regulate work practices through standardization as much as is practicable. In addition, this system factor allows technology, the physical environment and human resources to interface. As a result, processes are the means by which entropic and residual risks associated with these other system factors are able to interact. Processes allow various risk factors to combine, for example, the practice of driving a car brings the driver, vehicle and environmental conditions, including road and traffic systems, together. The levels of risk associated with each of these system factors determine the total level of risk.

Table 3.1 summarizes the sources of process risks and strategies with which they may be managed. It isolates interfaces such as process/ technology from the other system factors to break risk down into its component parts. An example is the best way of explaining the reasoning behind separating the system factors. Assume the process is 'climbing a ladder'. This activity involves a residual risk because of the height gained during the practice. The height defines the level of process residual risk, so therefore, climbing 30 meters has a much higher residual risk than climbing 3 meters. Other system factors then need to be considered. Firstly, the type of ladder will make a difference to the total level of residual risk. A long, open ladder, for instance, will be more dangerous than one that is caged and has regular rest platforms. The design of the technology therefore affects the level of residual risk associated with its use. Secondly, the physical environment is also important. There is a greater inherent danger in undertaking this process in a confined, poorly ventilated space than in a more open, controlled environment. The fit between the process and the environment thus affects the level of residual risk. Finally, the competencies and attributes of the climber have a bearing on inherent threats. The dangers will be greater for example, if the climber is a risk-taker or is unfit. A person in poor physical condition has a higher residual risk for this particular process than someone who is strong and healthy.

Table 3.1 shows that residual risk stems from the interaction of processes with each of the other system factors. The process/technology residual risk, for instance, has to be analyzed independently of human behavior

**Table 3.1** Strategies to manage the risks associated with processes

| Source of risk | Risk management strategy |
|---|---|
| *Residual risk* | Job safety analysis<br>Workplace inspections<br>Hazard inspections<br>Audits<br>Incident investigations |
| Process/Technology interaction | |
| Process/Physical environment interaction | |
| Process/Human resources interaction | |
| *Entropic risk* | |
| Process/Technology interaction | |
| Process/Physical environment interaction | |
| Process/Human resources interaction | |

because it is not possible to develop work practices that eliminate the risk associated with the use of technology. In other words, no procedure that involves technology is 100 per cent safe, particularly as it is impossible to predict the long-term consequences of use with certainty. For instance, a task carried out by remotely operated equipment will still have some residual risk due to unforeseeable residual risks. Each time the process is undertaken there will also be minor deviations that may cause a negative impact to accumulate on organizational systems and there is also some wear and tear on the equipment.

The process/physical environment interface has a residual risk because the workplace cannot be modified to the extent which allows work practices to occur in an ideal situation. For instance, outside work involves exposure to the elements while inside work leads to exposure to artificial atmospheres and lighting. The fact that processes cannot be developed to the extent that all risk is eliminated means that there is always a level of residual risk. Some processes, for example underground mining, have a high level of danger because the physical environment is not a good fit with the process. It can lead to destabilization which in turn prevents the process from being undertaken safely and efficiently. Finally, work practices cannot be developed to match perfectly the KSAs of individual employees. The greater the complexity of the task, the more demands are placed on the competencies of the worker to manage any inherent risk.

As shown in Table 3.1 process entropic risks stem from these same interactions. When other system factors degrade unchecked, the firm has to modify work practices to compensate for this. It is often the failure to modify processes in time that leads to an incident. An example is the death of a mechanical fitter in 1994 at an open-cut mining operation. He was trying to locate a fault in the rear swing gearbox of a rope shovel (large mining excavator). The deceased was looking through an inspection opening on the side of the gearbox,[4] being unaware that at the same time the shovel was being swung around using a bulldozer to push the bucket of the shovel. The deceased was trapped by the head in a nip point which was formed by a protruding component of the slowly rotating hoist drum and the rope guard. When the swing brake was released, a master switch caused the hoist drum brake to also be released. It was found that this tragedy was largely attributable to the lack of procedures and control of the operation being performed. There were no standard safe working procedures in place for undertaking this potentially hazardous task. Procedures were also lacking to ensure effective communication by radio, voice or visual contact, and the control of the operation was not solely in the fitter's hands. The hazards associated with the process had not been identified prior to its being undertaken.

A number of risks involved in this process can be identified using the entropy model. There were residual risks: firstly, at the nip point where the employee was trapped, and secondly, in the master switch which released both the hoist drum and swing brakes concurrently. The process itself had a high level of inherent risk because it had not been standardized and had high degrees of uncertainly and complexity. It was evident that the process/technology interface was degraded because the dipper bucket controls were no longer operational and a bulldozer was required to move it. Normally, the operator of the rope shovel would have controlled this process. Entropy was also present in the process/human resources interaction as is evident by the lack of communication. Under the circumstances, it was impossible for the deceased to anticipate all the risk factors present. Breaking down these risk factors into component parts – the system factor interfaces – provides a better understanding of the hazards involved and a more comprehensive approach to remedying these hazards.

## Risk management practices

Process risks can be identified and managed using four risk management practices, and followed up using post-incident practices such as incident investigations, which will be discussed later in this chapter. These risk management practices are JSA, workplace inspections, hazard inspections and audits. The purpose of these activities is to anticipate or identify risks before an incident occurs, and to eliminate, compress or manage the hazards to which workers are exposed using the four-fold strategy. These practices also contribute to the safety of persons who are not directly employed by the company such as visitors and contractors' workers.

### Job safety analysis

Work practices make up the total activity chain that turns inputs into outputs in the business. For example, a steel fabrication firm uses inputs such as labor, machinery and materials to manufacture building products. Work practices can be further broken down into tasks and these are comprised of steps. For instance, the changing of a vehicle's grease and oil is a task undertaken in a maintenance facility, and involves sequential steps including draining the existing lubricants, cleaning the engine and replacing the grease and oil. JSA is the review of each step in a task to identify the risks involved and is a critical practice in risk identification and management. Under productive safety, JSA is used to identify entropic and residual risks, not just 'risk' in a broad sense, and remedies are developed using the four-fold strategy provided by the entropy model.

Job safety analysis can be carried out at any time but is particularly required prior to new work practices being introduced, for example, when a firm starts manufacturing an additional product line that involves the use of new technologies. These changes affect system factor risk levels and also the risks associated with the interaction of these system factors. For example, the new technology may have a lower degree of residual risk which results in a change in the overall level shown in the model, however, it is also likely to cause a short-term rise in entropic risk through the technology/human resources interface. This is because it takes time for employees to become competent in the operation of the new production line. Entropy may also increase if the new technology is a poor fit with the physical environment, for instance, if it allows a greater rate of output.

In the short term, the company may have insufficient storage capacity to stockpile this product safely.

Job safety analysis can also be carried out as part of the auditing process. When there are obvious changes in the quality of system factors, for example, when hazards are identified or when safety concerns are raised, JSA should be used to identify the risks involved and to develop remedies. In firms that actively promote safety consciousness, any member of the organization may initiate a JSA. Workers are encouraged to take ownership of their own safety and the safety of others involved in or affected by the operations of the business.

A team of employees from the relevant workplace usually undertakes the JSA led by the supervisor and/or an OHS specialist. They make a list of the component steps within the workflow and analyze each of these for hazards. Traditionally, the risk assessment code (RAC)[5] or the risk scorecard (RSC),[6] shown in Fig. 3.3, have been used to evaluate the risk. (An alternative is to use the productive safety methodology for quantifying relative risk levels, presented in Chapter 7.) From these results, risk management strategies are formulated and incorporated into standard, safe work procedures.

The RAC evaluates the likelihood of an accident occurring and links it with the potential severity of the consequences to determine the urgency of response required. For example, if a hazard is 'likely' to result in 'moderate' damages and/or injuries, then it needs to be corrected using a level 2 intervention – immediate substitution or isolation. The RSC is more comprehensive and assesses the variables – probability, exposure and consequences – to determine a risk score. A line is drawn linking the probability with the exposure, for example, a hazard that is 'quite possible' to result in an incident to which the worker has 'occasional' exposure reaches the bottom of the tie line. The line is then extended to cut the relevant consequences point, for instance 'serious injury' with a resultant score of approximately 250. A hazard of this nature requires 'immediate' rectification by 'management'.

The entropy model provides an alternative approach to hazard control and reduction using the four-fold strategy. This involves eliminating entropic risk using corrective action in the short term, developing maintenance systems to prevent future degradation as a long-term strategy, managing residual risk in the short term, and compressing residual risk in the longer term. Appropriate on-going monitoring processes are also devised. Once the JSA form has been completed and approval granted by the manager, changes can be implemented. It is necessary to get managerial approval where changes require resourcing above the authorization level of the team leader or where such interventions affect more that a single work unit. This is to ensure internal strategic alignment of capital resources and consistency in the organization.

Tools like the RAC and RSC help the assessor identify how significant the process hazards are and how quickly they need to be addressed. These tools allow hazard management strategies to be prioritized. This is important given the limited resources that the company faces, as explained by the channel in Chapter 2. For example, a water leak that causes slippery floors in a high-activity work area would have an 'almost certain' likelihood

| Risk Assessment Code | | | | | |
|---|---|---|---|---|---|
| Likelihood | Consequences | | | | |
| | Insignificant | Minor | Moderate | Major | Catastrophic |
| Almost certain | 2 | 2 | 1 | 1 | 1 |
| Likely | 3 | 2 | 2 | 1 | 1 |
| Moderate | 4 | 3 | 2 | 1 | 1 |
| Unlikely | 4 | 4 | 3 | 2 | 1 |
| Rare | 4 | 4 | 3 | 2 | 2 |

Hierarchy of control for immediate action
Level 1 eliminate
Level 2 substitute or isolate
Level 3 engineering controls
Level 4 administrative controls or safe work procedures or personal protective equipment

Risk scorecard

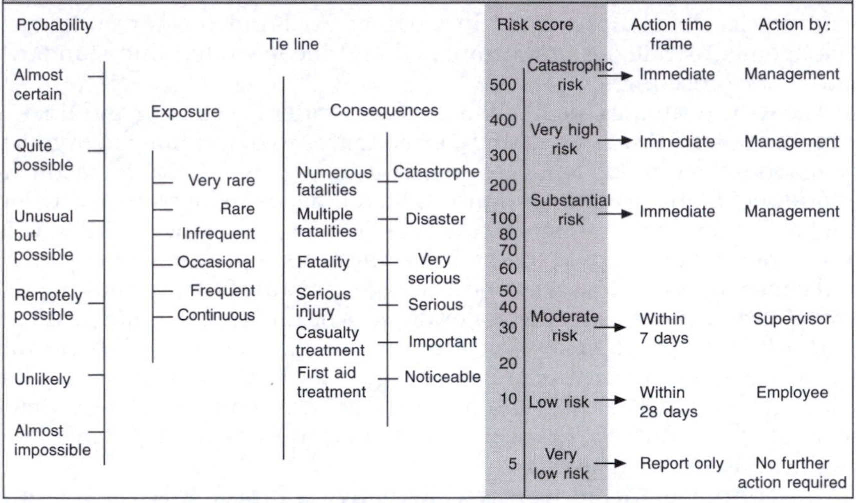

Probability: Almost certain; Quite possible; Unusual but possible; Remotely possible; Unlikely; Almost impossible

Exposure: Very rare; Rare; Infrequent; Occasional; Frequent; Continuous

Tie line

Consequences: Numerous fatalities – Catastrophe; Multiple fatalities – Disaster; Fatality – Very serious; Serious injury – Serious; Casualty treatment – Important; First aid treatment – Noticeable

| Risk score | | Action time frame | Action by: |
|---|---|---|---|
| 500 | Catastrophic risk | Immediate | Management |
| 400, 300 | Very high risk | Immediate | Management |
| 200, 100 | Substantial risk | Immediate | Management |
| 80, 70, 60, 50, 40, 30 | Moderate risk | Within 7 days | Supervisor |
| 20, 10 | Low risk | Within 28 days | Employee |
| 5 | Very low risk | Report only | No further action required |

Risk assessment code and risk scorecard adapted from resources given in refs 5 and 6, respectively.

**Figure 3.3** Risk assessment tools – risk assessment code and risk scorecard

of leading to an incident with 'moderate' consequences. The RAC and hierarchy of controls indicate that this hazard would need to be 'eliminated'. Work should cease until repairs are undertaken and the clean-up completed. In contrast, a 'minor' isolated leak in a nonwork area would have an 'unlikely' consequence of causing an accident in the short term and 'moderate' consequences. This could be corrected with 'engineering controls' or repairs with a lesser degree of urgency according to this approach.

The hierarchy of controls can similarly be used to manage residual risk and develop strategies for compression. The entropy model supports the application of this risk management tool and indicates that, independently of the likelihood or consequences, higher interventions are preferable to lower ones. Elimination of risks as far as is practicable is always preferable to applying lower order controls such as administration. The productive safety firm takes a longer-term perspective and pursues risk reduction by choosing higher-level interventions wherever possible and by applying the four-fold strategy for on-going hazard control. A JSA form for productive safety management is provided in Fig. 3.4.

At the top of the JSA form the process is identified. This involves describing what is done (the process), where it is done (the physical environment), what equipment is used (technology) and who undertakes the work (human resources). After this a list detailing the sequence of job steps from start to finish is generated so that each step can be analyzed for the hazards involved. These are described and identified as either residual or entropic risks. The RAC is then used to evaluate the level of risk, or alternatively, the method described in Chapter 7 can be applied. Control measures are developed using the four-fold strategy and transferred into the action plan.

Job safety analysis is the most important process in managing workplace risks. It should be undertaken whenever new work practices are introduced, when new technologies are installed and when the physical environment is significantly modified. New employees should also be taken through the process during induction to increase their awareness of workplace risks and to give them these analytical skills.

The failure of companies to carry out JSA prior to undertaking nonstandardized tasks which involve complexity or uncertainty has had serious repercussions. There are cases that illustrate this point. For example,

| Process description: | | | | |
|---|---|---|---|---|
| Work location: | | | | |
| Assessed by: | | | Date: | |
| Sequence of job steps<br>List the steps from start to finish | Hazard identification<br>What are the hazards for each step | Type of risk<br>Entropic and/or residual | Risk assessment<br>Use the risk assessment code | Risk control<br>What can be done to eliminate entropic risk?<br>What can be done to prevent future entropic risk?<br>What can be done to manage residual risk?<br>What can be done to compress residual risk? |
| | | | | |
| | | | | |
| | | | | |
| | | | | |
| | | | | |
| | | | | |
| Recommended action plan | | | Date of completion: | Persons responsible for implementation: |
| | | | | |
| | | | | |
| | | | | |
| | | | | |
| | | | | |
| Manager's comments: | | | Date: | Manager's signature: |

**Figure 3.4** Productive safety job safety analysis form

eight refinery employees were exposed to ammonia gas from three separate incidents during the modification and recommissioning of a primary flash tank.[7] In the first incident, seven employees suffered exposure when bolts were being undone before removing a temporary relief line. The second incident involved an explosion when the main relief line was cut. Later that day, when a fan was being fitted to the relief line, further exposure occurred. The workers suffered from shortness of breath, headaches, nausea and vomiting. The primary cause of these incidents was the lack of safe work practices for conducting these tasks. The contributing factors were numerous and some had accumulated over months. Most notable were the failure to recognize the significance of process changes and incidents that had happened in previous months. Other issues were:

- The failure to identify possible gaseous hazards;
- The failure adequately to investigate fuming, fires and explosions in the area which had occurred in the period leading up to the incidents;
- The failure to ensure that the system had been correctly isolated and purged;
- Lack of adequate supervision;
- The lack of adequate gas testing procedures, protective equipment and personal gas alarms;
- The absence of written procedures;
- The failure to report initial incidents to management;
- The lack of return to work procedures after exposure.[7]

A comprehensive JSA at the times when process changes were being introduced and thorough investigation of previous incidents would have greatly reduced the probability of these three events.

## Workplace inspections

Workplace inspections are also used to monitor and manage risks. Firms committed to safety undertake these reviews habitually to identify any changes in the condition of system factors since the last inspection. There are three types of changes that need to be identified, which are:

(1) A rise in entropic risk;
(2) The potential release of residual risk that may lead to an injury or damage;
(3) Those risks that may have previously been overlooked.

Supervisors and team leaders have a responsibility to carry out regular workplace inspections. It is also essential that employees do a pre-start check of the condition of equipment and the physical environment and again after the main meal break, as a minimum, in hazardous workplaces. To formalize these responsibilities, such practices should be included in position descriptions or workplace procedures. The pre-shift and post-main break checks are particularly important because they also psychologically prepare the employee for work by 'switching on' the operator's safety consciousness.

The entropy model indicates that these inspections should focus on the condition of the four system factors. Changes in operating procedures, the condition of plant and equipment, the safety, cleanliness and tidiness of the physical environment, and the health, competence and alertness of employees should be evaluated. The firm can take a strategic approach to the scheduling of additional workplace inspections by reviewing past incident statistics. From these it can determine the times in the work cycle with the highest probability of entropy as evident by the prevalence of accidents and near misses. The entropy model can be used to track these data as shown using the hypothetical example in Fig. 3.5.

Part (a) illustrates the number of undesirable events that the company has experienced over a year against the time that these incidents occurred. It shows that over a period of 24 hours, comprising three 8-hour shifts, system factors have experienced variable levels of entropy. As a result, certain times have higher incident rates than others. Part (b) applies the entropy model to illustrate how the probability of an accident has varied

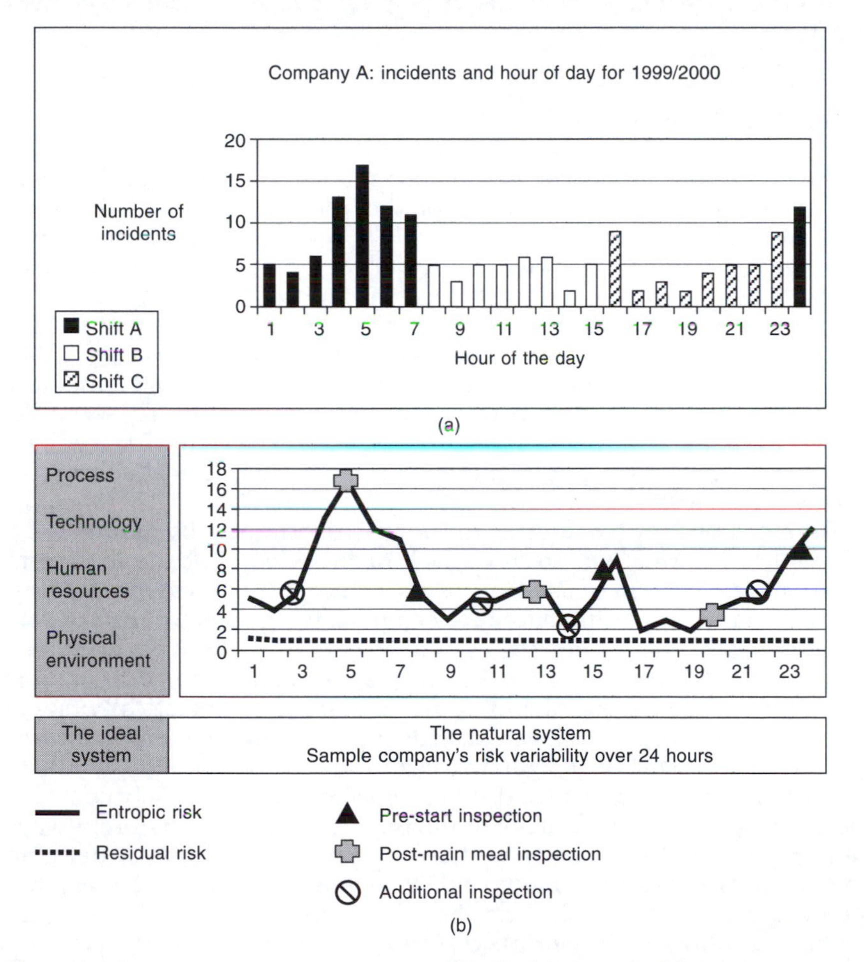


**Figure 3.5** Using the entropy model to track entropic risk and to schedule inspections

historically in this company over a 24-hour period. This information is used to determine the best times to carry out workplace inspections in order to prevent incidents. Additional inspections should be carried out before the peak incident times in each shift. On shift A, inspections are required at the 11.00 p.m. pre-start, at 2.00 a.m. prior to the tendency for rising degradation, and at 4.30 a.m. after the main meal break. The causes of these peaks should also be investigated so that work practices can be modified to prevent this degradation. For example, it may be found that fatigue is the main contributing factor to the incidents occurring between 3.00 and 7.00 in the morning. The company can modify work practices to prevent this entropic risk by, for instance, giving operators a 10 minute break prior to this rise. The figure also shows when the pre-start, post-main meal and additional inspections should be carried out for shifts B and C. Clearly, in this example, shift A, which occurs during the night when physical environment residual risks are higher and workers have a greater tendency towards degradation due to fatigue, has a higher susceptibility to incidents and requires more intensive risk management practices.

### Hazard inspections

Workplace inspections involve a general review of the four system factors in a work area. Hazard inspections, on the other hand, specifically focus on a particular risk source. Traditionally, these reviews have been concerned with the condition of either technology or the physical environment, for example, inspections of the crushing plant and slope stability at a mine site, respectively. The basic underlying premise of hazard monitoring is that there is an 'acceptable' level of risk. There are measures to quantify this level such as maximum acceptable concentration (MAC), permissible exposure limit (PEL) and short-term exposure limit (STEL). These cutoffs are based on the assumptions that exposure to levels below the defined value will have no long-term effects on the worker and that short-term effects are reversible.[1]

As explained in Chapter 2, firms are constrained by limited and incomplete information. Society's lack of knowledge of the long-term consequences of time-related exposure to hazards is evidence of this. There is, therefore, a difference between perceiving something to be safe and it actually being safe. The current debate over the risks of mobile phone use, which was discussed earlier, is an example. In addition, it is important to be aware of individual differences; specifically, some employees have lower tolerances than do others to environmental hazards. With the rising levels of allergies in western societies, variability in human resources capacity due to individual susceptibilities will become an increasing significant workplace health issue. It may not, therefore, always be appropriate to take a standardized or generic approach to defining hazard limits as harm may be dependent on individual physical or psychological responses. While the firm should still apply practicable measures for hazard control in the workplace, OHS practitioners in particular, need to be aware of the variability in tolerances among individual workers.

On an organizational scale, the primary purpose of hazard inspection is to monitor changes in the entropic and residual risks associated with the hazard. An example of the former is the level of emissions from a chimney-stack. If the scrubbers in the stack are failing to clean emissions effectively, this represents an increase in entropic risk; the scrubbers are below optimal safety, performance and system factor quality. High winds at an airport, which affect the safe landing and departure of aircraft, represent an increase in residual risk. When hazardous conditions rise above the industry standard 'acceptable' level, the probability of an incident or the threat to human health and safety has also risen. The greater this rise, the greater the need for urgency to correct the condition or to modify work practices to avoid the risk.

Traditional approaches to hazard inspection have focused on unsafe conditions in technologies and the physical environment. The entropy model suggests that hazards are also present in processes and human resources. There is a need, therefore, to consider also these system factors as potential sources of danger. For example, in the case of the mechanical fitter who was trapped while working on the rope shovel, the investigation revealed that the process was extremely dangerous and a pre-operational hazard inspection should have been undertaken before the work was attempted. Any work practice that is complex or that is not familiar to the worker, coupled with heavy equipment and/or a dangerous physical environment, is high risk. In addition, human resources can represent a hazard, for instance, a new group of trainees entering the workplace. Research has shown that young workers have a higher rate of injuries than older workers.[8] Accordingly, this group should be treated as a potential hazard and regularly monitored to ensure their safety and to prevent their limited competencies impacting the safety of others. Under productive safety management, hazard inspections apply to all four system factors, not just unsafe conditions.

## Audits

The third type of workplace inspection used to monitor risk is the audit. Audits are major reviews of the business' operations and are one of the most important processes in OHS management. The traditional definition of auditing is:

> Auditing is the process designed to improve safety performance by identifying deviations from what is safe.[3]

The prevailing approach is to look for deviations that represent unsafe acts and/or unsafe conditions. Current auditing practices cover such areas as: document control, purchasing criteria, workplace design, the condition of plant/equipment, emergency plans, hazard inspections, incident investigation, training and contractor management.[9] Productive safety management takes a more strategic view of auditing. It covers all the areas that are currently analyzed and breaks them into a more structured set of processes. Auditing is redefined as follows:

Auditing is an information gathering and evaluation activity involving three processes:

(1) Identifying the sources of risk that the organization is exposed to as a natural system;
(2) Evaluating the health of the organizational culture; and
(3) Measuring the achievement of strategies in the productive safety management plan.

The purpose of auditing is to uncover weaknesses in the total system, not just unsafe acts or conditions. These weaknesses are sources of entropic or residual risk as well as organizational management variables which hinder the achievement of optimal safety, performance and system factor quality. The first point of the audit definition involves separate reviews of each of the four system factors. This means that processes, technology, the physical environment and human resources audits are undertaken separately. (Each of these will be discussed in Chapters 3 to 6, respectively.) The reason for this approach is more effectively to identify and manage the risks within business activities.

The second point of the definition is covered by the behavioral audit. It examines those indicators of organizational culture which determine the extent to which OHS has become part of the total management system. The final point audits the effectiveness of the productive safety management plan including the appropriateness of strategies, action timetables and measures used to lift system factors towards the optimal state. Under productive safety management, therefore, auditing is not limited to what is currently unsafe. It evaluates the OHS management system in its entirety and results are used to forecast trends and set future goals. (This is discussed in Part 4.) Audits are used to gather information and to provide direction for strategy development and for continuous improvement.

## Process audits

The process audit begins with the identification of the work practices undertaken by the business. There are two options for gathering this information. The first is to compile a list of outputs of each business unit/workplace and determine the processes involved in producing this output. The second option is to look at the position descriptions for each workplace to identify the tasks that are undertaken. It is also important to speak to relevant supervisors to ensure that these provide a complete profile of work practices. The processes can be further classified into those that are currently standardized and those that are not. Figure 3.6 illustrates these steps.

When conducting the process audit, the list should be used for guidance only and should not restrict the auditor's ability to identify additional work practices which require analysis. The check includes the following questions:

(1) What processes are currently undertaken in the workplace?
(2) Which processes are standardized and which are not?

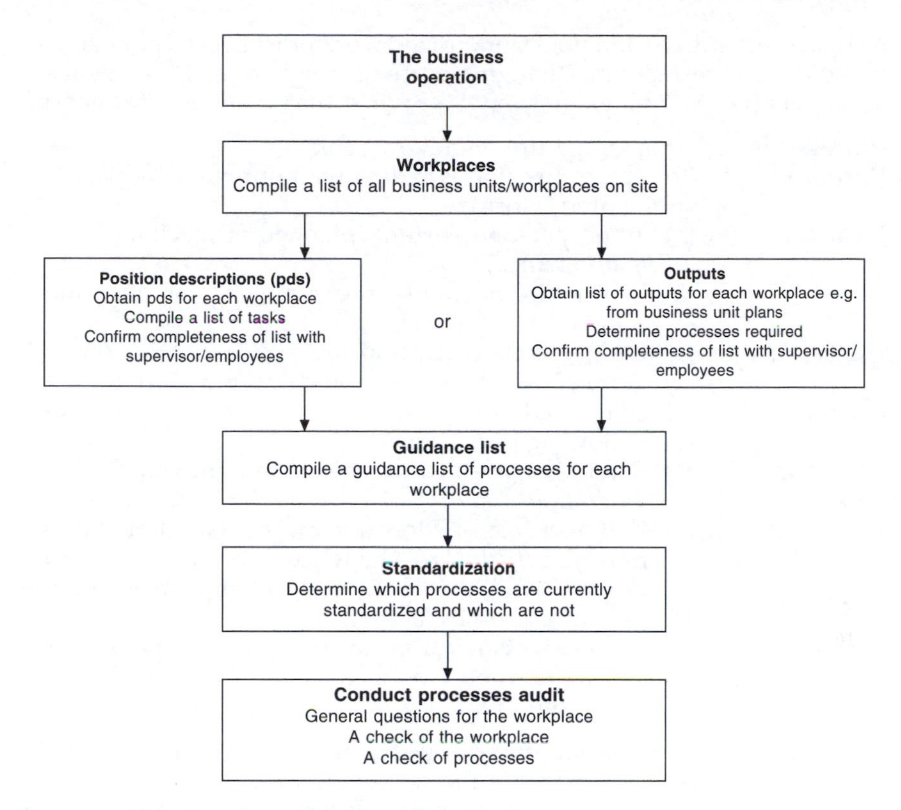


**Figure 3.6** Preparing for and conducting a process audit

(3) For those that are standardized, do any steps lead to entropic risk and what are the residual risks involved?
(4) For those processes that are not standardized, should they and can they be standardized?
(5) What is the potential for entropic risk and what are the residual risks associated with these nonstandardized tasks?
(6) Are standardized tasks adhered to?
(7) What have been the outcomes of each process? Have there been any negative consequences from them including lost time injuries (LTIs), near misses, poor productivity or poor quality?

The process audit uses JSA techniques and also applies the seven process investigators (The seven Ps). These are:

| | |
|---|---|
| Process: | What is being done? |
| Purpose: | Why do it? |
| Person: | Who is doing it? |
| Place: | Where is it being done? |
| Period: | When is it being done? |
| Procedure: | How is it being done? |
| Performance: | What are the outcomes – safety, productivity and quality? |

Both standardized and nonstandardized work practices can be analyzed using these investigators. Changing a tire on a light vehicle can be used as an example. Applying the seven Ps provides the following information:

| | |
|---|---|
| Process: | Changing a tire on a light vehicle. |
| Purpose: | To replace tire that is in less than optimal condition (punctured/worn). |
| Person: | (1) Tradesperson assistant (planned or reactive maintenance).<br>(2) Vehicle driver (unplanned maintenance following a puncture). |
| Place: | (1) Light vehicle workshop.<br>(2) Place chosen by driver following puncture. |
| Period: | (1) Scheduled into maintenance workshop activities.<br>(2) Following puncture. |
| Procedure: | (1) Refer standardized procedure in light vehicle workshop.<br>(2) Pull over to a safe location off the road. Use the jack to raise vehicle. Use the wheel brace to remove tire. Replace with the spare tire and tighten. Place the punctured tire in the vehicle. |
| Performance: | (1) No incidents reported for this process in past year.<br>(2) One injury reported when the driver injured his back lifting the tire. |

The auditors look at each procedural step and identify areas of risk. In the above example, the review of the workshop practices has identified another circumstance when this process is undertaken – the replacement of a punctured tire by nonmaintenance personnel. The process hazards, which include the risk of injury when lifting the tire, do not only occur within the workshop but also occur when employees are traveling between operations or while commuting to and from work.

Work practice risks can be further evaluated using 'what if' analysis. This involves questioning the factors that may cause deviations from optimal process safety. 'What if' analysis links the process to other systems factors by evaluating how they might affect how the practice is carried out and the consequences of deviation. For the tire change example, questions include:

(1) What if the vehicle driver does not know how to change a tire safely?
(2) What if the workshop is fully occupied, are tire changes carried out elsewhere?
(3) What if the vehicle driver is stranded on a major road? What addition risks are involved in the process?
(4) What if the vehicle driver has a back condition?

'What if' analysis is a method of stimulating thought about possible risks and appropriate remedies. For example, a response to question (1) above may be to introduce basic maintenance training and manual lifting techniques for those employees who regularly drive a company vehicle. Once risks have been identified for each process, these are documented and recommendations made to correct them using the four-fold strategy.

Strategies can be prioritized using either the risk assessment code given in Fig. 3.3 or the systems risk profile in Chapter 7.

## Visitor safety

The risk management practices discussed so far have focused primarily on protecting the health and safety of direct employees. The company is also responsible for the welfare of other individuals including visitors. Visitors are any persons who are not an employee or a contractor and include students on work experience programs, family members, children who are in the workplace for any reason, clients and customers.[10] Having people on-site who have no or limited knowledge of the risks which are present leads to additional hazards. Wherever there is a dilution of the average level of competency in a work area according to the number of people in attendance, there is an increase in residual risk in the human resources system factor and greater susceptibility to entropic risk.

The entropy model indicates that entropic risk should ordinarily be eliminated and maintenance strategies developed to prevent future degradation. In the case of visitors, however, it is not possible to eliminate this risk completely because they are only on-site for a short period of time. It is, therefore, important to develop strategies to minimize the risks that they introduce. These strategies are of three types. The first is the imposition of restrictions and controls on visitor access and behavior such as the tight regulation of arrivals and departures and restricting visitors from high-risk areas.

The second strategy is to provide constant supervision, particularly in hazardous worksites. The firm can establish accountabilities for visitor safety through its visitor policy, detailing the conditions under which outsiders may access the site, and through its 'sign-in' procedures. Visitors should be required to come to a designated point of entry, for example, the front gate or reception, to nominate the company officer they are to meet, sign in, and thus thereafter, the visitor becomes the responsibility of that officer until departure.

The third strategy is to develop communication systems that minimize this risk. Information is circulated in two ways. The first is by providing visitors with instruction that lifts their level of safety consciousness. This is to counter the entropic risk they introduce and should include a written statement documenting any restrictions that apply while on site[11] and an outline of emergency procedures. The second issue of communication is to advise relevant employees that visitors will be in their workplace during specific hours of work.

There are occasionally circumstances where visitors are at the operations for a number of weeks or months. In this situation it is essential that these people are treated like employees and put through induction. The failure to do so can result in an accident, for example, a serious incident resulted at an open pit mine site when a visiting researcher was driving too closely behind a haul truck. The truck was overloaded and materials began to fall from the tray due to the change in the center of gravity when the truck was going up a ramp. Ordinarily, mine workers are required to obtain a pit permit to drive in these environments, which specifies safe distances between trucks and other vehicles, but this was not done in this case. A rock hit the researcher's vehicle and he sustained serious facial injuries. Wherever possible, visitors should be prevented from contact with high-risk situations, and if this is not practicable because of the nature of their visit, they should be provided with the information, training and supervision to minimize the probability of them negatively interfacing with other system factors.

## Contractor management

Visitors and contractors' employees are groups for whom the firm has responsibility even though they do not come under the direct employ of the company. The relationship between the principal company and the contractor's employees is secondary. Legislation, however, extends the principal's duty of care to cover these workers. For example, the Occupational Safety and Health Act 1984 (Western Australia) states:

> 19.
> (4) For the purposes of this section, where, in the course of a trade or business carried on by him, a person (in this section called 'the principal') engages another person (in this section called 'the contractor') to carry out work for the principal –
> (a) the principal is deemed, in relation to matters over which he has control or, but for an agreement between him and the contractor to the contrary, would have had control, to be the employer of –
> (i) the contractor; and
> (ii) any person employed or engaged by the contractor to carry out or to assist in carrying out the work; and
> (b) the persons mentioned in paragraph (a) (i) and (ii) are deemed, in relation to those matters, to be employees of the principal.
> (5) Nothing in subsection (4) derogates from –
> (a) the duties of the principal to the contractor; or
> (b) the duties of the contractor to persons employed or engaged by him.[2]

To comply with this legislative requirement, principal firms have to take practicable measures to protect the health and safety of the contractors' employees, at each of the five stages in the life of the contract. These are writing tenders and contract specifications, evaluating tenders, pre-commencement planning of work, on-site control and monitoring, and review and improvement of the contractor management system.[12] At each stage, the principal makes decisions which affect its and the contractor's internal operations and the relationship with the contractor. If, for example, the principal selects contractors primarily on price, this constrains the contractor's ability to develop and implement effective risk management strategies. As safety issues arise during the life of the contract there will be an increased potential for conflict between the parties because of these resource constraints. The criteria upon which contractors are selected, therefore, have implications for the principal–contractor relationship. In particular, it determines the degree of alignment in this agreement that affects how well the firms can work together to achieve common goals, particularly those related to safety.

Price and the resultant quality of service delivery are key factors in this relationship. A further factor is proximity, which refers to physical distance and also, more importantly, to the extent of communication between the parties. The greater the interaction between them, the greater the likely impact on each other's cultures. In some circumstances, this contact is not limited to the work context. In remote single-industry townships where there is a principal company with owner-operated and contractor-operated sites, it is essential that the principal ensures that both types of operation have similar standards for health and safety because employees at both firms have the opportunity to interact professionally and socially. Any disparity in these standards may erode the credibility of the principal's OHS management strategy and thus have a negative impact on its organizational culture.

Those firms that conscientiously fulfill their duty of care provide their contractors with assistance in OHS management. This includes: verbal advice, training, written guidance material, examples of compliance documentation, proformas and checklists, feedback from audits and on-going support.[12] In effect, they increase the level of proximity in the principal–contractor relationship. Studies have found a number of benefits where governing firms increase communication and support. Firstly, principals feel more confident in their ability to demonstrate due diligence.[11] In addition, contractors understand risk management better and there is enhanced awareness of the need for effective OHS management along the contractual chain – from principal to contractor to sub-contractor. Research indicates that standards among contractors are gradually improving with reductions in LTIFR and also with some improvements in productivity. Contractors have indicated that initially additional time and resources are required to establish an OHS management system, however, this is generally a one-off cost[12] with benefits on-going thereafter. Generally speaking, the larger established mining contractors in Australia have well-structured systems. The smaller operators tend to be the firms that require most support from regulatory authorities.

In terms of organizational decision-making, the principal controls many of the variables that affect contractors' safety standards. The governing firm is able to fix the minimum standards of safety required of its contractors through a legally binding agreement. The principal also has the authority to define key parameters in the relationship such as the reporting structure and the degree of autonomy of the contracting company. For example, the principal can have one or more of its own employees such as a project manager or technical staff permanently on-site where the contractor operates to establish a direct reporting structure between the principal's designated manager and the contractor's manager. Under these conditions, the degree of accountability and of information sharing is greater than when the relationship is more at arms length. There are also more opportunities for the integration of cultures that particularly affect contractors' decision-making processes. In the first instance, the contractor has the benefits of increased assistance, but secondly, it faces a set of guidelines additional to the legislative requirements imposed by the external environment. These are the operation-specific standards defined by the principal. Through contract pricing, however, a large proportion of this financial cost may be borne by the principal firm.

Unfortunately, in current practices there can be a dilution of standards the further the relationship is from the governing company. This is because the contractor experiences conflicting pressures – legal requirements which demand 'acceptable' levels of OHS management and market pressures which focus on pricing. Contractors are often required to have an OHS management system as a precondition of tendering on major projects,[13] however, the same standards are not necessarily applied between the contractor and sub-contractors. Sometimes contractors require evidence of an OHS management system but fail to follow through with the assistance and training to establish a consistent approach amongst sub-contractors.[13]

Additionally, there is limited incentive for the principal to impose its own standards of OHS management on contractors and sub-contractors because of a number of issues. The first is that, as mentioned earlier, the cost of such systems is factored into the tender price and therefore the principal bears most of this cost. The second is that the greater the contractual distance between the two parties, the more difficult it is for any third party that sustains injury or loss, to claim damages against the principal. These secondary relationships make it harder for the claimant to establish negligence in a court of law using the 'reasonably foreseeable' and 'proximity' tests. A third factor is the length of the principal–contractor relationship. Some contracts are short and therefore it becomes impractical to transfer OHS systems between the two parties. In addition, there is a lack of standardization in industry with respect to contractor OHS management requirements.[13] Other barriers that hinder a consistent approach to the implementation of these systems in the contractual chain include lack of understanding of these legal responsibilities. Further problems include inadequate training for contract management staff, and lack of accountability, supervision and monitoring.[13] The contract industry is also characterized by high turnover. This is sometimes used as a justification for lower investment in training.

The external environment has also created a number of barriers to a consistent approach. As stated previously, these include a lack of standardization in OHS practices generally, which is exacerbated when competition based on price causes safety to be considered a cost rather than compatible with organizational performance. The priority for the contractor is invariably to win the tender and this process operates on short-term cycles accompanied in some cases by short-range, nonstrategic thinking. This approach has previously led to entrenched attitudes which hinder the implementation of effective OHS management systems in contractor firms.[13] As a result, this sector tends to have higher levels of risk tolerance and more reactive safety management practices than governing firms.

How can these barriers be overcome? The solution lies primarily with principal firms and the legislative framework. Competitive forces are the main drivers of contractor behavior. Price-based competition can lead contractors to tailor their submissions with the minimization of resourcing for OHS management. Where the principal requires a legally compliant and socially responsible risk management system, however, contractors will ensure that these criteria are met. Principal firms are, therefore, ultimately accountable for the standards of health and safety that occur within contractor operations. OHS legislation, in Australia, formalizes these responsibilities using general rather than prescriptive provisions so that the upper limit of accountability is not defined as such. A prescriptive approach in contrast could define the boundaries where the principal's responsibilities end and where the contractor's responsibilities start. The enabling approach allows for overlaps of accountability and the avenue for regulators to test each case on its merits. The on-going growth of the contract sector indicates that standards of compliance are an issue of some urgency for OHS legislators and professionals. Interestingly, research has associated this growth with a disorganizing effect on employer–employee relationships.[12] There appears to be a shift towards a greater degree of 'chaos' in these business partnerships and in work organization. Clearly, a more strategic approach to organizational decision-making encompassing contractor management is required.

The current intent of legislation is that both the principal and contractor are legally culpable in terms of the level of compliance of contract firms. How does this accountability affect the decision-making processes of the principal? The discussions thus far have highlighted the importance of proximity in the principal–contractor relationship. The more interdependence and communication between the parties, the more likely the two firms' organizational systems, particularly OHS management standards, will be aligned. The strategic alignment channel can be adapted to illustrate how these decision-making processes may be applied to affect positively this arrangement, as illustrated in Fig. 3.7.

The interdependencies and feedback arrows at the foot of the figure show that forces operating in the external environment influence the principal and the contractor. Both parties are required to be legally compliant and socially responsible. Inclusively, as described by legislation, for example, section 19 (4) of the Occupational Safety and Health Act 1984 (Western Australia), the principal is deemed to be the employer of the contractor and any of the contractor's workers. The principal's duty of care therefore includes a responsibility to select and manage contractors with due diligence. Consequently, the principal has to ensure that its contractors maintain system factors to an 'acceptable' level of risk.

The principal has to make strategic decisions to ensure these standards are met. It does this, firstly by including sound OHS management as a criterion in tender selection, and secondly by overseeing the contract to monitor compliance. As a result, both firms can achieve external strategic alignment. In addition, the principal can make resourcing decisions to balance the need for legal compliance against the cost of the contract. It

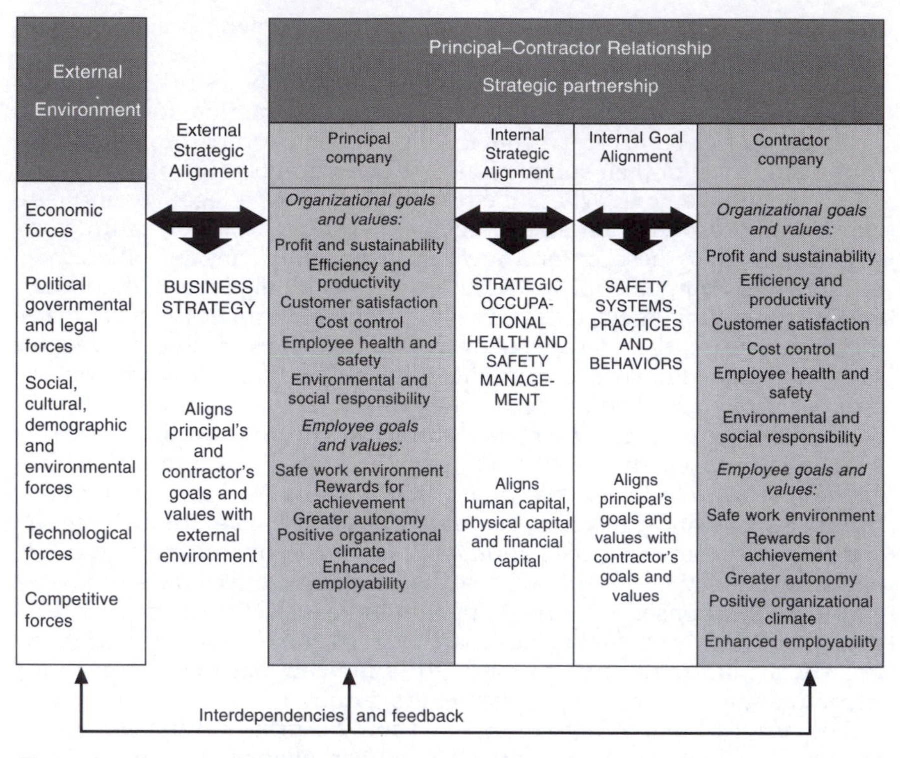


**Figure 3.7** Creating alignment in the principal–contractor relationship

can make allowances for the capital resources that the contractor has to outlay in order to implement 'acceptable' risk management practices. When standards in the final agreement reflect the standards applied by the principal in its own operations, with legal compliance as the minimum benchmark, there is congruence in the exchange of resources between the two organizations. These standards prevent the contractor from being risk tolerant while the principal is risk adverse, and also ensure that the price paid by the principal is sufficiently fair and reasonable to allow the contractor to achieve compliance. If there are differences in standards and attitudes to risk in the firms' respective management styles, their ability to pursue mutually beneficial outcomes and their communication may be affected. Alignment within the principal–contractor relationship may therefore be prevented.

Figure 3.7 shows that as socio-economic entities the respective parties have common goals. Both want to achieve profit and sustainability, efficiency and productivity, customer satisfaction, cost control, employee health and safety, and environmental and social responsibility. The workforces of the two companies also share common goals including a safe work environment, rewards for achievement, greater autonomy, a positive organizational climate and enhanced employability. Each organization consequently operates under the same decision-making rules as was shown in the channel in Chapter 2.

Within this shared framework, the principal–contractor relationship can lead to the development of compatible strategic OHS management systems which align the capital resources of both firms. This alignment occurs through the economic principle of 'utility'. 'Utility' is the basis of choice – the satisfaction or reward a product yields,[14] that fundamentally implies that the principal gets what it pays for and the contractor delivers according to how much it is compensated for. The price paid by the principal reflects the quality of service or product received from the contractor. There is therefore, an exchange of the principal's financial capital for the human, physical and financial capital of the contractor. This is the basis of the principal–contractor relationship that leads to internal strategic alignment so that the internal strategies and the resourcing balance of both firms are compatible.

If, on the other hand, the principal tries to off load responsibility for safety to the contractor and imposes price-based tendering, it cannot achieve alignment because of the disparity in standards. External strategic alignment cannot be achieved in the relationship if there is a failure to maintain legal compliance and social responsibility along the contractual chain. There is also lack of alignment in circumstances where principal firms have standards below those of the contractor. In such cases, the governing company may perceive advantages in outsourcing their operations. They are able to gain the benefits of the contractor's expertise without incurring the full cost of developing the necessary systems to manage risk. This is effectively an attempt to transfer responsibility for OHS from the principal to the contractor with the cost factored into a higher contract price. Under such circumstances, the principal may again fail to achieve alignment with external forces by neglecting to take ownership of its duty of care.

According to Fig. 3.7, alignment of goals and values allows the development of concordant safety systems, practices and behaviors. Common objectives minimize conflict between the two parties and ensure legal compliance and social responsibility. The principal–contractor relationship achieves internal goal alignment based on both parties fulfilling their responsibility for the health and safety of employees and the general public.

How can the principal assess potential contractors to determine whether an effective strategic partnership can be developed? This process begins with the preliminary stage of contract development. Firstly, the principal must include OHS management requirements in tender and contract specifications to ensure compliance with OHS legislation and with additional 'special conditions' tailored to meet the needs of the operation. General provisions for reporting, communication, notification of incidents and penalties for noncompliance should be included,[12] together with details of the OHS management system.

Secondly, as indicated by the entropy model, the principal should request information about the condition of the contractor's system factors, for example, the quality of the resources such as the technology and personnel that the contractor intends to employ. A third issue is the maintenance of system factors including whether the contractor applies safe work practices using JSA, its systems for the upkeep of technology and the physical environment, and its processes for the management of human resources, such as the use of sub-contractors. Additionally, it must be determined whether prospective contractors have systems in place to take appropriate corrective action including workplace inspections, hazard inspections, audits and incident investigations.

To ensure due diligence, the principal needs to evaluate the contractor's ability to manage risk for the life of the contract and to carry out any restorative work such as site rehabilitation required at the close of the operation. This latter issue is a prerequisite for the eventual discharge of the contract. The contractor's ability to manage risk depends on the processes it has in place to assess risks and to allocate resources for risk management. These matters should be addressed through formal planning systems that include risk management strategy development, and measures and methods to evaluate the results of its interventions. In situations where the operation is in a high-risk industry, the governing firm needs to evaluate thoroughly the contractor's strategies for the management and compression of inherent dangers.

The second stage of the contract is the evaluation of tenders. The principal should use the criteria above to evaluate the submissions. During the following stage – pre-commencement planning – monitoring is required to ensure that the selected provider puts systems in place to manage hazards effectively. Once the operations are running, on-going management and control practices are required to evaluate performance and safety, using regular site visits and formal structures for the reporting of results. The communication and feedback cycle helps to clarify and reinforce requirements and to address any obstacles to production, safety and system factor quality.

Applying the entropy model, at this contract implementation stage, the contractor's ability to prevent the degradation of system factors and to manage residual risk should be a primary concern. Matters to be monitored therefore include the induction and training of employees, practices to increase awareness such as toolbox meetings, inspection and monitoring of technology and the physical environment, the implementation of safe work practices and the contractor's self-evaluation methods such as record-keeping and internal reporting.

The final stage of the process is to review and improve the principal's contract management system. This system should therefore be accompanied by clear measurable objectives against which performance is evaluated. Any weaknesses identified during the life of the contract should be addressed to improve the quality of contract management. This may include better development of tender documentation, improved selection techniques, and more effective controls in the pre-commencement, operational and post-operational phases of the contract.

The principal–contractor relationship is formalized by legal agreement. The governing firm has to ensure due diligence by including provisions that define the contractor's responsibilities. The productive safety management approach recommends the inclusion of the elements shown in Fig. 3.8. These are divided into the four system factors, history and cultural issues, planning and communication.

The figure shows that the principal's exercise of due diligence begins

## System factors

### Compliance with:

Occupational health and safety legislation
Industry-specific legislation e.g. Mines Act
Australian standards
Special conditions required by the principal

Environmental protection legislation
Industrial relations legislation
Other relevant statutes, regulations and by-laws

#### Processes

Safe work practices

Risk management practices
- Job safety analysis
- Workplace inspections
- Hazard inspections
- Audits
- Visitor safety
- Sub-contractor management

Post-incident practices
- Incident reporting and investigation
- Evacuation procedures
- Emergency procedures
- First aid
- Workers' compensation
- Rehabilitation

Record keeping

#### Technology

Technology quality, output and reliability

Equipment maintenance

Parts availability and supplier relationships

Technological hazard management

Safety equipment

Monitoring equipment

Personal protective equipment

### History and cultural issues

Production and safety record

Fit between principal's and contractor's cultures

Management commitment and leadership

Performance management system

### Planning and communication

Safety management plan

Environmental management plan

Operational plans

Policies

Strategies and initiatives

Measures and targets

Formal reporting to principal

Reporting of special circumstances e.g. incidents, hazards and breaches

Communication between principal and contractor

Dispute resolution between principal and contractor

Discharge of contract

Employee reward systems

Commitment to training and development

Employee participation

#### Physical Environment

Environmental planning

Environmental monitoring

Nuisance management

Housekeeping

Prevention of unauthorized access

Residual risk hazard management

Accommodation and amenities

Post-operation clean-up and rehabilitation

#### Human Resources

Competence of personnel

Use of sub-contractors

Appointment of specialist personnel e.g. registered manager

Emergency response personnel

Clarity of responsibilities and accountabilities

Induction

Training

Toolbox and safety meetings

Fitness for work programs

Joint consultative committees

Industrial relations practices

Safety issues resolution

**Figure 3.8** Contractor selection criteria

with ensuring that the contractor is able to comply with OHS, industry-specific, environmental, industrial relations and other relevant legislation and standards. In addition, the contractor should meet the special conditions required by the principal that are specific to the operation.

Once these broad compliance issues have been assessed, the contractor's capacity to fulfill the elements of an effective OHS management system in the areas of processes, technology, the physical environment and human resources, should be evaluated. Primarily, these elements focus on taking appropriate corrective action, developing effective maintenance systems, the minimization and compression of residual risk, and the management of such risk by developing employees' safety consciousness and by implementing contingency plans.

The contractor's history and culture are also determinants of their suitability as strategic partners. In particular, it is advisable that the principal assesses the contractor's production and safety history, the quality of its leadership, its performance management and rewards systems, commitment to training and level of employee participation. Additionally, factors such as planning and communication, which affect the parties' working relationship, should be considered. To achieve alignment, the contractor's management and operational plans should be driven by the objective of meeting the customer's needs, as well as their own organizational goals. Further, the contractor must have a measurement system to evaluate its effectiveness in meeting the standards of compliance, output and quality required by the principal, in addition to its own financial objectives. Alignment is therefore developed and sustained through formal contractual structures and also through communication and feedback systems. Communication systems cover such issues as the formal reporting of results and other measures such as OHS deviations, procedures for dispute resolution and criteria for discharge of the contract.

Contractor management is an important risk control practice, and as is the case with owner-operated sites, the costs of getting it wrong can be substantial. For example, in 2001, the New South Wales Industrial Relations Commission in Australia fined a power supply company and an electrical contractor a total of $A118 000 for an accident in which a technician suffered electrical shock and burns, while another worker was also placed at risk. The men were installing mains cables from the transformer to an electrical switch room at the Wardell Sewage Treatment Works in March 1997 when one of them came into contact with a live 240-volt cable. The principal had granted access to unauthorized persons and the contractor had failed to provide safe work systems and appropriate training.[15] There were inadequate standards for the management of risk in this workplace and both the principal and contractor were found liable.

In contrast, the benefits of getting it right can be significant. For example, when the principal, Woodside Offshore Petroleum, set tough guidelines for contractors on the North West Shelf off the coast of Western Australia in 1989, it could not foresee the enormous benefits this would have. The company established Australia's most comprehensive safety program for sub-contractors to date.[16] In two years of potentially high risk not a single lost time injury was incurred. Since then a number of former Woodside contractors have attained best practice in OHS standards.

Part of this company's strategy was to demand that all sub-contractors demonstrate an acceptable level of safety performance on previous construction projects and it also applied consistent safety standards along the contractual chain. In addition to excellent LTIFR results, the company found a reduction in the frequency of unplanned events. The results suggest that Woodside had a sound strategy for the management of the residual risks involved in these projects, and also effective strategies for corrective action and maintenance to prevent degradation of system factors. The company established high standards and facilitated proximity so that the contractor's OHS systems and culture could be aligned with its own. Increasing the extent of alignment therefore results in on-going benefits for both contractual parties where legal compliance is used as the minimum standard of safety performance.

## Post-incident practices

All reasonable attempts should be made to prevent incidents from occurring. When an event occurs, however, the firm has to be prepared to respond appropriately to curtail further damage/injury and to take corrective action to restore system factors to an acceptable condition. The processes required to circumvent consequential damages and to reduce risk include incident investigation and reporting, emergency procedures, first aid, workers' compensation and rehabilitation. The types of incidents requiring a response include near misses, lost time and minor injuries, damage to technology and/or the environment, and the identification of new or previously undetected hazards.

### Incident investigations

Incidents can result from the degradation of system factors or from residual risk or a combination of these. An example of entropic risk leading to an incident is the case of a runaway remotely operated load-haul dump machine in an underground mine.[17] The machine was being reversed downhill out of a stope drawpoint (an area in the mine where gravity fed ore from a higher level is loaded into hauling units[18]) when it suddenly and unexpectedly increased speed towards the operator. The employee attempted to stop the motion by selecting 'full forward' and steering the machine away, however, it continued to roll back. The operator hit the emergency stop and got out of the path of the machine. The equipment came to a halt against the roadway side wall without any injuries being incurred.

The cause of the incident was that the engine had stalled because it was in low idling position and the revolutions were insufficient to prevent it from rolling back. This took the machine into a state of rapid entropy – a condition of operation that was no longer safe and productive. An outcome of the investigation was the need for employers and manufacturers to be mindful of their duty of care obligations. Specifically, they have to ensure that technology design changes are effectively communicated to employees and safe work practices are implemented following such changes.

Incidents can also arise from residual risks. An example is the case of an underground miner working with an airleg rock drill, who died when a 'hang up' (a void in a pile of broken rock) collapsed underneath him, in a shrinkage stope.[19] The shrinkage stope method of mining involves making a cavity by excavating out in successive flat or inclined slices and working upwards. After each slice is drilled and then blasted, enough broken ore is removed from below to provide a working space between the top of the pile of broken ore and the top of the cavity. As the rock is broken it increases in volume so that the excess volume has to be removed or drawn out of a series of drawpoints. The miners stand on this broken ore pile and drill into the solid ore slice.[18] If the rocks underneath the miner are not stable there is a significant danger that the materials he is standing on will collapse, as occurred in this case.

This accident resulted from a lack of awareness of the high inherent risk associated with this method. Often there are also loose rocks in the mine roof that can fall due to their own weight or due to movements in the earth. Generally, these hazards are monitored and remedied using practices such as scaling. This involves levering off thin layers of loose rock. Residual risks therefore have to be managed through continuous monitoring. Alternatively, companies should consider using mining methods which have a lower level of residual risk than shrinkage stoping.

When incidents occur prompt responses are required to prevent consequential damage and to obtain valuable information about risk levels. Such events may indicate that one or more system factors have degraded and/or that residual risk has reached an unacceptable level. Incidents also provide a warning to take corrective action to prevent repetitions. In some cases, these events indicate that there are underlying problems within the organizational system that have much broader implications than the incident itself. Investigations should, therefore, focus on the underlying systemic weaknesses that cause the event not the symptoms of the event. For example, in cases where operator fatigue is identified as a major contributing factor, work rosters and excessive overtime may be the root causes of safety deviations.

The entropy model provides direction for effective incident investigation. It indicates that facts should be gathered on the condition of each of the four system factors prior to, during and immediately after the incident. This involves considering the presence and effects of both entropic and residual risks. The information can be used to build up a chain of events as shown by Table 3.2 which illustrates the steps involved.

The chain of events identifies the risk factors or potential contributing variables that were present in each system area prior to the incident. The results of quantitative risk analysis from the systems risk profile (to be discussed in Chapter 7) can provide additional clues about these conditions. The profile allocates a numerical value to each of the system factors for a given activity to calculate the total risk score.

When an incident is being investigated, the assessor is concerned with three time frames – the time leading to the incident, the actual incident itself, and the time after the incident. These provide information about the three important dimensions of the event. The first is the condition of system factors prior to the event that affected the probability of the event occurring. The second issue is whether there were any triggers or catalysts that caused a rapid rise in entropic risk or the sudden translation of

**Table 3.2** Building a chain of events to determine the condition of system factors

| | Leading to the incident | The incident | Immediately after the incident |
|---|---|---|---|
| Processes | What process was being carried out? What risk factors were present? | What actions occurred? | What was the impact on operational activities? Where response procedures appropriate? |
| Technology | What technology was being used? What risk factors were present? | What technology was involved? | What damage was incurred or may occur if not rectified? |
| Physical environment | What were the workplace conditions? What risk factors were present? | What environmental changes occurred at the time? | What damage was incurred or may occur if not rectified? |
| Human resources | Who was carrying out the task? Who was in the vicinity? What risk factors were present? | Who was directly involved? Was anyone indirectly involved? | What harm was sustained? |

residual risk into imminent danger. An event may therefore occur when the deterioration of system factors has reached a critical point or alternatively, the event may happen in a seemingly safe workplace if system factors are triggered into a rapid state of decline, as also explained by Reason's model of resident pathogens. Extreme examples of this include criminal activities such as burglary or terrorism where a relatively safe environment becomes dangerous. Failure of the organizational system to deal with catalysts increases the risk and also potential consequential damages.

These catalysts cause severe changes in the condition of system factors. An example of this was the injury to Ray that was described in the Introduction. When the support pins broke this caused the ripper from the bulldozer to fall towards him and as a result he sustained serious injuries to both wrists. In this case a technological structural weakness was the catalyst of the accident. Triggers can also result from the physical environment. A case that illustrates this involved the inflow of storm water into a decline mine.[20] (A decline mine is one that has an inclined shaft usually following the downward slope of the orebody or vein.[18]) The area was already saturated by rain when a very heavy storm caused a rapid inflow down the decline that almost covered the face of the lowest heading. (A lower heading is a passage off a main cavity running along the bottom of the orebody or vein.[18]) The lack of sump capacity or pumping equipment with sufficient dewatering capacity resulted in immersion of the mining equipment at the mine face.[18] Fortunately, in this incident no one was injured, however, there have been other incidents of this nature that have resulted in the death of mine workers and extensive damage to mining machinery. For example, in the Emu mine disaster in Western Australia, six men lost their lives in a decline driven underground from the bottom of an open pit, as a result of sudden flooding of the pit following sustained and exceptionally heavy rainfall.[21]

Human resources, like the physical environment, can also rapidly deteriorate, which in turn can trigger an incident. For example, when an employee loses alertness and cognitive ability as a result of taking drugs the probability of an accident increases rapidly. Other matters that can cause sudden changes include extreme emotional responses to changes in conditions, such as fear or anger that cause the worker to lose the capacity to evaluate objectively risk factors.

The second phase of incident investigation therefore focuses on the variables present at the time of the accident. The final phase of incident investigation is the analysis of the consequences of an incident in the time immediately after and following the event. At this stage, the assessor needs to determine whether the systems were in place to respond effectively to the incident and to minimize the resultant damages and/or injuries. This covers OHS processes such as emergency procedures and first aid as well as contingencies for taking the required corrective action to prevent further degradation of system factors.

Productive safety management allows structured analysis of incidents by considering risk in these three time frames, by dividing the investigation into the four system factors, and by identifying sources of both the entropic and residual risk which contributed to the incident. The incident investigation report form is designed accordingly to ensure that investigators focus on the condition of the system throughout the chain of events. This differs from traditional approaches which focused on unsafe acts and unsafe conditions.

Figure 3.9 provides an extract of the incident investigation report form designed for productive safety management. In the form, system factors are explored in turn to identify any areas of entropy that may have been present prior to and at the time of the incident. This is followed by a summary of systems degradation in the incident analysis section. In some circumstances, it may be found that incidents are not entirely a result of entropy, for example, when high winds cause unsecured objects to become projectiles that cause secondary damage. In these cases, investigators need to determine the sources of residual risk that have resulted in breakage, loss or new hazards.

Incidents are therefore often caused by a combination of entropic and residual risks. For example, a car accident may be initiated by a loose rock rolling from the embankment on to the roadway. If the vehicle had poor brakes this may have affected the driver's ability to stop the vehicle in time to avoid the collision. To allow the investigator to consider both types of risk, the form leads him firstly to assess entropic risk in each system factor before analyzing residual risk. The format therefore follows the steps of the four-fold strategy. These steps help the investigator to identify entropic risk sources that indicate a failure in risk assessment and/or maintenance practices, then residual risk sources that identify lack of management controls to address inherent dangers, or alternatively, the presence of unforeseen hazards. Once all contributing factors have

| System Factors Analysis | | | |
|---|---|---|---|
| **1. Processes** | | | |
| 1.1 | Are procedures/standard practices set down for this task? If no, state what procedures could be implemented to reduce the risks involved. | Yes | No |
| 1.2 | Were procedures followed? If no, state why. | Yes | No |
| 1.3 | Is there a system for monitoring adherence to standardized work practices? | Yes | No |
| 1.4 | Are procedures in place to warn personnel of the hazards of this task? If yes, were they followed? If no, what can be done to rectify this issue? | Yes | No |
| 1.5 | Has this task been reviewed in the past? If yes, what were the conclusions of this review? | Yes | No |
| 1.6 | Were there any previous incidents related to this task? If yes, provide details. | Yes | No |
| 1.7 | Were employees advised of these incidents? | Yes | No |
| 1.8 | Were operating controls and instructions clearly visible? | Yes | No |
| 1.9 | Was specified protective clothing required to be worn whilst carrying out this task? If yes, was it being worn at the time, was it in good condition and was it worn correctly? | Yes | No |
| 1.10 | Were there any other factors related to processes which may have contributed to the incident? If yes, state them, including any residual risks. | Yes | No |
| **2. Technology** | | | |
| 2.1 | Was the technology in a safe operational condition i.e. without faults, design/quality or installation problems prior to the incident? If no, state the areas of fault/degradation and the cause. | Yes | No |
| 2.2 | Were adequate control mechanisms in place and operational e.g. guards/protective devices? If no, would controls have reduced the consequences or prevented the incident? | Yes | No |
| 2.3 | Had the technology been inspected regularly prior to the incident? If yes, when was the last inspection and what findings were made and action taken? | Yes | No |
| 2.4 | Has the technology been maintained on a regular basis? If yes, when was the last maintenance carried out and what was involved? | Yes | No |
| 2.5 | Was the correct technology used for the task? | Yes | No |
| 2.6 | Were there any other factors related to the condition, operation, maintenance of this technology which may have contributed to the incident? If yes, state these factors including any residual risks. | Yes | No |
| **3. Physical Environment** | | | |
| 3.1 | Did any of the following environmental factors contribute to the accident? If yes, tick and describe the conditions.<br>Rain ☐ Ice/snow ☐<br>Glare ☐ Fog ☐<br>Cold ☐ Humidity ☐<br>Heat ☐ Noise ☐<br>Fumes ☐ Restricted/confined space ☐<br>Vapor ☐ Radiation ☐<br>Gas ☐ Surface condition of ground/floor ☐<br>Insufficient lighting ☐ Ineffective physical barriers ☐<br>Other please specify ☐ | Yes | No |
| 3.2 | Was there free, safe access/egress to the location? If no, state the cause of blockage/restriction. | Yes | No |
| 3.3 | Are regular inspections carried out of the environment? If yes, when was the last inspection and what were the findings? | Yes | No |
| 3.4 | Were there any other factors related to the condition or maintenance of the physical environment which may have contributed to the incident? If yes, state these factors including any residual risks. | Yes | No |
| **4. Human Resources** | | | |
| 4.1 | Was the person concerned carrying out a task that was part of his/her normal duties? If no, explain the circumstances. | Yes | No |
| 4.2 | Was the person involved in an activity associated with work but not directly related to the task? If yes, explain the circumstances. | Yes | No |
| 4.3 | Was the person's immediate supervisor present in the area at the time of the accident? If no, state the location of the supervisor and any instructions given prior to leaving the area. | Yes | No |
| 4.4 | Could instruction have been given to prevent this incident? If yes, state the nature of instruction. | Yes | No |
| 4.5 | Was the task within the capability of the person concerned? If no, what training and/or experience were lacking? | Yes | No |

**Figure 3.9** Extract of incident investigation report form

| | | | |
|---|---|---|---|
| 4.6 | Had the person concerned been trained to carry out the task safely? If no, what additional training could have been given to prevent the incident? | Yes | No |
| 4.7 | Were any of the person's senses obscured? If yes, how? | Yes | No |
| 4.8 | Were there any other factors related to training, supervision or competency that may have contributed to the incident? If yes, what were these factors, including any residual risks, and how could they have been prevented? | Yes | No |

**Incident analysis**

| | | |
|---|---|---|
| Were system factors degraded at the time of the incident? If yes, specify which system factors and the nature of degradation. | Yes | No |
| If no, could the incident be attributable to residual risk? Provide reasons. | Yes | No |

**Post-incident analysis**

| | | |
|---|---|---|
| Were effective responses implemented after the incident to minimize consequential damages, for example, immediate reporting, corrective action, emergency procedures, evacuation procedures and/or medical treatment? Describe the actions taken and evaluate the appropriateness and timeliness of these actions. | Yes | No |

**Risk scorecard (use to evaluate risks present at time of incident and any remaining risks)**

Probability: Almost certain; Quite possible; Unusual but possible; Remotely possible; Unlikely; Almost impossible

Exposure: Very rare; Rare; Infrequent; Occasional; Frequent; Continuous

Tie line

Consequences: Numerous fatalities – Catastrophe; Multiple fatalities – Disaster; Fatality – Very serious; Serious injury – Serious; Casualty treatment – Important; First aid treatment – Noticeable

Risk score: 500, 400, 300, 200, 100, 80, 70, 60, 50, 40, 30, 20, 10, 5

| Risk | Action time frame | Action by: |
|---|---|---|
| Catastrophic risk | Immediate | Management |
| Very high risk | Immediate | Management |
| Substantial risk | Immediate | Management |
| Moderate risk | Within 7 days | Supervisor |
| Low risk | Within 28 days | Employee |
| Very low risk | Report only | No further action required |

**Corrective action plan/recommendations (to be completed by supervisor)**

| Action required | By whom | Due date | Date completed |
|---|---|---|---|
| Immediate action to eliminate entropic risk: | | | |
| On-going system factor maintenance to prevent future degradation: | | | |
| Strategies to manage residual risk: | | | |
| Long-term strategies to compress residual risk: | | | |

**Figure 3.9** *(Contd)*

been identified, the investigator assesses the effectiveness of post-incident responses, which includes whether the incident was reported immediately and by whom, the appropriateness and effectiveness of corrective action such as medical treatment, and where necessary, the implementation of emergency and evacuation procedures.

The incident investigation form contains the risk scorecard,[6] which may be used to evaluate the time frame for action and to manage the risks involved in and resulting from the event. Using this method, each risk is analyzed according to its probability, exposure and consequences of recurrence. The results of the evaluation are then used as a guide for change implementation.

Under productive safety management, the entropy model provides the framework for action using the four-fold strategy. Accordingly, the corrective action plan/recommendations section of the form is structured on this remedial method. Entropic risks are corrected in the short term and maintenance strategies are developed to prevent future degradation. Residual risks are addressed by managing them in the short term, for example, by isolating an unstable work area or providing employees with additional training to increase awareness of these hazards. Recommendations are also developed for the compression of residual risk in the longer term. Incident investigation, therefore, serves two purposes. The first is to trigger effective short-term responses to hazards and the second is to deliver long-term improvements in safety and performance through systems-based planning and implementation. The chain of events analysis therefore provides for strategic risk management responses.

How does the chain of events relate to the entropy model? What implications can be drawn from this relationship? The chain of events indicates that the assessor should consider the types and levels of risk present prior to, during and after the incident, which indicates that for an OHS system to be effective it has to address three issues. The first is the reduction of the baseline level of risk (the level of risk that the firm has when it undertakes its first risk assessment and evaluation) to minimize the probability of an incident. Secondly, the firm should identify and remove, or anticipate and plan for, catalysts that may lead to the rapid decline of system factor conditions. Finally, it should develop procedures to ensure timely and appropriate responses following an incident to minimize consequential damages. Figure 3.10 illustrates these three components of the chain of events.

On the left-hand side of the figure, it indicates that to increase the level of safety in a workplace, the firm has to firstly apply the four-fold strategy to reduce the baseline level of risk. The baseline is a combination of the static level of residual risk and the static level of susceptibility to entropy, as calculated later in Chapter 7. Secondly, potential catalysts that lead to the severe degradation of system factors (a substantial rise in entropic risk) or which trigger residual risks into imminent danger have to be anticipated, planned for and where possible prevented or managed. Fatigue and/or drug abuse are examples of catalysts which cause rapid deterioration of the human resources system factor that greatly increase the probability of an incident. For instance, there was a case of a fatality in an open cut mine in August 1994 that illustrated this. It occurred when a large haul truck overturned after the operator apparently lost control.[22] The driver suffered massive crush injuries to the chest and abdomen, which

| The Entropy Model | The Chain of Events | | |
|---|---|---|---|
| | Leading to the incident | The incident | Immediately after the incident |
| | **Baseline risk management** | **Catalyst risk management** | **Consequential damages management** |
| | Reduce the baseline level of risk in the firm using the four-fold strategy: | Identify and eliminate potential catalysts which: | Develop and implement strategies to minimize the consequential damages resulting from an incident: |
| | 1. Eliminate entropic risk<br>2. Implement maintenance strategies to prevent future entropy<br>3. Manage residual risk in the short term<br>4. Compress residual risk in the longer term | ■ Lead to rapid degradation of system factors<br>■ Trigger residual risks into imminent danger | ■ Evacuation plans<br>■ Emergency plans<br>■ Corrective action<br>■ First aid<br>■ Workers' compensation<br>■ Rehabilitation |

**Figure 3.10** The entropy model and the chain of events

resulted in death. A quantity of marijuana was found in the cabin of the truck during the investigation and as a result of the inquest, the coroner made the following recommendation which was supported by the jury:

> We recommend that on-going consideration be given to the issue of tiredness and the issue of drugs and alcohol within the industry.[22]

Catalysts are potentially disastrous workplace hazards that may or may not be readily apparent. Under the Mines Safety and Inspection Act 1994, for example, mining companies have a duty of care to control the risks of these hazards. The driver, in this case, also had a duty of care for her own health and that of other workers. Every employee working at the site had the responsibility to report any circumstances they deemed to be hazardous, including the use of substances that affect operator safety consciousness and competence.

Reiterating, catalysts that affect the level of degradation have to be anticipated and managed. There are also examples of triggers which send residual risk into extremely hazardous circumstances, for instance, flooding is a catalyst that triggers the residual risk of decline mining into imminent danger. Explosions and fires in underground coal mines are other catalysts of this type. The entropy model suggests that a rapid rise in entropic risk or the triggering of residual risk into severely hazardous conditions indicate that there is a loss of control within system factors, as illustrated in Fig. 3.11.

In part (a), the catalytic risk involves degradation of the human resources system factor, which correlates with a substantial rise in the probability of an incident, while the residual risk level remains stable. An example of this is when a driver takes drugs that reduce alertness and physical response times. Part (b) shows a catalytic risk triggering residual risk into imminent

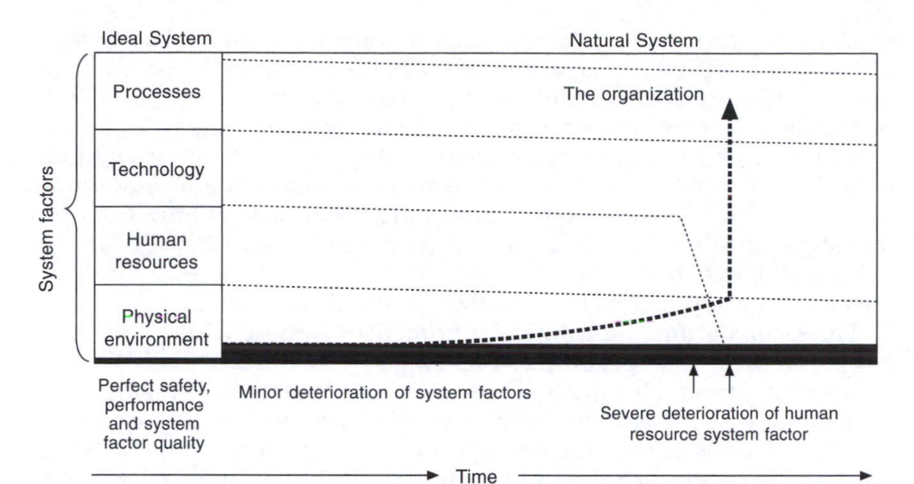


(a) Catalyst risk involving severe degradation of human resources system factor

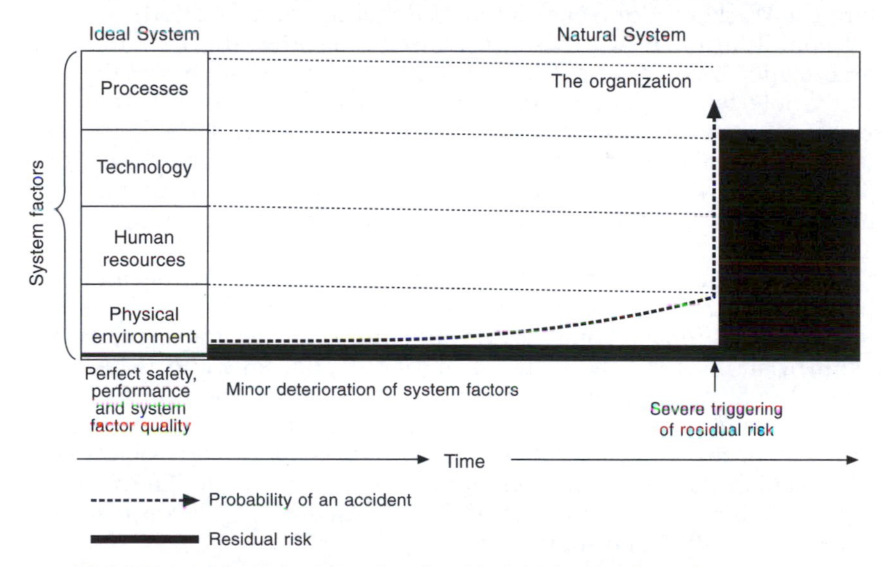


(b) Catalyst risk involving triggering of residual risk into imminent danger

**Figure 3.11** The effect of catalyst risks on the probability of an incident

danger, which also causes the probability of an incident to rise to severe and potentially uncontrollable levels. The case involving the heavy rain flooding the decline mine is an example of this. These illustrations highlight the importance of anticipating potential triggers and where possible, preventing them.

## Minimizing consequential damages

The first two areas of risk management shown in Fig. 3.10 are baseline risk management and catalyst risk management. In the event of an incident, the firm also has to have systems in place to contain consequential damages. This is illustrated in the final section of the figure as 'consequential damages management'. It involves limiting the extent of destruction, injury and/ or the development of new hazards. The incident investigator needs to consider the effectiveness of post-incident responses such as evacuations, emergency procedures and corrective action to determine whether these organizational systems were adequate after the event. Incident investigation, therefore, is a post-event practice that assesses the risk prior to the event, any catalytic risks that triggered the event and the effectiveness of practices to minimize resultant consequences.

The second group of post-incident practices is those which reduce the long-term harm to the health and wellbeing of the injured worker. Workers' compensation and rehabilitation processes ensure that the company meets its legal and social responsibility towards these individuals and their families. This is a requirement imposed on the firm by the external environment with each developed country having its own legislation to cover this responsibility. In Western Australia, businesses have to comply with the Workers' Compensation and Rehabilitation Act 1981.

Rehabilitation is a process that demonstrates the extent of the organization's commitment to health and safety, as such practices are compatible with common goals and values. These goals include the restoration of productivity for the individual and the company. Rehabilitation also considers psychological factors such as the preservation of the employee's self-esteem and the minimization of negative impacts on morale.[23] It is therefore a means of reinforcing internal goal alignment.

These practices also have strategic implications. The channel indicated that internal strategic alignment is concerned with the balance of physical, financial and human capital. Where an employee sustains an injury as a result of an incident, the value of human capital falls and is less than optimal in terms of safety, productivity and quality. The firm therefore, has to expend financial resources to facilitate the person's early and safe return to meaningful productive work, which is the primary objective of rehabilitation. It focuses on benefits for both the employee and the company. Again, the emphasis is on maintaining the quality of system factors in accordance with the entropy model.

## Summary

In this chapter the three types of processes – work practices, risk management practices and post-incident practices – have been discussed. These are the key areas in the processes system factor. In the following chapters, the risks associated with the other system factors – technology, the physical environment and human resources – will be discussed before presenting a method in Chapter 7 to combine these factors into business activities and quantify relative risk levels.

## References


1. Mathews, J. (1993) *Health and Safety at Work – Australian Trade Union Safety Representatives Handbook*. Pluto Press Australia, New South Wales, Australia.
2. *Occupational Safety and Health Act (1984) (Western Australia)*, Western Australian Government Printers, Perth, Australia. Available on website: http://www.austlii.edu.au/au/legis/wa/consol_act/osaha1984273/
3. Clark Hummerston Organization and Management Consulting Group (1990) *Managing Safety Participant's Guide*, Melbourne, Australia.
4. Department of Mineral and Petroleum Resources Western Australia (2001) *Significant Incident Report No. 52*. Available on website: http://notesweb.mpr.wa.gov.au/exis/SIR.NSF/6b390eea5649d21c48256097004aacb7/1550b866159bb3e14825616a0025e8af?OpenDocument
5. Flinders University (2001) *Topic 5: Managing Hazards: Control Hazards Now*, Hierarchy of control information. Available on website: http://wwwfp.cc.flinders.edu.au/Web_SupTrng/topic5/riskControl/Welcome.html
6. Kinney, C.F. and Wiruth, A.D. (1976) *Practical Risk Analysis and Safety Management Naval Weapons Center (Paper)*. China Lake, California, June 1976.
7. Department of Mineral and Petroleum Resources Western Australia (2001) *Significant Incident Report No. 107*. Available on website: http://notesweb.mpr.wa.gov.au/exis/SIR.NSF/6b390eea5649d21c48256097004aacb7/1fa42f29b0656309482568ee0030ca54?OpenDocument
8. Worksafe Western Australia (2001) *State of the Work Environment No. 26: Injuries and Diseases to Young Workers, Western Australia*, 1994/95. Available on website: http://www.safetyline.wa.gov.au/PageBin/injrstat0052.htm
9. National Occupational Health and Safety Commission Worksafe Australia (1994) *Positive Performance Indicators for Occupational Health and Safety*. Available on website: http://www.nohsc.gov.au/OHSInformation/NOHSCPublications/fulltext/docs/h2/ppio/ppio1a.htm
10. Worksafe Western Australia (1999) *Worksafe Plan – Western Australia's Assessment of Occupational Safety and Health Management Systems*. Available on website: http://www.safetyline.wa.gov.au/PageBin/bestplan0001.htm
11. Department of Education, Training and Industrial Relations (1999) *'Tri Safe' Management Systems Audit*. Available on website: http://www.whs.qld.gov.au/subject/auditing.htm
12. National Occupational Health and Safety Commission (1999) *Evaluation of Contractor OHS Compliance Initiatives – Managing your Contractors' Health and Safety: A Guide*. Available on website: http://www.nohsc.gov.au/ohsinformation/ohssolutions/97-98/09_concom_exec.htm
13. National Occupational Health and Safety Commission (2001) *Occupational Health and Safety Management Systems – A review of their effectiveness in securing healthy and safe workplaces*, April 2001. Available on website: http://www.nohsc.gov.au/Pdf/OHSSolutions/ohsms_review.pdf
14. Case, K.E. and Fair, R.C. (1989) *Principles of Economics*. Prentice Hall, New Jersey, USA.
15. National Safety Council of Australia (2001) *Electrical Accident results in $118,000 Fine*. Available on website: http://www.safetynews.com/ohs_email.php
16. Worksafe Western Australia (2001) *Work Practices: Woodside sets the Trend*. Available on website: http://www.safetyline.wa.gov.au:81/
17. Department of Mineral and Petroleum Resources Western Australia (2001) *Significant Incident Report No. 99*. Available on website: http://notesweb.mpr.wa.gov.au/exis/SIR.NSF/6b390eea5649d21c48256097004aacb7/2cebb7d66a875155482566c900027760?OpenDocument
18. Thrush, P.W. (ed) (1968) *A Dictionary of Mining, Mineral, and Related Terms*. US Department of Interior, US Government Printing Office, Washington, USA.
19. Department of Mineral and Petroleum Resources Western Australia (2001) *Fatal Accident Report dated 26/03/92*. Available on website: http://notesweb.mpr.wa.gov.au/exis/FATAL.NSF/0c932c8fa767542f482567550016f40c/584fa4e5bf76482567f200203312?OpenDocument
20. Department of Mineral and Petroleum Resources Western Australia (2001) *Significant Incident Report No. 76*. Available on website: http://notesweb.mpr.wa.gov.au/exis/SIR.NSF/6b390eea5649d21c48256097004aacb7/f6f5b73b1e199519482564c600158995?OpenDocument

21. Department of Mineral and Petroleum Resources Western Australia (2001) *Significant Incident Report No. 11*. Available on website: http://notesweb.dme.wa.gov.au/exis/SIR.NSF/85255e6f0052056085255d7f00607219/4c74be127e9743034825609d004bd03f?OpenDocument
22. Department of Mineral and Petroleum Resources Western Australia (2001) *Safety Bulletin No. 12*. Available on website: http://notesweb.mpr.wa.gov.au/exis/SBULL.NSF/85255e6f0052055f852558ac00697645/907596e5f03166dc482562f8000ee06b?OpenDocument
23. Chinnery, D. (1991) Consultation in multidisciplinary work in rehabilitation, *Psychological Perspectives on Occupational Health and Rehabilitation*. Harcourt Brace Jovanovich Group, Marrickville, pp. 426–453.

Chapter 4

# Technology

In the firm, technological risk management begins in the design, planning and purchase stages during the initial set-up of the business operation. In these early phases, the business has to make a number of decisions about capital resource usage. The external forces shown by the channel affect these decision-making processes and drive the company to act responsibly, for example, by providing and maintaining, as far as practicable, working conditions in which employees are not exposed to hazards. Accordingly, operations have to be designed with the lowest practicable levels of residual risk. In addition, plant and equipment purchased should be accompanied by manufacturers' guarantees that the technology, when used under normal operating conditions and with recommended precautions taken, does not pose a hazard to employees. The entropy model implies that what is meant by 'does not pose a hazard' is that the level of risk is 'acceptable' given the current state of technological knowledge and skills in society. It is not possible to completely eliminate residual risk, and therefore, all technologies are hazardous to some degree.

## Legal provisions

In the community, technological risk management also begins at the design and manufacturing stage. In Australia, the duties of manufacturers and suppliers are clearly set out in legislation, for example, the Occupational Safety and Health Act 1984 (Western Australia) states:

> 23. (1) A person who designs, manufactures, imports or supplies any plant for use at a workplace shall, so far as is practicable:
>   (a) ensure that the design and construction of the plant is such that persons who properly install, maintain or use the plant are not in doing so, exposed to hazards;
>   (b) test and examine, or arrange for the testing and examination of, the plant so as to ensure that its design and construction are as mentioned in paragraph (a); and

(c) ensure that adequate information in respect of:
  (i) any dangers associated with the plant;
  (ii) the specifications of the plant and the data obtained on the testing of the plant as mentioned in paragraph (b);
  (iii) the conditions necessary to ensure that persons properly using the plant are not, in doing so, exposed to hazards; and
  (iv) the proper maintenance of the plant.[1]

For a number of years this provision or its equivalent had not been tested in court, however, in 2001, a landmark decision was handed down that now sets a precedent for future cases. The full bench of the New South Wales Industrial Relations Commission tested the equivalent provision in the Occupational Health and Safety Act 1983 (New South Wales). The commission overturned the decision to acquit a woodchip machine manufacturer after an accident in which a worker lost both arms.[2] In its appeal against the acquittal, WorkCover, the government regulator, successfully argued that the company had supplied the machine knowing that the use intended was inappropriate. The machine would cause regular blockages at the waste transfer station where contaminated green waste was being processed.[3] Now that this provision has been legally tested and proven to be enforceable, a clear message is sent to manufacturers and suppliers to take all reasonable measures to ensure that their products are safe and used for the purposes intended.

In addition to provisions related to the duties of manufacturers, OHS legislation covers the duties of employers in relation to technology. These include providing and maintaining plant such that, so far as is practicable, employees are not exposed to hazards. Employers must also:

19. (1) (e) make arrangements for ensuring, so far as is practicable, that:
  (i) the use, cleaning, maintenance, transportation and disposal of plant; and
  (ii) the use, handling, processing, storage, transportation and disposal of substances, at the workplace is carried out in a manner such that his employees are not exposed to hazards.[1]

Subsection 19(1)(ii) above covers the use of substances or chemicals in the workplace. Such substances are included in the technology system factor of the entropy model.

## Risk and new technology

The reference to 'so far as practicable' in the legislation alludes to two constraints that affect the level of residual risk of a new technology and also the rate at which it degrades. Firstly, at the design and manufacturing stage, cost–benefit decisions have to be made concerning the level of safety built into a unit of equipment. For example, how much would a consumer be prepared to pay for a car that was 'guaranteed' to prevent injury in a collision at 100 km per hour? How much time and cost would a company be willing to dedicate to research and development of such a car? Limited resources cause the community to accept a level of risk that is noncompressible in the short term. The consequences for the firm are that plant and equipment have a residual risk. This residual risk is illustrated by the entropy model and indicates that organizations must learn to live with residual risk and therefore, establish strategies to manage it. Further, these cost–benefit decisions affect the life of a technology. Inputs that lengthen the period of return of a technology are more expensive. The willingness of the consumer to pay for this extended utility (benefit obtained) influences the quality of the product that, in turn, has an impact on the tendency of a technology to degrade. Cost–benefit decisions, therefore, affect the levels of entropic risk that plant/equipment has over its product life.

The second constraint is that companies have to make decisions about resourcing in the presence of scarcity as explained by the strategic alignment channel. Human, physical and financial capitals are limited with little flexibility to increase these resources in the short term. An increase in one, therefore, requires a decrease in another. For example, the decision to replace old technology (physical capital) requires expenditure of financial capital. With limited finances, the firm may need, in the short term, to accept the higher levels of residual risk inherent in the old compared to the new equipment. It may also have to compensate for higher rates of entropic risk using more rigorous, planned maintenance strategies.

Properly designed and well-manufactured technology has a lead-time before it begins to degrade and before this deterioration causes an 'unacceptable' hazard. From the cost point of view, however, it may still be more effective to continue to operate the old technology in the short term, even though it is less efficient, less safe and more costly in terms of maintenance than the new technology. Such decisions to continue operating suboptimal technology can be made on two bases. The first is that the technological risk is still 'acceptable' as determined by standards set by regulatory bodies and/or by agreement between management and the workforce, and that strategies are in place to manage this risk. The second is that the benefits of continued use outweigh the costs, factoring in the probability of this technological risk translating into an incident and the losses caused by suboptimal productivity, which are hidden costs.

These hidden costs can be explored further. The entropy model indicates that as system factors degrade the probability of an incident rises. Technologies with high levels of entropic and/or residual risk, therefore, potentially have a number of negative impacts on productivity and safety. These include equipment downtime – the machine has to be withdrawn from service for unplanned maintenance to rectify its deteriorated condition. Assuming the technology remains operational, entropy can lead to suboptimal output, for example, an electric shovel (large-scale mining equipment) may not operate as efficiently if its hydraulic components are in a degraded condition. In turn, this also places higher demands on the operator, particularly if management is measuring output as an indicator of employee or team efficiency. Degradation may also lead to suboptimal interaction with the physical environment, for example if the blade of a grader is not in good condition then the roadways can not be maintained to the required standard. The costs of degradation and inherent risk are therefore not limited to those that are easily quantified, such as loss of machine availability. There are also the nonaccounted costs of suboptimal output and increased demands on other system factors.

Additionally, there are the costs incurred when technological risk translates into an incident. These expenses can result from injury to an employee and/or damage to technology or the physical environment. New hazards may also be introduced. In OHS literature there have been a number of attempts to itemize these costs. The productivity model identifies a number of outlays associated with work-related injury.[4] These include loss of productive hours worked comprising the reduction in output and the cost of wages for these lost hours. It also includes costs associated with short-term absences, such as administration and workers' compensation insurance. Further, poor safety often causes a negative impact on workforce stability with the resultant turnover (resignations and departures of employees from the business), leading to loss of investment on training. Absence due to injury also forces the firm to allow more overtime, over-employment and the use of replacement workers. To compensate for lost productive time, the work rate may be increased, for example reduction of lunch breaks, which, in turn, may overload existing personnel. The determination of the costs associated with technological failures is, therefore, not simply lost time in man-hours and machine downtime. There are further ramifications that affect the firm's performance and which have implications for its culture.

In summary, the difficulties in cost–benefit analysis are how to factor in added safety benefits and how to factor in all the costs of losses caused by entropic and residual risks. In an underground mine, for example, the costs of structural supports for cavities and for the installation of equipment can easily be underestimated at the planning stage to the extent that the cost of mitigating residual risk in the physical environment and technologies can have a significant impact on viability. In addition to the constraints associated with estimating costs and benefits, the firm faces incomplete information about risks. Green-field sites are particularly exposed to unknown hazards and firms have to rely to a large extent on the claims made by manufacturers regarding the safety of their products. They also have to depend on the technical expertise of their employees, including mechanical and process engineers, in the selection of plant and equipment. Despite the difficulties in fully accounting for safety and risk, these internal strategic alignment decisions have to be based on cost–benefit analysis as it is the most workable method of determining appropriate usage of limited resources. Using this method, however, resultant strategies are 'practicable' rather than 'ideal' in both an economic and legal sense, and it is important for managers and OHS practitioners to understand these limitations.

The discussions thus far have implied that new technology is intrinsically safer than old technology. This may not always be the case for it is possible for the reverse to be true, particularly in the short term. The introduction of new equipment into the workplace changes the risk to which employees are exposed because of its interaction with other system factors. Firstly, there is a negative impact on human resources. Current employee KSAs may be inadequate to allow them to work safely and efficiently using the new equipment. Initially, employees require time to become familiar with technological change. These alterations, in effect, cause degradation of the human resources system factor. Consequently, it is essential that firms, when introducing system modifications, implement effective change control strategies to address such risks.

The time needed to acquire technology-specific competencies can be accelerated using pre-operational training, however, exposure and practice are still required. Figure 4.1 shows how the introduction of such change causes a decrease in the KSAs needed to perform safely and efficiently. It shows that with the degradation of human resource competencies there is a concurrent increase in the probability of an accident. As employees become familiar with the technology, the human resources system factor improves and the likelihood of an accident also declines. This continues until the employees are as knowledgeable and skilled with the new technology as they were with the old technology.

Equipment changes also introduce additional risks through the interaction with processes. Technological change requires new work practices to be developed. These can lead to unforeseen hazards, for example, modern plant that can be maintained through partial shutdown, may have more complex isolation and tag out procedures than old equipment with one centralized control panel. It is important to undertake JSA and adhere to manufacturer's instructions prior to and during the commissioning of new technology. The case of the refinery workers, discussed in the previous chapter was an example. It was the failure to recognize the impact of process changes required as a result of the technological changes, which contributed significantly to these incidents.[5] The problems were exacerbated by the poor fit between organizational competencies and the demands of the task. Employees were not equipped with the knowledge to understand the risks involved or the skills to carry out the work safely.

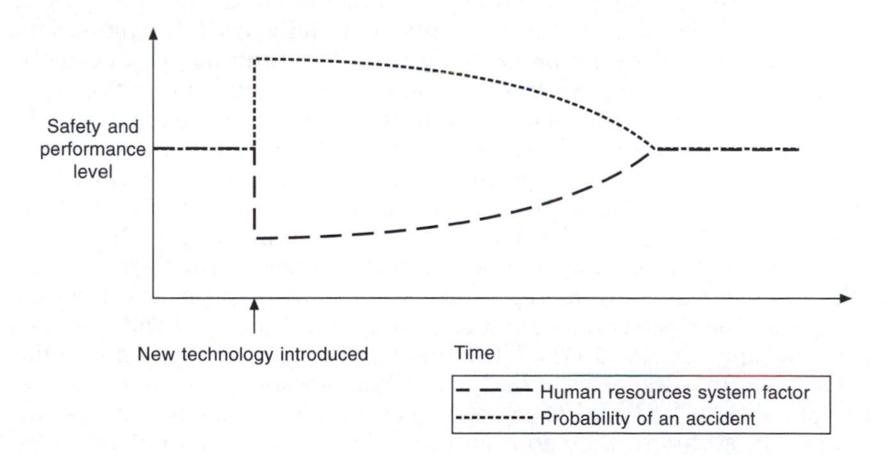


**Figure 4.1** The effect of introducing new technology on the human resources system factor

The introduction of new technology may also have repercussions for the physical environment. When firms commence their operations, the physical environment is usually designed to accommodate the types of technology used at the time. The later introduction of new technology can cause current infrastructure conditions to be unsuitable. An example is the introduction of computers into the workplace. In the paper office there were few power outlets. With technological change many offices have had to be modified. In buildings that were not designed for computerized operations it was not uncommon to see cables running across the floor. This new technology caused a degradation of the physical environment in which people worked and introduced tripping hazards. In addition, old office furniture and lighting systems were not necessarily suited to the use of computers. Consequently, with this technological change there has been an increasing emphasis on the ergonomic design of workplaces to create a fit between the types of technology employed, the work environment and human physical capacity.

New technology therefore often requires the physical environment to be modified. For example, modern plant may require additional ventilation, climate control or vibration protection. The impact of introduced technology on existing system factors should be evaluated prior to purchase and installation in terms of any changes in residual and entropic risks that may result. A further consideration in the purchase of new equipment is parts availability and supplier support. When parts are not available this has a direct impact on machine downtime and places constraints on the use of equipment. Where entropy causes a machine to become unsafe, it is important to take corrective action. This often involves dependence on the supplier to provide parts and services to remedy the degradation. Machine downtime has also a negative impact on productivity. Effective relationships with suppliers are thus an important part of the organizational strategy to manage residual risks and counter entropic risks in the technology system factor.

The strategic partnership between the firm and the supplier should also involve the provision of up-to-date information to assist the business to develop risk management strategies in relation to the technology that it employs. The firm should use a number of criteria to select its suppliers. These include the safety and reliability of the equipment, production capacity, cost, back-up parts, service and training, and the provision of information to assist risk management. Information sharing between the firm and its suppliers should flow in both directions. Where the firm experiences difficulties as a result of technological risks it should advise the manufacturer. The supplier can use this information to compress risks further through design modifications and enhanced quality.

The importance of information sharing becomes particularly evident where similar events are repeated such as two comparable aviation incidents. The first such incident occurred on 21 June 1999 and involved a Beech King Air 200 aircraft.[6] The findings of the investigation into this event were that the aircraft had lost cabin pressure and the passenger oxygen system had malfunctioned, placing those onboard in extreme danger. In addition, the cabin altitude alert system was not provided with an aural warning to indicate that the altitude had exceeded 12 500 feet and therefore, the pilot was not fully aware of the change in condition. Consequently, there were two technological system failures that contributed to the incident and fortunately, on this occasion, there were no fatalities. In response, the Australian Bureau of Air Safety Investigation recommended that the Civil Aviation Safety Authority issue a directive for an immediate check of the fitting of passenger oxygen system mask container doors on all such aircraft, and all other aircraft similarly equipped.[6]

The second incident involved the golfer, Payne Stewart, whose plane crashed in South Dakota on 25 October 1999, 4 months after the previous 'near miss', killing all six people on board.[7] The chartered, twin-engine Lear 35 appeared to have lost cabin pressure with the cause of such loss being unknown. Fighter jets were sent after the plane and the pilots noticed no structural damage. The windows, however, were frosted indicating the temperature inside was well below zero degrees. Although, as a result of the lack of evidence of the cause of this latter accident it is not possible to draw direct correlation between the two events, loss of cabin pressure appeared to be the cause in both cases. The earlier incident should have prompted greater attention to such matters in the industry.

These incidents illustrate the need for quick and effective responses when risk factors are identified that potentially can affect a whole industry or many operators. The above cases pinpointed a number of design faults that increased the risk of accidents as a result of depressurization. The first was the installation of passenger oxygen-supply container doors. It was found that incorrect fitting could lead to passenger incapacitation if masks were not deployed effectively. The second design fault was the lack of an aural warning in the cabin altitude alert system.[8] Had this information been available to industry members and had it been acted upon by the relevant companies, the death of Payne Stewart and his colleagues may have been avoided. Information-sharing is a very important part of risk management and begins with the strategic relationship between the firm and its suppliers.

## Technological hazards

All technologies pose some degree of risk that is either entropic or residual in nature. Residual risks stem from four sources. Firstly, technologies have design and manufacturing short-comings which affect the level of residual risk and also the rate at which the technology degrades as explained earlier. This means that plant and equipment cannot be made 100 per cent safe and this can be attributed to limited resources that make marginal safety improvements cost-restrictive. The level of safety is also constrained by the current state of technological competencies in the community. In addition, out of practical necessity some equipment is hazardous because of the purpose for which it is designed, for example, compactors and chainsaws.

The remaining three residual risk sources stem from the interaction of technology with other system factors. As identified earlier, there is never a perfect fit between technology and the physical environment. The latter is usually in a state of change and this can cause escalation in residual risk during the life of a business operation. An example of this is when the population of residents or other businesses increases in the region around a manufacturing or storage site with the effect that the severity of consequential damages resulting from a technological failure can rise significantly. This is particularly evident in countries lacking city planning and infrastructure management, but has also occurred in developed countries. An example is the explosion of a fireworks site in Enschede, The Netherlands, on 14 May 2000. At least 20 people were killed and a further 600 injured when a warehouse storing fireworks exploded near a residential area.[9] The original permit to store these goods was granted in 1977, when the surrounding area was not so heavily populated, and subsequently renewed three times. The issue of these permits and the storage of the explosives are under investigation by Dutch authorities and of particular concern is the assessment of changes in the physical environment surrounding technological operations during the life of such facilities.

The most notable example of these residual risks contributing to a major disaster is Bhopal. On the night of 2 December 1984, in Bhopal, India, a gas leak from a small pesticide plant owned by Union Carbide Corporation led to the death of 2500 people and injuries to a further 200 000.[10] The incident resulted from the largely unrealized dangers associated with the manufacture of the highly toxic chemical, methyl isocyanate (MIC). The immediate cause of the MIC leak was an influx of water into the chemical storage tank, but there were a number of catalysts adding to this including drought, agricultural economics and the gross mismanagement of entropic and residual risk factors. One of the most significant factors was an excessively large inventory of hazardous substances causing the inherent dangers to be extremely high. The severity of the mismatch between the technological design of the plant and storage conditions with environmental factors therefore largely determined the extent of this disastrous event.

Changes in climatic conditions can have a significant impact on the fit of technology with the physical environment. An example of this is the case in Romania involving a state company and an Australian company. These firms, according to the Australian company's press release, were operating an environmental clean up facility by treating and removing waste from various toxic mining sites, whilst extracting residual gold and silver from abandoned tailings.[11] On 30 January 2000, following the heaviest snowfalls in 50 years, there was an overflow of tailings from the dam causing 100 000 $m^3$ of water to escape containing traces of cyanide. The spill was caused by a combination of bad weather and inadequacies in the design of the system to cope with such conditions. This case highlights that changes in fit between technology and the physical environment need to be anticipated. This is to ensure that the residual risks associated with this interface are prevented, through effective design and monitored, from becoming a threat to the environment and public health and safety.

Residual risk is also affected by the interaction of technology with processes. Substitution of old for new technologies does not necessarily reduce the level of residual risk involved in company operations. Economies of scale potentially mean that the extent of consequential damages can be much greater than for smaller operations. This is particularly evident in underground mining:

> Mechanizing has taken the miner away from the immediate face but, at the same time, it has introduced large and fast moving mobile machinery into a relatively confined environment, and has required the opening up of larger size excavations which are harder to support. What you tend to get when you go from close quarter mining to large-scale mining is less of the individual, one miner being hit situation, but an increase in the potential for organizational accidents where you might have a big event.[12]

There is therefore a direct relationship between the size of the technology employed and the extent of residual risk associated with it. This will be explained further in Chapter 7 when such risks are quantified in relative terms.

The final residual risk source is the interaction between technology and human resources caused by the demands technology places on the operator. In this regard, the level of exposure is important. The greater the duration of operation and the severity of the hazard, the greater the likelihood that the risk will translate into harm to the health of the operator.[13] Use of computers is an example that shows the importance of the contact time between human resources and technology. Sedentary work such as typing involving the frequent, repetitive use of the same muscle group throughout the day can lead to musculoskeletal trauma. The physiological effect of such activities, over long periods of time, can cause swelling and decreased lubrication within the synovial sheaths that support the tendons.[14] Consequently, extended durations of such tasks allow the residual risk to translate into injury. According to the four-fold strategy, such residual risk needs to be managed. In this case, appropriate interventions include using job redesign, ergonomics and suitable corrective exercises.

A further example of the importance of duration in the human resources/technology interaction is noise and vibration. Noise is one of the most widespread hazards facing Australian workers.[15] Noise-induced occupational hearing loss depends on two factors – the intensity of the noise and the length of exposure. The greater these two factors, the greater the damage. Short-term exposure can lead to a temporary threshold shift that is essentially reversible whilst longer-term exposure can lead to permanent damage to hair cells in the ear and industrial deafness. It is important to manage these residual risks in the short term to prevent this damage using ear-muffs, and to implement strategies to compress this risk in the longer term by the installation of technologies with lower noise levels or, as a secondary measure, using suppression methods. OHS practitioners and others involved in OHS management need to be aware that the consequences of residual risks associated with the human resources/technology interface are often only apparent when the worker has had longer-term exposure. Nevertheless, such risks have to be managed as standard operational practice to prevent these consequential damages from occurring.

As discussed earlier, equipment quality determined at the design and manufacturing stage affects the rate at which technology degrades. Product quality therefore affects the level of entropic risk over time. The second variable determining the rate of degradation is the operation of the technology. There are three parameters that determine this rate which are the length of operating time, the quality of utilization and the implementation of maintenance. The greater the operating time the greater the deterioration of components. In addition, to this utilization rate, the quality of operation of the machinery affects the rate of wear and tear. 'Thrashing' a machine – rough gear changes, excessive acceleration and hard braking – will increase wear and tear, whilst operation according to manufacturers' instructions or within the intended capacity of the machine will minimize it.

The sources of decay for technology are, therefore, wear and tear and the technology/operator interface. Wear and tear is a combination of the interaction of a unit of technology with other technologies, the physical environment, human resources and processes. For example, a tracked bulldozer suffers wear in the engine during operation, in the tracks as a result of contact with the roadway, and of the bucket during excavation. Whether the process involves shifting sand or rock will also affect the rate of degradation. Harder materials will increase the rate of deterioration and greater weights of materials being shifted increases the workload on the engine. There is therefore a direct relationship between the key variables of the process and technological entropic risk.

Wear and tear is particularly significant as a risk factor because it leads to suboptimal output and safety, as suggested by the entropy model. It also affects the interaction between the operator and the equipment. Degraded equipment places greater demands on the operator, for example, cutting a tree branch with a blunt saw is more strenuous than doing the same task with a well-maintained sharp saw. In addition, when equipment fails, the worker's reaction has an influence on the consequences of the failure. The employee's competencies, state of health and vigilance determine his ability to either avoid damage/injury or minimize the extent of such damage/injury. Where, for example, a vehicle tire blowout occurs, the driver's skills have an influence on whether a collision results or whether further damage is averted. If the interaction between technology and the physical environment is also placing additional demands on the operator, for example, the vehicle is not suited to the road conditions, then the tendency towards entropic risk is greater still.

These interactions between the driver, vehicle and physical environment, have been illustrated using chaos theory.[16] Figure 4.2 shows these relationships. In part (a) the vehicle is in a poor condition and the road is in a fair condition which results in stable oscillations in the driver's vigilance. The operator varies in attentiveness without any serious consequences and is able to manage the demands associated with the poor condition of the vehicle. In part (b) the vehicle is in good condition and the road in a bad state. The effect on the operator is chaotic oscillations in vigilance and the possibility of an accident. This situation puts very

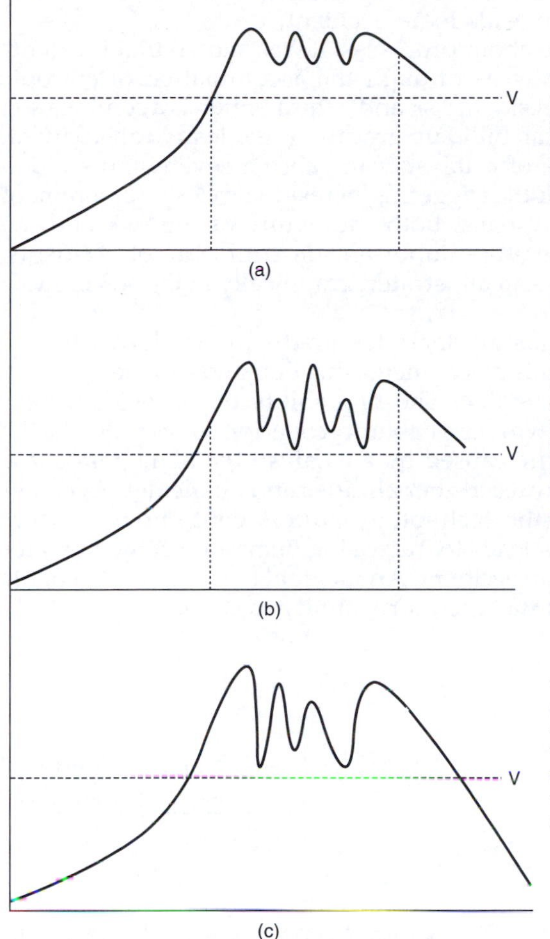


Adopted from Fig. 5 on page 103 'Chaos theory of occupational accidents' by B. Hacking and C. Thompson, *Journal of Occupational Health and Safety – Australia and New Zealand 1992,* 8(2).

**Figure 4.2** The effect of road and truck conditions on the driver's vigilance

high demands on the driver and the inconsistency of these demands makes it more difficult for him to make the necessary adjustments, for example when the road has a pothole or uneven surface. The risk of an accident is contained by the driver's alertness, but this depends on his physical condition and competencies. Finally in part (c) the vehicle and road are both in bad condition. Chaotic oscillations in vigilance occur and fall below 'V' which is the cut-off point where overload results in a certain accident. The driver is not able to cope with the combined effect of these demands.

Figure 4.2 shows that technology and its interaction with other system factors has a significant effect on the operator's ability to manage risk. The greater the severity of degradation and the greater the process duration, the more likely it is that the technological demands will exceed the worker's ability to cope. This in turn leads to an accident.

In the earlier discussions about processes, it was shown that incidents occur in three ways. The first is through the accumulation of entropic risk. The second is through residual risk and a third option is a combination of these. In addition, risks can build up gradually and lead to an accident. Alternatively, the presence of catalysts can cause a severe and sudden rise in entropic risks and/or the triggering of residual risks into imminent threats to organizational systems. Both the nature of the risk and the severity of the risk are therefore important determinants of accidents. These time-related variables are illustrated graphically in Fig. 4.3 to show how they apply to technologies.

In Fig. 4.3(a), the technology degrades gradually as shown by the dashed line. Concurrently this places increasing demands on the operator, as shown by the continuous line. The probability of an incident rises accordingly, however, the worker is able to cope for some time. At the point of technological failure caused by a catalyst, the demands on the operator reach a point that exceeds her abilities and an incident becomes imminent. In Fig. 4.3(b), the technology suffers continuous gradual degradation, which in turn, leads to increasing demands on the operator and a rising probability of an accident. An incident becomes unavoidable when the demands exceed the operator's ability to cope.

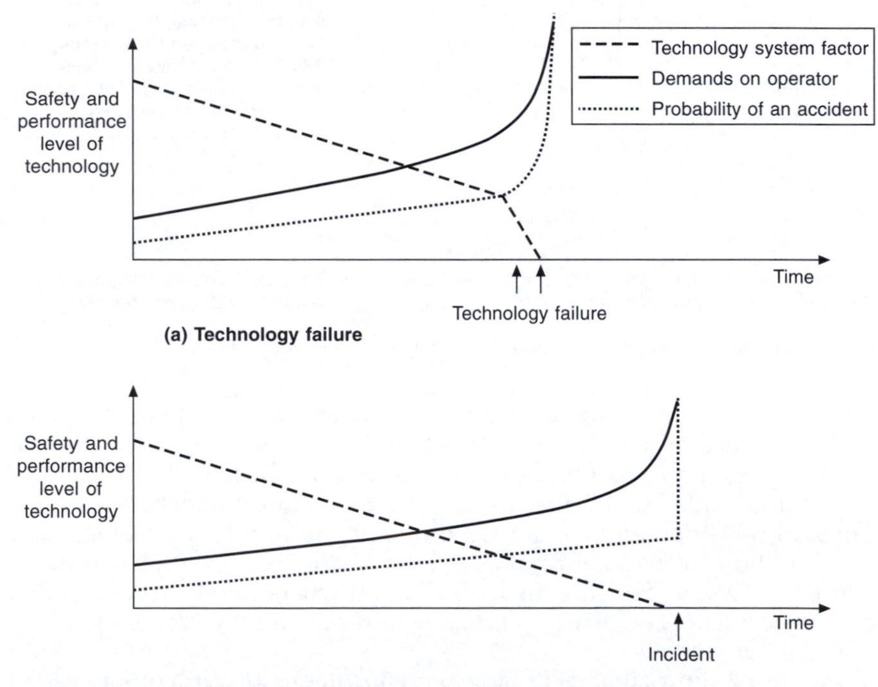


**Figure 4.3** Technological degradation and the probability of an incident

## Managing technological risks

How can technological risks be managed? Residual risk management begins with the supplier at the design and manufacturing stage. For the purchaser, OHS legislation requires them to provide plant so that, as far as is practicable, employees are not exposed to hazards. The implications are that the onus is on firms to evaluate technology to minimize these residual risks prior to purchase which requires them to establish pre-purchase evaluation criteria that consider both the safety and performance of the technology. The evaluation stage uses the information that is currently available from the supplier and industry sources as the basis for equipment selection.

The firm should also require its suppliers to provide advice about the risk factors associated with the technology, during the operational life of the equipment, so that these risks can be managed effectively. Reiterating, the criteria for technology selection include the safety and reliability of the equipment, production capacity, cost, back-up parts, service and training, and the provision of supplier information. In addition, the firm needs to establish techniques for the monitoring of technological performance and safety to identify unforeseen residual risks. These risk management strategies are summarized in Table 4.1.

In Table 4.1 there are a number of strategies described for both residual and entropic risks associated with the interface between technology and other system factors. The residual risk resulting from the interaction of technology with the physical environment can be managed using a number of risk reduction techniques. The first is to identify potential hazards and design the work environment to maximize the fit with the technology to reduce such hazards. For example, in large earthmoving equipment there is a vision shadow that prevents the operator from having a full view of the surroundings. This is a residual risk associated with this type of technology. The physical environment can be adapted to compensate for this. In open pit mining operations, for instance, earthen barriers are built at the edge of roadways to reduce the likelihood of vehicles being driven over the edge of the level/bench. Additional lighting can be provided to working areas to increase visibility at night. Wherever the firm uses mobile plant, the physical environment has to be designed, as far as practicable, to minimize the residual risks associated with the interaction of these two system factors.

Fixed technology also introduces residual risk. To mitigate this, it must be installed in accordance with the manufacturer's instructions and relevant employees trained in its safe operation. It may also be appropriate to develop additional workplace-specific safety standards to manage any residual risks, for instance, gas appliances, such as hot water systems and stoves, have a residual risk because a delay in ignition can cause a flash flame to exit the appliance.[17] When lighting these appliances, particularly for the first time or after an interruption of the gas supply, it is important to follow the manufacturer's instructions. In its procedures, the firm will also have to specify which of its employees are authorized to carry out the task. The case involving the electrocution of a technician (Chapter 3) which resulted in the principal firm being penalized for

**Table 4.1** Strategies to manage the risks associated with technology

| Source of risk | Risk management strategy |
|---|---|
| *Residual risk* | |
| Design and manufacturing short-comings | • Pre-purchase evaluation of safety and performance record of technology options<br>• Using supplier relationships to obtain up-to-date information about safety performance<br>• Monitoring of technology's performance and safety |
| Technology/physical environment interaction | • Effective design and planning of work environment<br>• Pre-installation evaluation of risk factors and potential hazards<br>• Installation of technology in accordance with manufacturer's instructions<br>• Development of workplace-specific safety standards to manage any residual risks |
| Technology/process interaction | • Evaluation of suitability of technology for the process<br>• Standardization of technology<br>• Modifications of process in the short term to manage residual risk<br>• Modification of technology to compress residual risk in the longer term |
| Technology/human resources interaction | • Purchase of ergonomically-designed, operator-friendly technology<br>• Modification of existing technologies to fit the operator<br>• Job design<br>• Modification of work practices |
| *Entropic risk* | |
| Wear and tear | • Proactive scheduled maintenance<br>• Regular monitoring of condition of technology<br>• Planned replacement of parts/equipment prior to rapid rise in entropic risk<br>• Reactive maintenance |
| Technology/operator interaction | • Pre-operation training<br>• On-going refresher courses where necessary<br>• Organizational culture which reinforces desired behaviors<br>• Systems to correct undesirable behaviors |

allowing access to non-authorized persons, is an example of the need for strict controls to manage the risks associated with technology.

When installing or using technology, potential catalysts from the physical environment which may trigger inherent risks into imminent danger, also need to be considered. For example, when moisture is present where electrical appliances are being used, there is a high risk of electric shock. A case is the death of a boilermaker who was electrocuted at a Western Australian ship-building company.[18] The deceased had been using a standard manual metal arc welding machine to repair the internal workings of a ship. He had been in a confined space for almost 3 hours and the workplace temperature may have been as high as 60°C. When found, his gloves and clothing were wet with perspiration and a new electrode rod had been fitted in the electrode holder. As a result of this case, the coroner handed down a requirement for the installation of voltage reducing devices (VRDs) on all alternating current welding equipment used for industrial purposes. This example highlights the need to anticipate activities that involve the combination of technological and environmental factors that translate residual risks into imminent danger. It would also have been appropriate for work practices to be modified, for example regular breaks from the task, given the extreme conditions involved.

A further source of inherent risk is the interaction between technology and processes. As summarized in Table 4.1, these risks can be managed in a number of ways. The first is to determine the suitability of the technology for the process. An example of this is dewatering pumps that do not have a cut-off device fitted to prevent them from cavitating (pumping dry) for extended periods.[19] Such pumps are not appropriate where the pump is not closely monitored by an employee. The result can be the build up of hot water in the delivery hose and if the worker is unaware of the continued operation without a flow of water through the pump, serious scalds can occur. To remedy this, firms can modify work practices using JSA to ensure that the correct shut-off procedures are followed. In the longer term, this residual risk may be compressed through the modification of the technology. The technology–process interface is managed primarily by following the manufacturer's instructions; specifically, only using the technology for purposes for which it is designed and enforcing the recommended operating procedures. Long-term solutions to these residual risks involve better technology designs such as an automatic cut-off when the pump starts cavitating.

To manage residual risk, firms should standardize technologies as much as possible. For example, mining companies usually purchase a fleet of the same make and model of haul truck rather than having a number of different suppliers or products. This standardization also applies to tools and other equipment. The advantage is that a consistent approach can be used to reduce the residual risks caused by the interaction of the technology with other system factors. Information gathering and dissemination processes, such as training, are also less complex when equipment is standardized and it is also often more cost-effective.

Using systemized technologies reduces the variety of processes that need to be undertaken in the workplace in terms of both equipment operation and maintenance. Residual risk is reduced because workers are able to develop specialized KSAs in the use and servicing of the technology. As a result, the demands on the employee are also minimized. For example, if a pool of light vehicles consists of all the same type of vehicle, drivers are more easily able to adjust to the minor variations in performance of each car. On the other hand, increasing the variety of vehicles used places greater demands on operators. In an activity where the driver is accustomed to an automatic transmission vehicle and on occasion drives a manual transmission vehicle, this leads to higher demands in the latter case. The driver has to remain conscious of the type of vehicle being operated, otherwise he may stall the car when braking and create a hazard. Reducing workplace complexity by having standardized technologies that allow defined procedures based on JSA to be developed, therefore, reduces residual risk.

There is also inherent risk at the technology/human resources interface. This has attracted considerable attention in recent decades with the concurrent rise in job simplification, such as positions that only involve data entry. As a consequence of reducing the variety of tasks undertaken in a given job, the times of exposure for certain work activities have increased. In data entry roles people are sitting in front of computers for longer and they have fewer opportunities to move around whilst performing productive work.

There are a number of strategies that can be employed to manage the residual risk at this interface. The first approach is to minimize the level of residual risk by purchasing equipment which best fits the needs of the operator. As far as practicable, therefore, technology should be ergonomically designed. It should also have minimal levels of negative characteristics, such as emissions, vibrations and noise that affect the worker's physiological or psychological health. In firms with established equipment, it may be possible to modify these technologies to make them more operator-friendly and to compress residual risk levels. Often simple solutions can be found to increase user comfort, for example, raising computer visual display units (VDUs) to eye level by placing old telephone books underneath them, or having a small cushion on the seat at the base of the spine to support the lower back.

Job design can also be used to minimize the potential consequences of residual risks. This includes rotating workers so that they undertake a variety of tasks, for instance, in a factory, employees on a food manufacturing production line may be shifted between jobs that require sitting and standing, or that require sorting and moderate lifting. Job rotation is an effective means of managing such risks. Job enrichment that involves building a variety of tasks into a job[14] can also be used to manage residual risk caused by the repetitive use of the same muscle and tendon groups. An example is the enrichment of typing pool positions into secretarial roles. This reduces the length of time spent continuously typing by providing other tasks including answering telephone enquiries, delivering documents and filing.

In addition to changing the types and variety of tasks undertaken, work practices can be modified to reduce the impact of residual risks on worker health and safety. Modifications include providing rest breaks and encouraging exercise that counters the physical stress caused by repetitive use of the same physiological systems. An issue that has become contentious in recent years is the monitoring of the pace at which employees work. In computerized offices, for example, some companies are using electronic systems to monitor the logging errors or keystroke rates of typists.[14] This in turn, puts the worker under additional pressures to work continuously at a higher pace, thereby increasing the stress and severity of the residual risk associated with computer usage.

In addition to residual risks, the firm has to control entropic risks associated with technology. These stem from two main sources – wear and tear and the equipment/operator interface. The management of wear and tear requires both proactive and reactive responses. Firstly, it can be addressed through scheduled maintenance. For plant and equipment this is done in accordance with the manufacturer's instructions, as a minimum, with additional upkeep geared to the specific demands of the firm.

Maintenance can also be triggered when changes are identified during routine inspections. Specifically, the supervisor should assess the current level of wear, anticipate the remaining life of components and evaluate the rate of degradation. An example is where a team leader uses his experience to evaluate the condition of truck tires and to forecast when they need to be changed. Subsequent routine inspections are used to ensure that the components are still in an acceptable condition and to revise this forecast if necessary. Maintenance is planned to maximize the utilization rate of the technology without compromising safety.

Regular inspections help to prevent wear and tear from rising to an inadmissible level of entropic risk. Equipment failures are inevitable unless influenced by the frequency, thoroughness and extent of maintenance activities.[20] A further two incidents involving small aircraft illustrate that routine inspections and maintenance, together with residual risk assessment and compression, are important means of reducing the probability of undesirable events. On 4 September 2000, eight lives were lost when a Beechcraft King Air 200 flying from Perth to Leonora in Western Australia crashed. The investigation determined that the pilot and passengers were most likely incapacitated due to hypobaric (altitude) hypoxia resulting from the cabin being under-pressurized and their not receiving supplemental oxygen,[21] suggesting that the mask deployment system had failed. Although the investigation could not be conclusive about the cause of depressurization, there did appear to be some parallels with the Payne Stewart case that occurred in October 1999. The Perth 'ghost plane' continued to fly on autopilot for 5 hours before crashing in Queensland.

In October 2001, an incident involving an aircraft of the same make and model occurred. In this case, a complex set of variables was involved. In the first instance, the aircraft's air conditioning system had failed to operate properly after a history of problems with six instances of maintenance recorded since January 2001. In response to the oppressive heat and humidity on the day of the incident, the pilot attempted to speed up proceedings and did not complete the pre-take-off and after-take-off cabin pressurization checks.[22] He had neglected to undertake the required regular inspections and adjustments of on-board systems. This reduced the effectiveness of the aircraft's cockpit warning system that would ordinarily have alerted the pilot to lack of cabin pressure. In addition to highlighting the need for routine system checks as part of standard work procedures, the case lead to the mandatory fitting of aural warnings to operate in conjunction with the cabin altitude alert warning system on all Beechcraft Super King Air and other similar aircraft. These incidents indicate that it is imperative that hazard information is shared to ensure that incidents are not repeated and that firms in the relevant industry take necessary remedial actions.

The latter case also raises the importance of reactive maintenance to counter wear and tear and to return technological systems to the required standard of operation. This is the least preferred approach because it means that the technology has been allowed to reach substandard safety and productivity. It is a major concern because the failure to anticipate or to monitor entropy often is the cause of accidents. An example is the case where a box assembly fell from a drilling rig at a mine in the Western Australian goldfields causing severe head and other injuries to a driller.[23] The box assembly was usually attached to the drill head and was designed to cope with the high air pressures and abrasive nature of the drill cutting process. The box, weighing approximately 300 kg, broke away from its welded-support fitting and the attachment bolts also sheared off from the drill head. The contributing factors that caused the box to fall were that one of the welded-support brackets had failed prior to the day of the accident and the remaining weld broke on the accident day. In addition, two of the attachment bolts to the drill head had sheared off the day before and had not been replaced. Further, the box had been substituted before the accident. The old design had a safety link that involved attaching a safety chain to the rig, whereas the new box did not have this attachment and also was intended for a different type of rig arrangement.

The case illustrates how entropic risk can escalate and lead to an accident. It highlights the importance of responding quickly and effectively to such risks. According to the entropy model, corrective action should have been taken in the presence of entropic risk. Maintenance strategies should also have been developed to prevent future degradation. These particular problems could have been identified through workplace inspections, specific hazard inspections and incident investigations, in addition to spot checks at any time during the work cycle.

Given that technology degrades, the firm has to decide when to take corrective action to remedy this condition. The key objective of these decisions is to ensure that risk is maintained below the 'acceptable' level and can be managed effectively. To meet the required duty of care, managers have to take the demands on the operator which are caused by this entropic risk into consideration, in addition to the impact on work processes and the physical environment. For example, if road conditions on a construction site are slippery as a result of heavy rain, it is more critical for truck tires to be in a sound condition to allow for good traction than in dry conditions. If the level of traffic on the site is high, the stopping capacity of the truck becomes a major safety issue to allow the driver to respond quickly to the actions of other vehicles. The definition of 'safe' in terms of equipment condition therefore has to take these secondary factors, such as physical environmental conditions and driver competency, into consideration.

The decision to take corrective action therefore depends on two issues. The first is the actual condition of the technology, for example, the air-conditioning and oxygen deployment systems in the small aircraft discussed previously required immediate corrective action because of the severity of the risk. The second is how critical this condition is when combined with the risks associated with other system factors – other technologies, processes, the physical environment and human resources. This relationship was shown in the example about the truck tires given in Fig. 4.2. When road conditions are less favorable the state of the tires becomes increasingly important as a risk control factor.

Technological entropy is primarily caused by wear and tear. A related issue is that it is also caused by the suboptimal interaction of technology and human resources. Specifically, the quality of operation – how well the worker uses the equipment – affects the rate of wear. Whilst the operator can increase the rate of deterioration of the equipment, degraded equipment can also place additional demands on the worker. The longer the duration of operation and the greater the level of wear, the harder it is for the employee to maintain vigilance and operate safely. There is therefore a reciprocal relationship between these system factors; the condition of technology affects the operator and vice versa. For this reason, pre-operational training is very important. It provides the worker with the competencies to use the technology safely and efficiently. When equipment is used appropriately this increases its productivity by maximizing output and optimizing its life.

Operator competencies and behaviors are addressed by the management system and influenced by the organizational culture. Behavioral management involves developing employee skills and attitudes that lead to the correct use of technology and appropriate interaction with other system factors. In Part 3 of this book, the elements of this management system, for example, defining accountabilities for the operation and maintenance of equipment, effective training strategies, and a measurement format that evaluates the condition of system factors, will be discussed. Such systems should also contain procedures for the correction of undesirable behaviors, commonly referred to as disciplinary actions, which usually include penalties for the intentional misuse of technologies or for the failure to maintain it adequately. Although punitive courses of action may be required from time to time, the emphasis of the productive safety culture is on education and reinforcement of positive behaviors rather than punishment.

Technology is mostly thought of as the equipment and tools used to generate output for the firm. Other equipment, that is not part of the production line, also needs to be considered. For example, technology can be used to monitor hazards, such as the concentration of toxic gas emissions and the intensity of noise. Also in this category are signs, emergency response equipment and personal protective equipment (PPE). Some devices, such as fire extinguishers, are needed to contain the consequential damages of an incident. This equipment, like the machinery used to produce output, has to be operated correctly and maintained in a safe and efficient condition. The failure to do so can expose employees to considerable risk. This has been shown to be the case in a number of failures of fire suppressions systems onboard excavators that have resulted in threats to the lives of drivers.[24] In addition to maintaining this equipment, the firm has to ensure that systems are adequate to deal with potential hazards. Other considerations are the location of system hardware, the level of protection afforded by the technology, for example, the fire resistance of the materials used, installation standards and activation time. An inventory of safety equipment should be drawn up and a schedule of regular inspections and maintenance carried out.

PPE has also to be maintained well as it is the last line of defense against hazards. It is at the bottom of the hierarchy of controls (refer to

Fig. 3.3). This equipment must meet the relevant national or legislated standard. It is intended to be used by a single wearer, fitted correctly to the individual to ensure optimum protection, and therefore, should not be shared. Legislation such as the Occupational Safety and Health Regulations 1996 (Western Australia) contains specific provisions concerning the supply and wearing of PPE. For example, Section 308 of the regulations requires the employer to provide a personal hearing protection to each employee exposed to workplace noise above the 'action' level.[25] Other PPE covered by this legislation include respiratory equipment, eye protection, fire-resistant aprons and gloves, safety helmets, footwear and safety belts/harnesses.

Broader provisions related to the supply and wearing of PPE are contained in the Occupational Safety and Health Act 1984 (Western Australia). These require the employer to provide employees with adequate personal protective clothing and equipment to protect them against hazards where it is not practicable to avoid the presence of these hazards in the workplace. This is covered by Section 19(1)(d).[1] The Act also indicates that an employee who fails to use such equipment provided by the employer and after he has been properly instructed in its use, contravenes the 'duties of employees' provisions. It is, therefore, the worker's responsibility to use PPE in the appropriate manner and to maintain it in a good condition. The supervisor's responsibility is to ensure that employees are provided with, use, wear and maintain their PPE and are made aware of procedures related to its use.

## The technology audit

One of the most effective means of identifying risks is the auditing process. Under productive safety management, audits that specifically target technological risks are carried out. To conduct the technology audit the information initially gathered during the process audit, discussed in Chapter 3, is used to identify the technologies employed by the firm. The business operation is broken down into workplaces and each area reviewed to identify the plant, equipment, tools and products used or stored there. A guidance list is compiled to assist investigators in the auditing process. The list, however, should not restrict the auditor's ability to identify additional sources of technological risk. The process of compiling a list is illustrated in Fig. 4.4.

The check includes the following questions:

(1) Does the manufacturer/supplier's information provide warning of hazards?
(2) When was the technology purchased?
(3) What has been the technology's performance and safety record at this site?
(4) What has been the technology's performance and safety record in this industry?
(5) Do the design, operation and storage method fit the current physical environment?

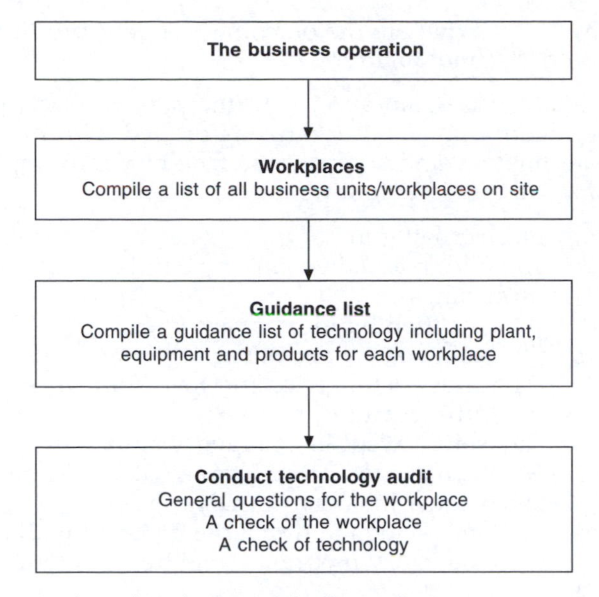


**Figure 4.4** Preparing for and conducting a technology audit

(6) Has the technology been installed correctly?
(7) Is the technology suitable for this process?
(8) Is the technology being operated or stored safely?
(9) Has a strategy for planned maintenance and inspections been devised?
(10) Have maintenance plans and inspections been carried out?
(11) Has reactive maintenance been required since the last audit?
(12) Have the hazards associated with the technology been identified, documented and quantified?
(13) Are any parts replacements/overhauls/decommissioning planned for the forthcoming period?
(14) Have all operators been given pre-operational training?
(15) Have any employee suggestions been raised in relation to this technology?
(16) Are any PPE required to be worn when using or in contact with this technology?
(17) Is the PPE adequate and well-maintained?

When undertaking the technology audit similar steps are applied as were used during the process audit. The latter applied the seven process investigators (the 7 Ps). These are adapted slightly so that the investigators are the T and 6 Ps, as follows:

| | |
|---|---|
| Technology: | What is being used? |
| Purpose: | Why is it being used? |
| Person: | Who is using it? |
| Place: | Where is it being used? |
| Period: | When is it being used? |
| Procedure: | How is it being used? |

Performance: What are the outcomes – safety, productivity and quality?

'What if' analysis is again applied to the person, place, period and procedure to identify potential hazards associated with the interaction between the technology and other system factors. For example, analysis of the use of a nail gun is as follows:

Technology: Electric nail gun
Process: To build/maintain company buildings, fittings and furnishings
Person: (1) Qualified builder or carpenter
(2) Maintenance personnel
Place: (1) Infrastructure sites and township workshop
(2) Infrastructure sites
Period: Day shift (except in emergency situations)
Procedures: No standardized procedures available (clarify steps taken using Job Safety Analysis)
Performance: (1) One accidental discharge reported with no injuries
(2) One injury reported – nail through foot

When 'what if' analysis is applied, potential questions may include:

(1) What if the maintenance personnel have not been given training to use the nail gun?
(2) What if the nail gun has to be used in a high activity work area – what additional risks would result?
(3) What if the nail gun gets jammed?
(4) What if the incorrect sized nails are used?
(5) What if the nail gun does not meet the required national standard?

These analyses and investigation tools help the firm to identify residual and entropic risks associated with technology. A nail gun would, for example, have a serious design short-coming or significant residual risk if it was not provided with a safety switch to prevent inadvertent discharge. The primary source of residual risk is that the equipment involves the pressurized release of a projectile that is capable of causing serious injury or damage. Entropic risks may be apparent if, for example, the equipment is corroded in parts due to exposure to the weather and its mechanisms are stiff due to poor maintenance. It is also important for the auditor to check whether the technology meets the national standard because, from time to time, unapproved equipment which is strictly illegal and dangerous, is offered for sale. This has been found to be the case with electric power tools.[26]

The auditor should document these technological risks and develop recommendations using the four-fold strategy. Reiterating, this involves eliminating entropic risk in the short term, establishing maintenance strategies to prevent future degradation, managing residual risk in the short term, and compressing residual risk in the longer term.

For the nail gun example, alternatives for the management of these risks include:

(1) Existing areas of corrosion on these tools are to be scrubbed back and

coated to prevent further deterioration. The equipment is to be cleaned and maintained to ensure that mechanisms work safely and efficiently.

(2) Housekeeping practices are to include a post-shift check to ensure that tools and equipment are returned to the assigned storage place out of the weather and in a safe location where they can not be accessed by unauthorized personnel.

(3) Relevant employees are to be briefed at the next toolbox meeting regarding the residual risks associated with the use of nail guns. Correct operating and maintenance procedures will be discussed and implemented.

(4) Future purchases will consider the safety features of the equipment, in addition to suitability for the purpose.

This audit provides a profile of the technological risks faced by the firm. From this, the investigation team can prioritize degradation prevention and residual risk management strategies to make best use of the firm's limited resources. The method for quantifying these risks will be presented in Chapter 7.

## Maintenance of technology

The entropy model indicated that maintenance is required to address degradation of the technology system factor. Who is responsible for this maintenance? The channel suggests that resourcing decisions are primarily management prerogative. The levels of human, physical and financial capital allocated to maintenance are determined at the upper- and middle-management levels. It is therefore, management's responsibility to ensure that sufficient resources are made available for maintenance to reduce risk as far as practicable and to provide a workplace where employees are not exposed to technological hazards.

The purpose of maintenance is to operate equipment as close to the optimal output and safety level over the life of the product as possible. This reduces risk and also maximizes the return on the investment in plant and equipment. As such, the technology should not only pay for itself by contributing to the production of outputs, but also start to give benefits beyond its initial cost. Without getting into the complexities of cost–benefit analysis, a day-to-day example is the purchase of a new or nearly new vehicle. When the owner operates the car in a manner that minimizes degradation and maintains it regularly, after a number of years, the owner continues to get satisfactory use (utility) from the vehicle, even though its market value may have depreciated considerably. For example, the vehicle may only be saleable for $2000, but because it is still reliable and safe, its worth to the owner is close to the cost of a nearly new vehicle – $20 000 – because this is what it would cost the owner to replace it. The longer the owner can keep using the vehicle and it continues to be safe, efficient and cost-effective to run, the more benefit is derived from it. The initial or fixed cost of the vehicle is spread over a longer life. In accounting terms, firms write off the fixed cost of capital purchases through depreciation over a predetermined number of years.[27] When it optimizes the life of its technology through planned maintenance, it reaches a point where the benefits nullify the initial cost of purchase, and returns continue to be attained until the costs of maintenance outweigh the productive benefit of the equipment. Accordingly, optimum use is made of limited resources.

What issues does management have to consider when making resourcing decisions for the maintenance of technology? The first is that resource allocation has to be aligned with the company's goals so that sufficient resources are committed to make these goals achievable. The options available in relation to these decisions, however, are not the same from firm to firm. Companies face varying constraints and therefore, their objectives will also be different. The mining industry can be used as an example. In some multinational mining companies that are very well resourced and have large-scale operations, the purpose of maintenance is to ensure high levels of output and safety in the short term. These firms may change their technologies regularly to pursue economies of scale. Large firms have the resource availability to regularly purchase bigger trucks, shovels and other plant which allow more output to be produced. The implications are that by having state-of-the-art equipment, they potentially have lower levels of inherent risk and less exposure to entropic risk as a result of their technologies.

Whilst most of the major mining contracting companies have high quality, well-maintained equipment and thereby are able to sustain the performance standards required by principal firms, smaller scale operators can struggle. As a consequence of less financial resource availability these firms are often forced to get a much longer life from their technologies thus increasing their exposure to entropic risk. In addition, pursuing economies of scale may not apply because their client market often requires them to operate small- to medium-sized mines. One of the purposes of maintenance, therefore, is to extend the operational productivity of their existing equipment. Such companies may also need to be more proactive to manage higher residual risks and to counter the degradation of aging technologies.

Regardless of the level of resources available to the firm, it is important that it evaluate the effectiveness of resourcing decisions to determine whether goals were achieved, whether sufficient capital was allocated and how these decisions can be improved. Consequently, the firm has to establish measurement criteria to compare actual outcomes against desired outcomes. For example, to measure the effectiveness of maintenance budgets there are a number of issues to be considered. The first is whether there were any equipment failures that resulted in damage or injury. If a number of such incidents occurs it may indicate that insufficient maintenance is being carried out or that the levels of technological risk are unacceptable. Secondly, the firm will also need to compare the level of output produced by the technology against targets, for instance, the volume of ore crushed at the ore processing plant. If the level of output is well below expected, lack of or poor maintenance practices, such as rework, may be a contributing factor.

A further matter to be measured is equipment availability. Planned maintenance reduces availability because it requires the plant to be removed from the production line. This is a reduction in operating time that is necessary to upkeep the equipment in a safe and efficient condition. On the other hand, reactive maintenance results in unplanned and disruptive reductions in availability and is considered to be undesirable. If, for instance, the firm experiences an increase in breakdown maintenance, this may indicate that insufficient resources have been allocated to preventative maintenance. Breakdowns warn that entropic risk has reached a counter-productive level and that more resources need to be allocated to strategic maintenance. Rework also affects equipment availability, because it means that the plant has to be returned to the maintenance workshop for further repairs. Managers should consider rework to be very serious for two reasons. The first is that it is costly. The second is that it is indicative of poor quality workmanship. The entropy model suggests that when technologies are not being maintained effectively, degradation is not prevented, and therefore, the level of entropic risk tends to rise. There are thus, two costs incurred by poor maintenance, which are the costs of rework and the hidden costs of uncontrolled degradation.

Strategic and budgetary decisions concerning maintenance practices have a direct impact on organizational achievement and can be evaluated using tangible measures, such as the number of incidents caused by equipment failure and the costs of these failures. Resourcing decisions can also have a less obvious impact on the firm's performance because of their effect on the organizational culture. Employees evaluate the level of company commitment to safety according to the resources allocated to OHS management, as well as other considerations, such as the quality and consistency of operational decision-making. In the example given previously where the maintenance crew requested the supply of two-way radios for work in the open-pit mine and management rejected the proposal, employees considered this to be indicative of a lack of commitment to safety. Management, therefore, has to be aware that resourcing decisions can reinforce or erode common goals and values.

The firm has to make decisions firstly about the amount of financial resources allocated to the maintenance of equipment and secondly, how this work will be carried out. This includes whether to have these duties undertaken internally by company employees or outsourced to contractors. The impact of contracting out technological maintenance has led to considerable debate in Australia and other western countries in recent years. The airline industry is an example. Unions claimed that the spate of incidents, in 2000 and 2001, in a major airline company were partly attributable to outsourcing.[28] Incidents included:

September 2: The engine of a 747 jet scraped along a Perth runway while it was landing in gusty winds.
September 23: QF1 careered off the runway at Bangkok airport, causing $100 million worth of damage to the aircraft.
November 2: Fumes entered the cockpit of a 767 Sydney–Melbourne flight.
November 14: An emergency exit chute inflated inside a jet flying from Brisbane to Auckland.
December 23: A jet carrying 70 people circled Canberra for almost an hour because of a wing flap malfunction.

December 26: An engine part fell from a 747 flight from Brisbane to Japan.

January 21: A flight on its way to Cairns turned back after smoke was detected in the cabin.[28]

The now defunct Ansett Airlines (an Australian domestic airline) also experienced problems in the area of maintenance prior to its demise in late 2001. The Civil Aviation Safety Authority intervened and Ansett was required to put forward a plan for the upgrade of its maintenance practices. This included a total overhaul of the maintenance division involving a review of staff, improved training, better management of spare parts, a review of tooling and the introduction of a tool management system. Significantly, the plan included introducing stronger maintenance planning to properly control and direct such work. The company intended to bring additional resources from Boeing for specific aircraft repair planning issues.[29] These maintenance deficiencies have since been found to be symptomatic of broader management problems which preceded Ansett's collapse.

Maintenance resourcing decisions are important not only in terms of dollars committed, but also how maintenance is carried out. The issue is not so much about whether the work is done internally or outsourced, but about accountability, the level of control, quality and the impact on the organizational culture. An article on dozer safety in open-cut mining operations indicates that in addition to applying maintenance strategies to manage technological risks, maintenance has to be considered as a strategy for keeping up the standard of all systems to the required level.[30] It therefore must extend to the management of risks that occur at the interface between technology and other system factors. This includes ensuring adherence to safe operating procedures, the monitoring and maintenance of physical environment changes that have an impact on the operation of technologies, and the maintenance of operators' workplace-specific knowledge. The article states that firms should make provisions for:

Adequate lighting of working areas should be provided at night. It is not sufficient to rely on equipment running lights. Fixed lighting or trailer-mounted lighting plant should be maintained at working locations.

Appropriate planned maintenance and fault repair, coupled with immediate, management-supported and mandated equipment shutdown in cases where problems cannot be fixed immediately.

Familiarisation of crews with pit geography and layout. This is particularly important where long distance commute systems of work are in place and crews may need to be updated on changes to their workplaces during their time off site . . .

Operating practices established at the mine for bulldozers must be adhered to and regularly monitored. The practices and procedures will be determined by the mine management in consultation with the equipment supplier and the plant operators . . .

> Strict enforcement of the wearing of seatbelts by all operators of and passengers in mobile plant, at all times when the equipment is operating.[30]

The article indicates that a comprehensive technological risk management strategy involves on-going preventative maintenance and before risks become unacceptable, corrective action must be taken. The entropy model indicates that these interventions are needed to counter the tendency of technology and other system factors to degrade. The final point to be considered in making resourcing decisions of this type is that preventative maintenance is good business. The entropy model illustrates that as technology degrades it becomes less productive and less safe. In this era of slower economic growth, better equipment utilization is becoming increasingly important.[31] It is a significant factor affecting the firm's sustainability. Company goals should therefore include shifting technologies towards zero breakdowns that, as far as practicable, requires the elimination of reactive maintenance due to degradation. Reductions in total machine availability should, thus, be attributable to only two factors. These are firstly, planned maintenance and secondly, breakdowns due to unforeseen hazards or events after all practical efforts have been made to evaluate and eliminate potential risks.

A number of companies have experienced the benefits of effective risk management on the 'bottom line'. When Woodside Offshore Petroleum implemented strict controls on the management of its contractors one of its findings was that sound planning in OHS management reduced the frequency of unplanned events. The division's general manager indicated that unplanned events, whether they caused danger or not, detract from the efficiency of the business. By planning to eliminate these deviations, the operations also eliminated incidents that reduced organizational effectiveness.[32] Preventative maintenance therefore helps firms account for unplanned occurrences such as equipment breakdowns that detract from productivity. These practices take into consideration the hidden costs of risks and factor these into resourcing decisions related to technological maintenance.

As stated earlier, the channel indicates that resourcing decisions are the prerogative of management and therefore, final responsibility for maintenance rests with them. It is at senior company levels that budgets for planned maintenance, equipment replacements and training are determined. It is also at these levels that the criteria for maintenance scheduling are authorized. For example, will the equipment be services every 5000 or 10 000 operating hours? At what stage of wear and tear will components be replaced? These are management decisions that have a direct impact on the safety and output performance of technology. Nonmanagement employees also have a major influence on the condition of this system factor. Their competencies and attitudes affect the quality of maintenance and operation of equipment. There is therefore a strong case for behavioral management linked to these technological maintenance practices, as discussed further in Chapter 6 and Part 3 of this book.

## Summary

This chapter has described how technology introduces both residual and entropic risks into the workplace. These risks can be managed using the four-fold strategy proposed by the entropy model. Practical measures include, firstly, the careful selection of technologies in the planning stage to minimize residual risk and the tendency of equipment to degrade. Secondly, at the operational stage, the firm has to implement corrective action and preventative maintenance strategies to address wear and tear. In addition, training is required to prevent the suboptimal interaction of technology and the operator. It was shown that technology introduces many additional risks when there is a poor fit with the workplace and with the worker. In the following chapter, the risks associated with the physical environment will be discussed. This chapter has particular significance for hazardous industries, such as mining, oil and gas, and construction in which the residual risks associated with the physical environment are high.

## References


1. *Occupational Safety and Health Act 1984 (Western Australia)*, Western Australian Government Printers, Perth, Australia. Available on website: http://www.austlii.edu.au/au/legis/wa/consol_act/osaha1984273/
2. Miller, G. (2002) *Trends in OH&S Prosecutions*. Available on website: www.workcover.nsw.gov.au/pdf/ohoseduc_prosecutiontrends.pdf
3. National Safety Council of Australia (2001) *Amputation leads to landmark judgement*. Available on website: www.safetynews.com/#amputation
4. Oxenburgh, M. (1992) How to evaluate the cost of sickness, injury and absenteeism. Chapter 1: *Increasing Productivity and Profit through Health and Safety*, CCH International, Sydney, Australia, pp. 13–51.
5. Department of Mineral and Petroleum Resources Western Australia (2001) *Significant Incident Report No. 107*. Available on website: http://notesweb.mpr.wa.gov.au/exis/SIR.NSF/6b390eea5649d21c48256097004aacb7/1fa42f29b0656309482568ee0030ca54?OpenDocument
6. Bureau of Air Safety Investigation (2001) *Safety Deficiency and Interim Recommendation No. IR19990084*, issues 21 July 1999. Available on website: http://www.atsb.gov.au/aviation/int_rec/ir19990084.cfm
7. CNN (1999) *Payne Stewart dead at 42 – U.S. Open champion killed in South Dakota plane crash*, posted: Monday 25 October 1999. Available on website: www.sportsillustrated.cnn.com/augusta/stories/102599/payne/html
8. Wainwright, R. and Roberts, G. (2000) *Warning linked to doomed plane*. Available on website: www.abc.net.au/news/2000/09/item20000905050030_1.htm
9. Comiteau, L. (2000) After the blast – more questions than answers remain in the wake of a deadly fireworks factory explosion. *Time Europe* May 29, 2000, 55, 21. Available on website: www.time.com/time/europe/magazine
10. Reason, J. (1991) Resident pathogens and risk management. *Safety in Australia*, 9, 3, pp. 8–15, Australia.
11. Esmeralda Exploration Limited (2000) *Media Release 13 June 2000*, Aurul to recommence operations at Baia Mare. Available on company website: www.esmeralda.com.au
12. Butterly, N. (2000) Mechanised mining creates new safety hazards for industry. *Gold Gazette*, 16 October 2000, pp. 1–2, Australia.
13. Kinney, C.F. and Wiruth, A.D. (1976) *Practical Risk Analysis and Safety Management Naval Weapons Center (Paper)*. China Lake, California, USA, June 1976.
14. Thompson, D. and Rempel, D. (1994) Designing and redesigning jobs, *Occupational Health and Safety*, 2nd edn. National Safety Council, Illinois, USA, pp. 164–172.

15. Mathews, J. (1993) *Health and Safety at Work – Australian Trade Union Safety Representatives Handbook*. Pluto Press Australia, New South Wales, Australia.
16. Hocking, B. and Thompson, C.J. (1992) Chaos theory of occupational accidents. *Journal of Occupational Health and Safety – Australia and New Zealand*, 8, 2, pp. 99–108, Australia, (figure from page 103).
17. Department of Mineral and Petroleum Resources Western Australia (2001) Safe lighting of gas appliances. In *Minesafe* Vol. 12 No. 1, May 2001 p. 10.
18. Department of Mineral and Petroleum Resources Western Australia (2001) Stop welding electrocutions NOW! In *Minesafe* Vol. 12 No. 1, May 2001, p. 4.
19. Natural Resources and Environment Department (Victoria) (2001) *Significant Incident Report No. 05/00 – Scalding Incident*, Minerals and Petroleum Regulation Division, East Melbourne, Victoria, Australia.
20. Kincaid, W.H. (2001) How maintenance contributes to poor safety performance. *Occupational Hazards*, August 2001, Penton Media, Inc. in association with The Gale Group and Looksmart. Available on website: http://www.findarticles.com/cf_dls/m4333/8_63/78364156/print.jhtml
21. Australian Transport Safety Bureau (2002) *Media Release 7 March 2002 – ATSB Releases Report on VH-SKC Burketown Fatal Accident*. Available on website: http://www.atsb.gov.au/atsb/media/mrel019.cfm
22. Australian Transport Safety Bureau (2002) *Accident and Incident Report Occurrence Number 200105188*. Available on website: http://www.atsb.gov.au/aviation/occurs/occurs_detail.cfm?ID=392
23. Department of Mineral and Petroleum Resources Western Australia (2001) *Significant Incident Report No. 109*. Available on website: http://notesweb.dme.wa.gov.au/exis/SIR.NSF/6b390eea5649d21c48256097004aacb7/db7776e88086d258482569b90081ea50?OpenDocument
24. Mineral Resources Department of New South Wales (2001) *Safety Alert – Update on Fire Suppression Systems*, prepared by Ray Leggett on 4 January 2001.
25. *Occupational Safety and Health Regulations 1996 (Western Australia)*, Western Australian Government Printers, Perth. Available on website: http://www.safetyline.wa.gov.au/sub3.htm#5
26. Department of Mineral and Petroleum Resources Western Australia (2001) 'Dodgy' power tools for sale – Take heed! In *Minesafe* Vol. 12 No. 1, May 2001, p. 8.
27. Hoggett, J. and Edwards, L. (1983) *Accounting in Australia*, 2nd edn. John Wiley and Sons, Brisbane, Australia.
28. Wainwright, R. (2001) Accidents will happen – Qantas defends its safety record but unions and pilots say its standards have dropped. Available on website: www.smh.com.au/news/0004/29/review/review10.html
29. Civil Aviation Safety Authority (2001) Ansett's reform package approved. *Media Release Friday 20 April 2001*. Available on website: http://www.casa.gov.au/hotopics/other/01-04-20transcript.htm
30. Department of Mineral and Petroleum Resources Western Australia (2001) *Safety Bulletin No: 63 Dozer Safety in Open Cut Operations*, issued 5 June 2001. Available on website: http://notesweb.dme.wa.gov.au/exis/SBULL.NSF/85255e6f0052055f852558ac00697645/1176ee3c8ec4a8d448256a620014d07f?OpenDocument
31. Imai, M. (1986) *Kaizen – The Key to Japan's Competitive Success*. McGraw-Hill Publishing Company, New York.
32. Worksafe Western Australia (2001) *Work Practices: Woodside sets the Trend.* Available on website: http://www.safetyline.wa.gov.au:81/

Chapter 5

# Physical environment

The physical environment is defined as infrastructure and locational factors. It includes company buildings, fittings and fixtures such as fencing, lighting, fans and storage systems. Items such as furniture and tools are part of the technology system factor because they are not fixed. The physical environment also includes the natural features of the company site such as topography, flora and fauna, climatic conditions and soil. The firm is legally required to manage the risks associated with the physical environment, as far as is practicable, to prevent injuries to employees and other persons on site. The management of the physical environment also takes into consideration the potential for consequential damages including injuries to members of the general public and damage to infrastructure and natural systems located outside the firm's boundaries.

The channel described in Chapter 2, explained that the firm is required to be both legally compliant and socially responsible. This is particularly the case in the management of the physical environment. There are stringent controls to ensure that firms operate responsibly to prevent negative consequences to the environment within the company site and also beyond the company gates. It is, therefore, in this area of risk control that there is the strongest relationship between OHS and environmental management.

## Linking OHS and environmental management

Case studies undertaken in The Netherlands and Denmark indicate that OHS and environmental management are linked in two ways. The first is on a technical basis stemming from the notion that hazard sources are similar and therefore these programs have synergies.[1] When managing physical environment risks using an OHS management system this concurrently contributes to the prevention of environmental damage. Likewise, environmental management strategies can have a positive impact on health and safety.

The second way in which they are linked is the effect on the organizational culture. Providing workers with the opportunities for improved participation allows the firm to achieve leverage that results in the flow-on of ideas to environmental management.[1] Accordingly, productive safety management not only proposes a new perspective on risk management which integrates safety and production as compatible goals using a multi-disciplinary approach, but also provides the systems for a supportive culture. Employee participation is fundamental to this culture because organizational achievement requires each individual to take ownership of safety. It also requires management systems that encourage and reward this ownership.

In relation to the multi-disciplinary approach, there has been some debate concerning whether OHS and environmental management should be integrated. Some authors propose that there is a threat of OHS management becoming relegated to low priority:

> Much stronger than is the case for OHS management, in environmental management there is an emphasis on external control, a requirement of supply chain management, and of responsibility for product effects during their entire life cycle. As a consequence, the material topics that are dealt with in environmental management may have stronger links to all pervasive risks – and thus to strategic company issues – than those on OHS management.[1]

This perspective on environmental management assumes that it is a very broad area of risk management that includes product quality control as a means of preventing injuries or damages which may be caused by the company's goods after they have been sold. This is an attempt to merge quality and environmental management. In addition, it has been suggested that attempting to integrate all these disciplines can have serious consequences for OHS processes. When quality and OHS management systems are amalgamated, OHS auditing becomes very complex:

> It is not enough for the auditor to establish that there are arrangements in place to address a particular element of the health and safety management 'standard' and that those arrangements are followed. The auditor needs to establish that the particular arrangements are adequate relative to the hazards and risks associated with the organisation's activities.[2]

The entropy model provides a simpler explanation of how these management areas are related. If entropic and residual risks are effectively managed and system factors are shifted towards optimal performance and safety, the firm's products should also become safer. This is because the productive safety culture involves the permeation of risk consciousness into all areas of business activity including product design and manufacturing. Employees develop an understanding of the nature of risk and the effect their competencies and attitudes have on safety, system factor quality and productivity. In manufacturing firms, therefore, they should also understand how these risk factors affect product quality, particularly at the production stage. As employees' competencies and risk awareness improve, an enhanced appreciation of how residual risks and degradation of components can be remedied or controlled to improve the quality of company goods should also result.

OHS, environmental management and quality assurance are also linked through the relationship between companies and their suppliers.

In Chapter 4 it was suggested that stronger interdependencies should be developed in the business sector between suppliers and customers. This should involve increased information sharing better to manage risks, to reduce product residual risks and the tendency of technologies to degrade. Through linkages in the supply chain there are opportunities to improve product development by responding to customer feedback, and thus to achieve higher standards of safety and health in workplaces with the availability of safer technologies and workplace design systems. In addition, such improvements should contribute towards the minimization of negative environmental impacts of technology use.

The relationship between the three management disciplines can be explained further using the entropy model. Each discipline aims to enhance the quality of system factors and is affected by residual and entropic risk levels. For example, the hazards associated with the physical environment have implications for both safety and environmental management with failures having the potential to create injury or damage. To manage environmental risks effectively therefore, the company's environmental and OHS management plans must be aligned. For instance, where there is a risk of hazardous run-off from a processing plant contaminating a nearby waterway, the control of such risks needs to be considered in both plans because it has both environmental and health implications. In addition, both disciplines need to be part of a total management system in which production and safety are compatible goals. The environment cannot be managed well where the company culture allows output to take precedence over safety because this compromises both safety and environmental care. In addition, these disciplines are concerned with maintaining the quality and integrity of the physical environment, as evident in legislation requiring the rehabilitation of company sites to ensure that, as far as is practicable, such sites are returned to their natural condition and are made safe. OHS and environmental management, therefore, both focus on maintaining the quality of the physical environment system factor to varying degrees. The former is concerned with maintenance from the safety and health point of view, while the latter is concerned with safety and natural sustainability.

Productive safety management facilitates quality control because it is driven by the maintenance of these system factors at as high a standard of safety and performance as is practicable, and the minimization of consequential repercussions of business activities. For instance, the quality of technologies is assured, as explained in the previous chapter, using purchasing selection criteria and planned maintenance strategies. Process quality is sustained by standardization where possible, and by making work practices safe and efficient using techniques such as JSA. The capability of human resources, in terms of competencies and levels of safety consciousness, are considered to be critical success factors in preventing degradation and managing residual risk. In addition, retention of quality human resources is a high priority. As a holistic quality approach, productive safety management also involves the use of measurement systems to ensure that company safety and performance objectives are achieved and that systems are maintained at optimal level. In a broad sense, the approach applies similar principles of accountability and feedback that are used in quality management systems.

Productive safety management also links OHS, environmental management and quality assurance using the channel. It explains the need for firms to be legally compliant and socially responsible. Firms that fail to meet enforceable standards in the areas of safety, environmental protection or product quality face penalties. They also risk claims of negligence through the common law system. In addition, the channel reinforces the need for employee participation. Worker involvement enhances safety consciousness, encourages environmentally responsible behaviors, and builds competencies that affect product quality and problem solving in all three of these management areas.

The entropy model and the channel help to explain the interdependence of OHS, environmental and quality management. The disciplines overlap but maintain their specialist integrity which prevents one system being relegated to the others. The purpose of conceptually proposing a set of boundaries for these disciplines is to minimize role conflicts in organizations that require a multi-disciplinary approach and to allow for high degrees of expertise together with collaborate problem solving. For example, management may elect to focus environmental management on the control of factors that do not have a direct bearing on the health and safety of employees. In this case, the risk elements within the physical environment, that potentially lead to injury or operations damage, may be addressed through the OHS management system. To ensure a balanced approach the input of environmental specialists should be sought during the development of the OHS plan in relation to any areas of overlap of risk management responsibility.

The risks associated with the firm's saleable product may be excluded from environmental management if the company takes a holistic approach to quality that integrates product quality through the maintenance of system factors. In other words, if the firm's strategies are driven by the maintenance of system factors at a high standard of quality, then as a consequence, its products should also improve in quality. In addition, when the company applies legal compliance and social responsibility as the benchmarks for firm behavior, then it must also ensure that its goods/services meet these standards. A conceptual relationship between these three management disciplines is illustrated in Fig. 5.1. It shows how system factors affect the total risk prevention strategy of the organization.

In the center of the diagram is the productive safety management system. It is an integrated, total management system and is used to regulate the internal environment of the company by focusing on the performance, safety and quality of the four system factors. Surrounding the firm is the immediate environment which has natural, infrastructure and social characteristics. In the event of a major incident the company's operations can have two types of impact on the immediate environment. It can cause damage to natural or infrastructure systems and/or it can cause injury to members of the community. The characteristics of the immediate environment affect the extent of this harm, for example, in Bhopal, failures in the internal environment led to the contamination of the immediate

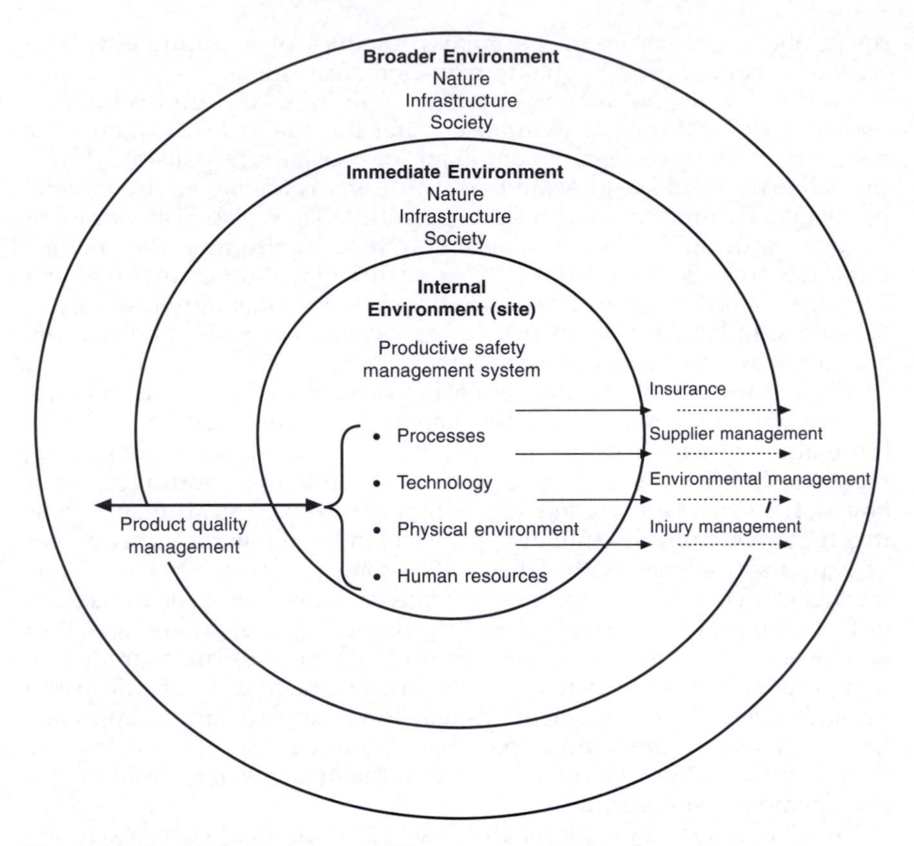


**Figure 5.1** The relationship between OHS, environmental and quality management

external environment. Many members of the general public were killed or injured because of the high population density around the operation.

Figure 5.1 indicates that firms should anticipate the consequential damages to the immediate surroundings when assessing risks to the internal physical environment. In some cases, such as the mining companies operating in Romania, in which the cyanide spill from their tailings dam allegedly polluted rivers many kilometers away, damage may extend to the broader environment. This is shown by the continuation of the environmental management arrow in the diagram. The potential for consequential harm explains the link between OHS management and environmental management.

As part of its OHS management strategy, the firm also has to manage the risks associated with technology. It does this by establishing a supplier management system, as discussed earlier, to minimize residual risks and to prevent entropic risks associated with technology. Its suppliers may be in the immediate environment, for example, in the same town, or in the broader environment. The supplier management arrow shows the link between the internal environment of the firm and its suppliers.

The control of human resource hazards is part of the firm's risk management strategy. These resources are the people the company employs, and as shown in the figure, workers may reside locally in the immediate environment or further afield. Any fatality or injury that occurs as a result of a failure of the firm's operations has consequences for both its workforce and for the society. For example, in small towns dependent on a few businesses where most of the employees live locally, the impact on the morale of the community following an incident can be serious. The broader community also bears part of the cost of workplace injuries through the public health system, through insurance costs and loss of national productivity. The business minimizes the consequential damages in these circumstances through its injury management system. The figure shows that injury management is related to OHS through the human resources system factor. The impact of such injuries reaches the immediate society and the broader society as shown by the injury management arrow.

Companies also take out insurance to protect themselves against damages resulting from failures in their OHS, environmental and quality management systems. This covers such circumstances as: injuries to workers, damage to internal infrastructure and technology, injuries to the general public, damage to the external environment and injury or damage caused by faulty goods produced by the company. Insurance is, therefore, a process to mitigate the financial repercussions of system failures in the areas within these three disciplines.

Finally, the figure shows that all four system factors have an influence on quality control as indicated by the product quality management arrow. This illustrates that quality, environmental and OHS management systems can be integrated as suggested by international standards such as ISO 9000. In the diagram, product quality affects both the immediate and broader external environment wherever customers are based. As explained by the entropy model in the discussions about technologies, the firm's products will have a residual risk and also, with time, will start to degrade. The firm controls these hazards to an 'acceptable' level by managing the risks within its system factors and by shifting these factors towards optimal performance and safety. The reciprocal relationship between the firm and the external environment, explained by the channel, indicates that forces in the external environment determine the maximum 'acceptable' risk of such products. Government regulators define this risk using specific standards prescribed in legislation or by testing cases before the courts. The community can also test the level of acceptable risk by pursuing common law remedies after incurring damages from product failures.

The value of illustrating these relationships between OHS and other company management strategies is that the benefits of improving and maintaining internal system factors can clearly be seen. The control of risk factors associated with the physical environment, for example, contributes to sound environmental management. The minimization of residual risk and degradation, so that system factors are shifted towards optimal performance and safety, leads to better product quality control. The objective of returning employees to productive work following an accident through the firm's injury management system ensures that the company fulfills its responsibilities to the worker and to society.

Given that OHS and environmental management strategies are linked through the quality of system factors, what choices are made and actions taken by responsible firms? These businesses use legal compliance as the minimum benchmark for their organizational systems. They also self-regulate to achieve industry best practice in OHS and environmental management. Self-regulation leads to socially responsible outcomes which reduce the probability that business activities will incur negative consequences for the workforce and the external environment. Figure 5.2 illustrates the quality standards applied to OHS and environmental management systems in this regard.

The responsible firm manages the impact of its operations on both the internal and external environments. It achieves legal compliance by fulfilling its duty of care to employees and the general public through its OHS management system. The firm also complies with its legal obligations in relation to the internal physical environment and the operation's surroundings through its environmental management system. The pursuit of socially responsible outcomes causes the company to self-regulate. It implements industry codes of practice and guidelines as well as setting workplace-specific safety standards. It develops rehabilitation programs that consider the needs of families of injured workers so that affected employees receive the support they need to return to productive work. This minimizes the negative impact of workplace accidents on employees and on the community. The responsible firm also self-regulates in the area of environmental management. This involves investing in research

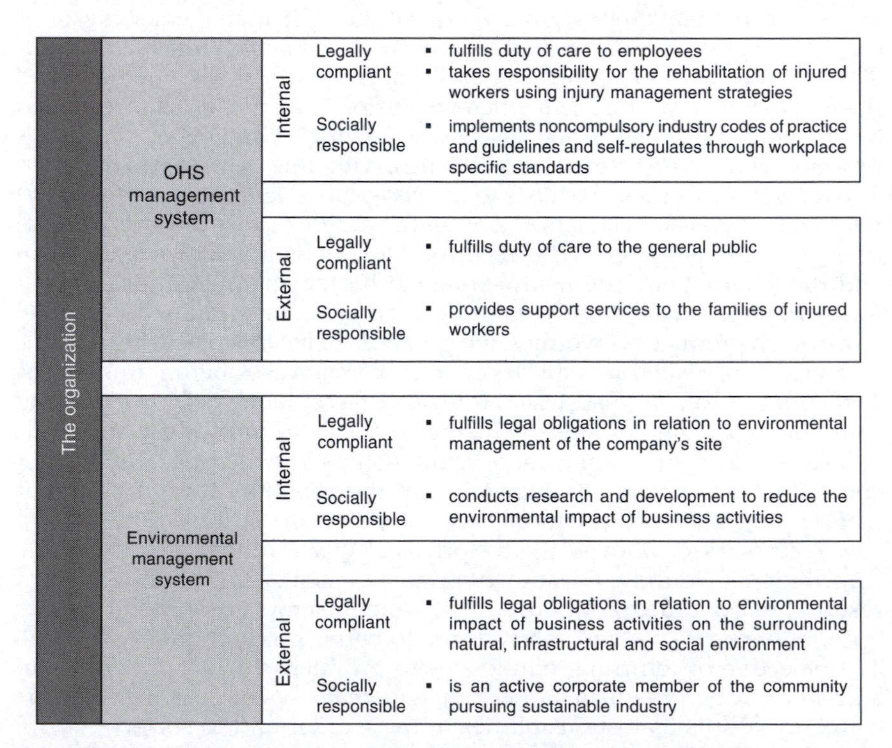


**Figure 5.2** Achieving legal compliance and social responsibility through OHS and environmental management systems

and development to reduce the environmental impact of business activities, and acting as an active corporate member of the community to build a sustainable industry.

The figure shows, therefore, that OHS and environmental management have common underlying principles. These are, firstly, that legal compliance is the minimum benchmark that the firm must achieve, and secondly, that self-regulation leads to socially responsible outcomes. Firms have to achieve legal compliance in the short term because they are held accountable by government regulators through legislation and by the community through the legal system. The time frame for social responsibility on the other hand, may be longer term. The commitment to self-regulation, community support, organizational development and investment in research and development, are all long-range strategies for those firms that do not currently have sophisticated systems or sufficient resources. The immediate implementation of OHS and environmental management systems, however, allows these firms to improve the quality of their system factors, attain legal compliance and establish a strategic direction that pursues socially responsible outcomes.

## Legal provisions

The discussions thus far have identified that the physical environment system factor and the maintenance of its quality provide the link between the OHS and environmental management systems. In addition, the firm is required to set legal compliance as its minimum benchmark in both these disciplines. In relation to OHS management, the firm has a duty of care to provide employees with a safe work environment. These responsibilities are set out in legislation, for example:

> 19. (1) An employer shall, so far as is practicable, provide and maintain a working environment in which his employees are not exposed to hazards and in particular, but without limiting the generality of the foregoing, an employer shall – provide and maintain workplaces, plant, and systems of work such that, so far as practicable, his employees are not exposed to hazards.[3]

Firms are also required to comply with laws covering the environmental impact of business activities. In Western Australia, for example, new mining projects are required to comply with: the Mine Safety Inspection Act 1994, the Mining Act 1978, the Environmental Protection Act 1986, the Occupiers Liability Act 1985, the Environmental Protection and Biodiversity Conservation Act 1999, the Wildlife Conservation Act 1950, the Aboriginal Heritage Act 1972, the Country Areas Water Supply Act 1947, the Conservation and Land Management Act 1984 and the Bush Fires Act 1954.[4] Legislative requirements vary from industry to industry, however, there are some Acts and Regulations such as the Occupational Safety and Health Act 1984 (Western Australia) which apply to all workplaces in the state. Most developed countries have legal frameworks based on this type of structure, having both industry-specific and general-industry legislation.

There is clearly an overlap in OHS and environmental management, particularly as failures of organizational systems have implications in both these areas. Risk management, as a broad concept, is therefore a common theme to both disciplines. The firm needs to establish some conceptual boundaries between professional fields as discussed earlier, for example, for the purpose of auditing, whilst drawing them under the 'umbrella' of an integrated management system. For this reason, it remains important for the risks that have OHS implications to be treated independently of nonsafety-specific risks using the auditing processes described for each system factor in Chapters 3 to 6.

## Physical environment hazards

What are the hazards associated with the physical environment? These hazards result from the natural environment and from man-made structures. Unlike other system factors, physical environment hazards can be categorized relatively simply. They are summarized in Table 5.1 under the headings of climatic, structural, chemical, biological and other. The categories given are all residual risks that may be associated with the physical environment and are not limited to the workplace. They also apply to nonwork and leisure activities, and therefore, are important in the management of public health and safety in addition to workplace safety.

Climatic hazards include extreme temperatures, which can apply excessive demands on business activities. Such conditions can have an effect on system factors, for example, processes that involve chemical or biological reactions such as fermentation are slowed when the temperature drops. Technologies may not operate as efficiently in such environments, for example, it is more difficult to start a car engine on winter mornings and machinery can overheat when it is very hot. The temperature also affects worker comfort, safety and the ability to maintain vigilance. The case involving the death of the boilermaker, given in Chapter 4, who was electrocuted as a result of prolific sweating, reinforces this point. Infrastructure design, or in this case, the ship's compartment, and the materials from which it was made affect the temperature by either retaining or deflecting heat. Ventilation is another important design consideration as it affects airflow that either allows hot air to be dispersed or causes it to accumulate.

Extreme temperatures can be fatal as is evident in a case involving the death of a field survey worker from exercise-induced heat exhaustion. The incident occurred approximately 30 kilometers from Coober Pedy in South Australia on 9 February 1992.[5] Two employees were traveling in a remote area when their vehicle became bogged down. After repeatedly freeing the vehicle and getting stuck again, they decided to set off for the Stuart Highway using maps and a compass. The distance was 20 kilometers over gently rolling, open terrain covered with small rocks. They took 5 liters of water and left the truck at about 1.00 p.m. The maximum

**Table 5.1** Physical environment hazards

| Climatic | Structural (natural or infrastructural) | Chemical | Biological | Other |
|---|---|---|---|---|
| Extreme temperatures<br>• Very hot weather<br>• Very cold weather | Elevation or heights<br>• Falls<br>• Falling objects | Particles and fumes<br>• Inhalation<br>• Contamination<br>• Accumulation of deposits on surfaces | Biohazards<br>• Bacteria<br>• Viruses<br>• Parasites | Noise (nontechnological origin)<br>• Storm activity and other natural conditions<br>• Human interaction |
| Solar radiation<br>• Sunburn<br>• Glare | Instability<br>• Earthquakes and tremors<br>• Loose rocks and earth<br>• Unstable infrastructure | Chemicals (naturally occurring)<br>• Inhalation<br>• Burns<br>• Irritation<br>• Chemical reactions | Dangerous animals<br>• Poisonous snakes<br>• Poisonous spiders<br>• Poisonous marine life<br>• Harmful insects<br>• Vicious animals<br>• Predators<br>• Nesting birds | Submersion<br>• Drowning<br>• The bends<br>• Cramp<br>• Irritation to eyes, nose and ears<br>• Loss of vision and hearing<br>• Fatigue<br>• Hidden objects in water and physical hazards |
| Precipitation and moisture<br>• Rain and hail<br>• Snow and sleet<br>• Fog and smog<br>• Corrosion<br>• Obstruction<br>• Weight | Surface conditions<br>• Irregular surfaces<br>• Slippery surfaces<br>• Loose surfaces<br>• Steep slopes<br>• Extremely hot surfaces<br>• Extremely cold surfaces<br>• Glare from shiny surfaces | Radioactive substances (naturally occurring)<br>• Radiation poisoning<br>• Exposure to carcinogen<br>• Chemical reactions<br>• Contamination | Toxic vegetation<br>• Poisonous plants<br>• Allergic respiratory reactions<br>• Allergic skin irritations<br>• Digestive reactions | Insufficient light<br>• Injury to worker/s<br>• Damage to equipment and/or infrastructure<br>• Damage to the environment |
| Lightning<br>• Injury to worker/s<br>• Damage to equipment or infrastructure | Confined spaces<br>• Restricted movement<br>• Poor ventilation<br>• Build up of heat | | | Disturbing odors<br>• Nausea<br>• Vomiting |

*(Contd)*

**Table 5.1** *(Contd)*

| Climatic | Structural (natural or infrastructural) | Chemical | Biological | Other |
|---|---|---|---|---|
| | • Psychological discomfort | | | |
| High winds<br>• Falling structures<br>• Air-borne projectiles<br>• Dust storms | Drainage<br>• Flooding<br>• Ponding of liquids<br>• Contamination<br>• Slippery conditions<br>• Erosion | | | |
| Atmospheric pressure<br>• Altitude sickness<br>• Severe oxygen deprivation<br>• Travel sickness | Vibration and movement<br>• Loss of balance<br>• Dizziness<br>• Nausea | | | |
| Fire (nontechnological origin)<br>• Bush fires<br>• Spontaneous ignition of flammable materials | Access/egress<br>• Injury to worker/s<br>• Higher consequential damages following incident | | | |

temperature recorded on that day was 35.6°C. They stopped every half an hour for a rest and drink. After completing 10 kilometers, they rested for 15 minutes near a stream. The deceased did not use the stream water to cool down. After a further 5 kilometers, he indicated that his legs were numb and he then collapsed. The surviving employee covered the remaining kilometers in 45 minutes, got picked up by a bus and taken to Coober Pedy where he affected a rescue, which was to no avail.

The factors to be taken into account for remote area work which were highlighted by this incident were that the onset of heatstroke can occur at ambient temperatures not necessarily considered to be excessive and is more likely on humid days. In addition, it can occur suddenly and without warning. In this case, the core body temperature of the surviving employee once he reached hospital was 38°C, even after his considerable exertion in the last 4.5 kilometers of the trek. This indicates that a considerable difference in the ability of each person to handle heat stress can exist. A further factor to be considered is the amount of time such employees are given to acclimatize to the conditions.

A case involving the death of a 19-year-old exploration worker in circumstances suggestive of heat exposure[6] illustrates these dangers further. The cause of death was not determined by a post-mortem examination but indicators were consistent with death due to exposure to extreme heat. The case may be used to illustrate how easily and rapidly climatic factors can contribute to loss of life. The deceased was a geology student who had little previous experience in these harsh Australian conditions in which the daytime temperatures were around 41°C.[6] The student had been suffering gastric symptoms and vomited several times on the first night in the camp. On the second day he was unable to continue work and around mid-morning began to walk back to the vehicle a short distance away. It is believed that his state of dehydration led to rapid deterioration and as a result he did not make it back to the vehicle.

These incidents highlight the need for awareness of the potential consequences of harsh climates and that each individual has different susceptibilities to such conditions. Contingency plans must also provide for medical interventions when employees become ill, and also to treat other conditions such as sunburn, that may result from working in harsh areas.

Solar radiation is a multi-risk hazard as it can lead to overheating, sunburn and also create glare that reduces worker comfort or visibility. When drivers are travelling from Western Australia to the eastern states, for example, the sun is in direct view at sunrise and this can prevent the driver from seeing objects or animals on the road. This glare can either have a climatic origin or it can result from reflection of artificial light from the shiny surfaces in workplaces.

Other climatic factors include precipitation and moisture which create hazards by making working conditions more difficult. In the longer term it can lead to structural damage such as metal corrosion or timber rot. Precipitation can result in the obstruction of roadways and prevent access to work areas. In combination with design faults, the accumulation of rainwater or snow can cause structures to collapse. Drainage design is therefore an important factor in preventing such risks. In addition, the natural topography or the design of infrastructure can allow flooding or ponding to occur which can lead to slippery or unworkable surface conditions. In some workplaces, such as chemical storage facilities, firms are required to erect bund walls to capture spillage. Poor design and maintenance of these structures can lead to site contamination and threats to the health of workers and the general public. Wherever there is an uncontrolled flow, either resulting from precipitation or leakage, there is also the potential for secondary damage including the erosion of slopes, roadways and other work areas.

Lightning is a hazard for workers who are outdoors during an electrical storm. This risk rises when the employee is in an open area such as a recreation field, on the sea or under trees. In fact, there were three deaths at work from lightning strikes in Western Australia in 1999 and 2000. Ten people are killed in this manner, on average, in Australia each year.[7]

High winds are also hazardous. They can destabilize natural and man-made structures causing them to collapse. They can also result in objects having kinetic energy – becoming projectiles – as is the case during cyclones/hurricanes, or they can generate dust storms that reduce visibility and in some cases, cause operations to cease. The key issue in relation to such hazards is under what conditions is it acceptably safe to continue to work and when is it not. In addition, firms can implement suppression systems to reduce the severity of such conditions. For example, some mining operators use water tanks to wet the surfaces of open-cut mines to reduce dust hazards.

Climatic conditions also include atmospheric pressure, which is a hazard when aircraft systems fail. This was evident in the previously described aviation accidents. Atmospheric pressure is also problematic for employees who have to work at high altitudes, for example, workers who commute between Antofagusta, on the coast in Chile to the Escondida mine located at the base of the Andes mountains. Variations in air pressures during flying cause travel sickness in some individuals.

The final hazard in this category is fire of a nontechnological source, such as bush fires, which are natural threats that need to be managed, particularly in the forestry industry and in areas near natural bush land. Some of these climatic hazards are associated with natural disasters, and firms need to be aware of their susceptibility to such threats to ensure that effective contingency plans are implemented during such events.

The second category of physical environment hazards is structural factors. These can be either natural or man-made, with a significant risk in this group associated with structural height. The severity of the residual risk relates to the elevation of either the land, for example, at the top edge of an open-cut mine or cliff face, or the height of infrastructure. In addition, residual risks stem from exposure to falling objects at the base of or at a lower elevation than a nearby height. In the period 1995–1996 to 1999–2000, falls from heights represented 9 per cent of fatalities in underground mines and 10 per cent in surface mining operations in Western Australia.[8] The figures for death from being struck by a falling object was 9 and 30 per cent, respectively. Height-related injuries are also significant in the manufacturing and construction industries.

Instability is another significant structural hazard and can be attributed to weaknesses in infrastructure or to geological characteristics of the workplace. Rockfalls caused 50 per cent of fatalities in the underground mines that were surveyed in the aforementioned period. Surface conditions in workplaces are a further physical characteristic in this category. These include the condition of workshop floors, for example, whether they are smooth, are nonslip or have irregularities that may be a tripping hazard, and the state of road surfaces and other areas where human resources or technologies undertake work. Some work environments have steep slopes that have to be inspected and may require barriers such as fencing to prevent rock falls from causing injury or damage. Surfaces can also become hazardous when they become extremely hot or extremely cold, for instance, railway sleepers become hot enough to fry an egg in the middle of summer in desert regions and contact with such surfaces can cause serious burns. Likewise, infrastructure close to technologies and processes that generate heat can become dangerously hot.

The chemical composition of the physical environment can be a source of hazards. Most environmental chemicals are inert, and therefore, are not hazardous. In some cases, however, they can generate particles and fumes, for example, in some metalliferous mines sulfur dioxide is emitted from deposits of pyretic shale when it comes into contact with the air. This is a hazard for workers if inhaled, causing irritation of the eyes, nose and throat. When it rains this sulfur dioxide becomes sulfuric acid. Acid mine drainage can have a negative impact on the environment.

Infrastructure and site design can be used to compress residual risks. For example, the layout of buildings affects ventilation and the dispersion of particles and fumes from processes and technologies. One of the areas of current investigation in OHS management is 'sick building syndrome' (SBS), defined by the World Health Organization:

> as an extra frequency of irritative symptoms from the eyes, nose, throat and lower airways, skin reactions, unspecific hypersensitivity reactions, mental fatigue, headache, nausea or dizziness while in a particular premises.[9]

More than 50 per cent of Australian employees work in office environments with approximately 10 to 30 per cent of inhabitants being affected. Research indicates that this does not necessarily increase absenteeism but it is believed to reduce productivity. Biohazards, such as Legionnaire's disease, are a further risk factor in these environments. The Legionnella bacteria may multiply in and are then spread by air-conditioning systems.[10] The disease results in chronic lung infection and can cause death.

There are a number of industries, including farming, abattoirs, clinical pathology and medicine, where biohazards are a major residual risk. Biohazards include bacteria, viruses and parasites. Within this category are also insect-borne diseases such as malaria, which are a health risk, for example, for workers commuting between western countries and Asia on a fly-in fly-out basis. Employees working in the natural environment can also be exposed to dangerous creatures, such as poisonous snakes and spiders, and vegetation that can lead to skin irritation or contact dermatitis. Some workers may also have reactions such as asthma and hay fever to air-borne allergens.

The remaining physical environment hazards do not fit clearly into the aforementioned categories. These include noise, which can result from natural processes or human activity, that cause the worker to be distracted from the task. More commonly, however, this hazard is associated with technology not the physical environment. Workplace design has an effect on the dispersion of noise. Confined spaces cause noise to reverberate whereas open spaces allow the noise to dissipate.

Water is a hazard in a number of ways – precipitation, as a flow, accumulation, as a conductor and through submersion. Some jobs require the worker to be submerged, such as lifesavers and swimming pool attendants, resulting in exposure to these residual risks. In the case of divers who inspect and repair deep-sea oil rigs, submersion can cause the 'bends', which is a potentially fatal condition resulting from the build-up of excess gases in the bloodstream.[11] In addition, submersion reduces the person's visual and auditory receptiveness. This hinders hazard identification and prevents effective communication, for example, people have died or been injured as a result of diving into shallow water without inspecting it for sub-surface objects. Other physical hazards such as aquatic weeds can cause people to become entangled and drown. Work activities carried out in water also lead to fatigue or physical degradation so therefore, the length of time that the activity is undertaken has an effect on the risks involved.

Table 5.1 also identifies insufficient light as a source of risk. This increases considerably the risks of injury and damage regardless of whether the activity is undertaken at work, at home or in a public place. Inadequate lighting occurs as a result of nightfall, unfavorable climatic conditions, the physical environment itself such as in underground mines, and from poorly illuminated workspaces. The problem is a significant issue in 24-hour operations, and firms have to make decisions regarding the extent to which resources are allocated to provide adequate illumination according to the tasks undertaken in an area. Artificial light has two cost factors. These are capital outlay for the installation of lighting systems and the on-going costs of electricity. The firm has to determine the level of lighting it provides to various workplaces and other areas, to ensure that employees have sufficient visibility to avoid hazards and to work safely. Lighting is a significant environmental factor in reducing risk and also affects the process of 'tricking' the body into maintaining function during normally inactive hours of the day. As a result, it is important to provide adequate lighting in operations that work shift rosters.

> Disruption of circadian rhythms through work schedules means not only that people are expected to be awake and active at the inappropriate time in the cycles, but also that environmental factors (like light and dark) that keep the individual's cycle on track are out of kilter.[12]

Accidents resulting from poor visibility happen very easily. For example, it was the primary cause of an incident involving an open-pit mine production operator who suffered a dislocated shoulder and bruising to his hip and back when he stepped into a three meter cavity in the berm.[13] (A berm is a flat portion or step in an open-pit mine wall left by design to catch loose rocks falling from upper levels of the wall.) The worker was running out levels near an excavator at night and could not see the cavity caused by blasting. Two light towers lit the area but were insufficient and the operator was not carrying a torch. He was not aware of the hazard because it was not visible. Applying the four-fold strategy this accident could easily have been avoided. Appropriate strategies could have included:

(1) Corrective action involving the erection of barriers around the cavity;
(2) Maintenance strategies including work practice improvements such as compulsory use of torches or helmet lights in poorly illuminated work areas;
(3) Management of residual risk in the short term through training that encourages vigilance in these hazardous work areas;
(4) Compression of residual risk using capital expenditure to improve lighting in such work areas.

Poor lighting reduces visibility, and in combination with other environmental factors, such as unfavorable weather and road surface conditions, can create a situation of even higher risk. For example, a case involving a mobile plant incident at an open-pit mine illustrates this combined effect. A truck driver was reversing his truck through a 1.5-meter high windrow (protection barrier made up of loose rock) at a waste dump when he became bogged down because he could not clearly see the windrow and road surface conditions. Fortunately, no damage or injuries resulted on this occasion. The incident occurred because the illumination from the light tower was obscured due to heavy rain and the shadow of two other trucks at the site.[14] Combinations of physical environmental hazards greatly increase the level of risk, placing additional demands on the operator to remain vigilant and to apply his competencies to manage these risks. When other risks are factored into the situation, the worker's coping capacity can be exceeded resulting in an accident.

The final hazard which may affect the worker is disturbing odors. These do not cause any long-term harm but may induce nausea and vomiting which causes temporary incapacitation and inability to carry out the task.

In high-risk industries such as mining and construction, the physical environment is often the major source of residual risk and therefore contributes to a significant proportion of incidents. For example, rock falls were the main cause of fatalities in Western Australian mines from 1995 to 1997,[15] as shown in Table 5.2. There were 11 fatalities attributable to this cause. A further two incidents – structural failure and fall from height – made a total of 13 or 52 per cent related to physical environment hazards for the period. These statistics indicate the need to develop strategies to manage residual risk and to enhance employee awareness. The report from which these statistics have been obtained also identifies lack of knowledge to be an issue, particularly in relation to management knowledge of geotechnical (earth science and mineral engineering) matters. A multidisciplinary approach to risk management and a greater emphasis on information-sharing and training is called for in relation to such hazards.

Statistical data on the 'agency of occurrence' of incidents in a number

**Table 5.2** Primary cause of Western Australian mining fatalities from 1995 to 1997

| Primary cause | No. of fatalities for year | | | |
|---|---|---|---|---|
| | 1995 | 1996 | 1997 | Total |
| Rock fall | 1 | 4 | 6 | 11 |
| Explosives detonation | | | 1 | 1 |
| Run over | 3 | 1 | | 4 |
| Tractor rollover | | | 1 | 1 |
| Heavy vehicle accident | | | 1 | 1 |
| Caught by/between (machinery) | 2 | | | 2 |
| Tire explosion | | 1 | | 1 |
| Structural failure | | 1 | | 1 |
| Caught by/between (not machinery) | 1 | | | 1 |
| Fall from height | 1 | | | 1 |
| Aviation | | | 1 | 1 |
| *Mining sector:* | | | | |
| Underground | 3 | 4 | 7 | 14 |
| Surface | 5 | 3 | 3 | 11 |
| Total for industry for year | 8 | 7 | 10 | 25 |

Reproduced with the permission of the Department of Minerals and Petroleum Resources Western Australia. Refer to Ref. 15 for source details.

of industries provide evidence of the significance of environmental factors in causing or contributing to incidents. These statistics separate environmental factors from other agents, such as machinery and chemicals. They include, for the purpose of the study, the outdoor, indoor and underground environments. The results were that 14.8, 8.8 and 13.2 per cent of incidents were triggered by environmental factors for the mining, manufacturing and construction sectors in Western Australia in 1998–1999, respectively.[15] The data are shown in Table 5.3. The impact of the environmental factors may have been even greater than these figures indicate depending on how these classifications were determined. For example, would the previous example in which a truck became bogged down as a result of insufficient light be counted as 'mobile plant and equipment' or 'environmental factors'? In addition, does the grouping 'animals, human and biological agents' include snakebites and other biological hazards which were included in the physical environment categories presented earlier in this chapter? The impact of the physical environment as a contributing factor may therefore be greater than the above results indicate.

The discussions thus far have explained that the physical environment has residual risks. The main causes of this risk in the work context are the presence of uncontrollable natural systems and the inability of the business to adapt the environment to meet its needs precisely. The operational site will, therefore, always have some design shortcomings and also residual risk stemming from the interaction between the physical environment and other system factors, particularly technology. This interaction was introduced in the previous chapter in the discussions on technological hazards. Failure to adapt the physical environment to match technology

**Table 5.3** Agency of occurrence in the mining, manufacturing and construction industries in Western Australia 1998–1999

| Agency of occurrence | Mining | | Manufacturing | | Construction | | All persons in WA |
|---|---|---|---|---|---|---|---|
| | No. | % | No. | % | No. | % | |
| Machinery | 74 | 6.7 | 859 | 14.1 | 222 | 7.4 | 8.3 |
| Mobile plant and equipment | 230 | 21.0 | 295 | 4.9 | 285 | 9.6 | 9.3 |
| Powered equipment and tools | 46 | 4.2 | 614 | 10.1 | 170 | 5.7 | 6.5 |
| Nonpowered equipment and tools | 247 | 22.5 | 1284 | 21.1 | 716 | 24.0 | 23.8 |
| Chemicals | 28 | 2.6 | 154 | 2.5 | 50 | 1.7 | 2.0 |
| Materials and substances | 214 | 19.5 | 1321 | 21.7 | 699 | 23.4 | 15.5 |
| Environmental factors | 162 | 14.8 | 532 | 8.8 | 394 | 13.2 | 12.6 |
| Animal, human and biological agents | 21 | 1.9 | 264 | 4.3 | 27 | 0.9 | 8.0 |
| Other and unspecified | 75 | 6.8 | 756 | 12.4 | 420 | 14.1 | 14.1 |
| Total | 1097 | 100 | 6079 | 100 | 2983 | 100 | 100 |

Reproduced courtesy of WorkSafe Division, Department of Consumer and Employment Protection, Western Australia (www.safetylinewa.gov.au)

or reciprocally, to purchase technology that suits the environment, increases the level of residual risk. As an example, the residual risk is higher driving a car on an unsealed road than on a sealed road. The options to manage this risk are either to seal the road or to operate a four-wheel drive vehicle that is better designed for rough road conditions. The severity of risk at the interface between system factors therefore depends on the fit between such factors. Consequently, the interaction of the physical environment with processes is also a source of residual risk. The greater the disparity between these system factors, the greater the risk. For instance, offshore oil and gas operations in the North Sea experience a higher level of residual risk than oil rigs on land in Texas. The process is the same – the extraction of oil and gas – but the physical environment risks are considerably different.

Inherent risks are also present in the interaction between human resources and the physical environment. The human body is not designed to withstand the severity of conditions that may arise. It is vulnerable to extreme temperatures, chemicals and the other hazards identified earlier in Table 5.1. In addition, although firms attempt to control the physical environment so that they can carry out business activities by modifying the natural surroundings, these have a tendency to degrade which leads to rising entropic risk. A perfect fit between the firm's activities, capital resources and the environment cannot be obtained and requires proactive, on-going maintenance to stabilize these interactions at an 'acceptable' level of risk.

While the physical environment affects workers, reciprocally, employees can also cause degradation of the environment either intentionally or inadvertently. If the operator drives too close to the edge of the road, for example, this tends to damage the bitumen edges. In the workplace, poor housekeeping is a major source of degradation, for instance, floors become increasingly unsafe when tripping hazards are introduced or spills are not cleaned up. The physical environment therefore degrades naturally and also as a result of its interaction with other system factors. Mining and manufacturing are processes that cause technology and human resources to have an intensive interaction with the environment, which causes it to degrade more quickly, and therefore such firms are exposed to rising levels of entropic risk. In open-cut mines, for example, as the pit gets deeper the risk of rock falls increases as does the potential severity of damage in the event of an incident. Industries which involve obtaining resources from the physical environment that leads to fundamental changes in that environment, concurrently cause variations in the levels of residual risk and potential for entropic risk. The operation's exposure to environmental hazards, including inherent dangers, therefore, is not static. The greater the intensity of interaction between the physical environment and other system factors, the greater the variability of risks.

## Managing physical environment risks

The strategies for managing the residual and entropic risks associated with the physical environment are described in Table 5.4. As is the case in the management of technological risks, the control of environmental risks begins in the planning and design stage. Poor design is a major contributor to fatalities and lost time injuries. For instance, in June 2000 at an underground gold mine in the northern goldfields of Western Australia, a fill barricade (barrier of rocks and material) at the base of a stope/cavity near the bottom of the mine gave way as a result of inadequate

**Table 5.4** Strategies to manage the risks associated with the physical environment

| Source of hazard | Risk management strategy |
|---|---|
| *Residual risk* | |
| Site design shortcomings | • Pre-development evaluation of physical environment factors<br>• Evaluation of other or similar sites of the company, competitors and industries<br>• Monitoring of impact of physical environment on performance and safety<br>• Documentation of residual risks |
| Physical environment/ Technology interaction | • Effective design and planning of work environment<br>• Pre-installation evaluation of risk factors and potential hazards<br>• Installation of technology in accordance with manufacturer's instructions<br>• Safe work practices/operating procedures |
| Physical environment/ Human resources interaction | • Modification of the physical environment to fit the needs of workers as far as practicable<br>• Flexibility in work practices to keep the demands on the worker to an 'acceptable' level<br>• Safe work practices<br>• Monitoring of worker health, safety and vigilance<br>• Training in residual risk management |
| Physical environment/ Process interaction | • Evaluation of suitability of the process for the physical environment<br>• Modification of existing processes in the short term<br>• Modification of physical environment (where possible) in the longer term |
| *Entropic risk* | |
| Natural degradation of the physical environment | • Regular monitoring of the condition of the physical environment<br>• Emergency response procedures and contingency plans<br>• Maintenance of natural environment and infrastructure |
| Physical environment/ Operator interaction | • Regular monitoring of condition of physical environment<br>• Induction and training<br>• Good housekeeping practices<br>• Organizational culture which reinforces desired behaviors |
| Physical environment/ Technology interaction | • Regular monitoring of condition of physical environment<br>• Proactive scheduled maintenance of the physical environment<br>• Reactive maintenance |
| Physical environment/ Process interaction | • Regular monitoring of condition of physical environment<br>• Safe work practices/operating procedures<br>• Proactive scheduled maintenance of the physical environment<br>• Reactive maintenance |

drainage.[16] This allowed approximately 18 000 $m^3$ of fill to enter the lower levels of the mine and the decline. The incident resulted in the death of three mine workers. The construction of stoping systems is a planning and design issue in underground mining. The findings of this incident report also highlighted the need to take physical environmental factors outside the immediate stope environment into consideration when evaluating residual risk at the pre-operational stage. The report states that:

> Even when fill is fully drained and consolidated, there exists the possibility of re-charge from unexpected groundwater inflows, flood or unexpected rainfall events. Systems need to be in place to quickly detect such recharge and to deal with it appropriately. Similarly, regular routine checks of the continued integrity of fill barricades and bulkheads (even in unfrequented areas of the mine) are required.[16]

Firms can adopt a number of strategies to address residual risks stemming from site design shortcoming. The first is to undertake a pre-development evaluation of the physical environment to identify factors that may affect safety and production output. Data may also be collected from other company sites to assist management in anticipating the residual risks inherent in the new site. Once operations have begun, the physical environment should be monitored, residual risks documented and action plans implemented to manage these hazards.

To control the risks inherent in the physical environment/technology interaction, effective design and planning is again important. The firm should carry out a pre-installation evaluation of risk factors and potential hazards. These hazards result from the imperfect fit between the technology and the environment. Technological improvements tend to make a unit of equipment safer, however, when it is introduced into the workplace and interacts with the physical environment, residual risks are not necessarily reduced. For instance, as a result of larger-scale mechanization instead of one person being affected by an incident, more injuries and damage are likely when an incident occurs.[17] This indicates that specific safety standards and operating procedures need to be developed to manage the residual risks in each workplace. Concurrently, emergency procedures have to be established and practiced to minimize consequential damages from such events.

The environment has a significant effect on the safety and productivity of workers. It puts demands on the worker physically and psychologically. For the firm to maximize its return on human capital investment these demands have to be minimized, so that the worker can focus on performing the task. Consequently, as far as is practicable, the workplace should be modified to fit the biological capacity of employees. Interventions can be developed to address the physical environment hazards that are detailed in Table 5.1. This includes, for example, temperature and humidity control in enclosed workplaces, protection from solar radiation when working outdoors, use of barriers and safety harnesses near elevations, safe work practices for confined spaces, protection against chemicals and noise and provision of adequate lighting.

The entropy model indicates that residual risks should be managed in the short term and compressed in the longer term. The inherent risk in the physical environment/human resources interaction should be managed using the hierarchy of controls. Accordingly, hazards need to be eliminated, substituted or isolated, controlled through engineering, or controlled through administrative/safe work procedures or by using personal protective equipment (PPE), with elimination being preferable over latter interventions. Where interaction cannot be avoided, for example, the employee has to work at a height, interventions have to prevent potential injury and damage. This involves anticipating possible failures, implementing safe work practices, supplying PPE and providing workers with training in these practices and the use of this equipment.

The fit between human resources and the physical environment is not static because the health and level of safety consciousness of an individual is variable. Illness can cause reduced concentration and productivity, thereby making the worker more vulnerable to workplace hazards. For this reason, it is important to allow some flexibility in work practices to keep the demands on the worker to an 'acceptable' level. In hazardous environmental conditions it is particularly important to monitor the health, safety and vigilance of workers. This is particularly evident in cases of exposure to extreme climatic conditions. The primary responsibility for monitoring the effect of these residual risks on employees lies with the supervisor. Workers also have a duty to report any condition that may be hazardous including circumstances where the physical environment causes them to be at risk. Employees develop the ability to determine what should be reported when they are provided with training that instils knowledge of residual and entropic risks, the sources of these risks and their potential consequences. They must also be willing to report such matters and this requires management to encourage employee involvement and to respond to legitimate worker concerns.

The final source of residual risk, shown in Table 5.4, is the physical environment/process interaction. Management of this risk begins with an evaluation of the suitability of the process for a particular site. This is an area addressed during feasibility studies and when obtaining government approval prior to the commencement of operations. Consent for a project to proceed indicates that regulators deem the fit between the process and the environment to be 'acceptable' provided that all reasonable measures are taken to manage risk factors.

After a company's operations have commenced, the physical environment tends to change during its life. When substantial changes occur, processes have to be modified accordingly and legislation often requires regulators to be notified. In open-cut mines, for example, in the event of a landslide, which is more likely to occur as the pit gets deeper, processes have to be altered to manage subsequent hazards such as isolation of the unstable section and cessation of further mining in that area of the pit. In addition, according to the Mines Safety and Inspection Act 1994 (Western Australia), the mines inspector has to be notified of any extensive subsidence, settlement or fall of ground.[18] In the short term, when environmental residual risks become detrimental to processes, business activities have to be adjusted. Many industries are aware of how these residual risks affect the safety of their operations. There are also a number of guidelines issued by government departments to address these types of risks. An example is Working Safely in Wet Weather – Construction Industry which states that:

> Provided work is arranged to minimise hazards associated with wet weather, and safe systems of work are followed, work at construction workplaces can continue safely. Taking steps to control these hazards will protect the safety and health of employees, and will benefit companies and enterprises through:
>
> - reduced injury and disease
> - higher levels of job satisfaction and reduced absenteeism
> - increased efficiency and productivity.[19]

The guideline identifies a number of strategies to manage this residual risk. These include:

- Working under sheltered structures or temporary shelters in wet weather;
- Working on tasks that are not made hazardous by wet weather;
- Making environmental changes such as ensuring good drainage;
- Using pumps to disperse flooding;
- Providing workers with warm dry shelter for dry clothing;
- Providing waterproof clothing, safety shoes and gum boots;
- Monitoring employees who have a medical condition that reduces their tolerance to cold or wet conditions.[19]

In the short term, such residual risks have to be managed and this often includes clearly defining the conditions under which it is acceptable to continue work and when work should be ceased. According to the four-fold strategy, residual risk should also be compressed in the longer-term. The channel indicates that this time frame depends on the resources available to the firm. As the internal environment changes over the life of the operation, it may need to be modified or adapted to better accommodate processes and to compress this risk. In addition, the forces in the business context are also evolving and there is a tendency for standards to become more stringent rather than looser. In effect, this means that 'acceptable' residual risk is set at lower and lower levels by regulators and therefore companies are forced to expend resources to comply or risk incurring penalties. For example, organizations are becoming increasing aware of the need to provide slip-resistant floors in reception and accessible areas to reduce their exposure to public liability. Two Australian standards were released in 1994 to reduce this hazard. They are AS 3661.1 Slip Resistance of Pedestrian Surfaces: Requirements and AS 3661.2 Slip Resistance of Pedestrian Surfaces: Guide for Reduction of Slip Hazards.[20]

The legislative framework, therefore, addresses physical environment residual risks in two ways. The first is that parameters are set that define the circumstances under which it is acceptable to operate in the presence of residual risks on the proviso that it is managed well, as was the case with the guidelines for working in wet conditions. The second is to define the standards for physical environment planning, design and construction to minimize residual risks, as contained in the guide for reduction of slip hazards. The guidelines detail the types of modifications that can be undertaken to reduce this risk.

Table 5.4 explains that the physical environment poses both residual and entropic risks. The latter arise because the environment tends to degrade naturally, which reduces the fit between it and the firm's other system factors. The key to preventing injury or damage is regular monitoring of its condition to allow timely corrective action. In cases of sudden degradation due to unforeseen circumstances, the firm needs to have contingency plans to contain consequential damages including emergency response procedures such as systems recovery, medical treatment, isolation of danger and notification of relevant authorities.

Failures in the physical environment can have very severe consequences, particularly when environmental conditions are altered by business activities. This was evident in the Aberfan disaster.[21] The village of Aberfan is at the base of a hill below the Merthyr Vale colliery waste dump. On the morning of 21 October 1966, the structure started sliding down the hillside resulting in a wave of rock, coal slurry and water rumbling down which buried houses and the school. One hundred and forty-four men, women and children lost their lives with 116 of the victims being school children. The parents of a deceased child requested that the cause of fatality on the death certificate read, 'buried alive by the National Coals Board'.

Incidents similar to the Aberfan disaster can also result from natural phenomena. The probability of occurrence increases drastically, however, when business activities also cause physical environment changes. The processes of the firm make the environment less stable than it would ordinarily be, for example, natural slopes tend to be vegetated and this reduces erosion and decreases water-logging. In contrast, man-made rock and soil structures are often not stabilized in this manner to the same extent and this makes them more hazardous. Man-made slopes tend to have a higher level of risk and can degrade more rapidly through processes such as adding a high content of fine material to the stockpile, not maintaining physical barricades and by directing water to the slope. In the mining and construction industries particularly, there are considerable hazards associated with fill materials. The combination of physical environment factors increases the severity of these hazards, for instance, as indicated earlier, stockpiles become increasingly dangerous as the fill becomes saturated. Drainage is therefore an important design consideration and barricades are usually erected to mitigate this risk. When workplace activities introduce entropic risks then these residual risks can be triggered into imminent danger, for example, flooding from a failed process such as pumping wastewater discharge close to the stockpile or leakage from a processing plant can saturate the fill causing a slide. The firm therefore has to anticipate the impact of its activities on inherent dangers and be aware that entropic risks can trigger residual risks into imminent threats. In operations that are inherently high-risk, it is therefore, extremely important to prevent catalytic hazards and severe degradation.

Entropic risks of the physical environment also include the deterioration of infrastructure. To prevent this, buildings, fences and other fixtures have to be maintained. Poorly kept infrastructure is potentially dangerous, for instance, rotten timbers can eventually cause a structure to collapse. When combined with residual risks, these hazards can lead to active threats. An example is when a 5-meter long section of roof guttering from a plant air compressor building fell to the ground during a period of high winds.[22] The combination of residual and entropic risks in the physical environment greatly increases the probability of an undesirable event and the consequences of these hazards.

The interaction between the operator and the physical environment can also result in degradation. The worker can accelerate the deterioration of the physical environment by performing his duties in a suboptimal manner, for example, not cleaning up spills or deviating from designated roadways whilst operating machinery. The environment can also be damaged by intentional safety violations. The primary means of managing the potential entropic risk at the physical environment–human resources interface is induction and training which provides employees with an understanding of risks, with hazard identification and control skills and safety awareness. Training also enhances the competencies of operators so that safe work practices are adhered to and good habits, such as housekeeping, are developed. The organizational culture can be used to reinforce desired behaviors.

The key to preventing human behavior from inducing deterioration of the physical environment, therefore, is education. The standard of housekeeping in a workplace is the clearest indicator of how effectively this interaction is being managed. Training, however, is not a substitute for providing a workplace designed to fit the capacity of employees and the tasks that are undertaken. For example, in a maintenance workshop it would be inappropriate to expect workers to keep the facilities tidy if there are no systems for the storage of tools. The design of the physical environment must support business activities and allow for safe and efficient operations.

As described in the previous chapter, the interaction between technology and the environment causes degradation of both these system factors. Rough road conditions increase the wear and tear on truck tires. Likewise, heavy machinery causes the roadway to deteriorate. Technology can also be a catalyst that causes entropic risk to rise rapidly or residual risks to result in imminent danger. A fatality resulting from a fallen tree branch is an example. In May 1995, an earthmoving company's supervisor was killed when a branch fell from a dead tree. The 20-tonne excavator was digging test holes for soil structure evaluation on a rural property prior to the construction of a water catchment dam, when the boom of the excavator struck a nearby dead tree. A branch, 12 meters up, broke off and struck the supervisor. The contributing factors were that the hazard had not been identified and that the operator's vision was restricted by the boom, window frame and hydraulic ram of the vehicle.[23] Lack of risk assessment prior to the task being undertaken, together with lack of visibility, introduced entropy into the activity. As a result, the residual risk caused by the presence of a dead tree was released and resulted in the accident. The case highlights the need for hazard inspections prior to the commencement of work. Specifically, the following questions need to be asked in any such circumstances:

(1) What are the residual risks present in the area?
(2) Is the environment currently degraded, to what extent and what can be done to mitigate this degradation?
(3) How is the use of technology, the worker's competency level and the process going to affect these risks?

In the above example, a significant residual risk was the lack of unrestricted space to undertake the work and that the tree was dead and therefore unstable. Had the tree been alive and strong the consequences of the impact would not have been as serious although the equipment may have been damaged. To mitigate this risk, the dead tree could have been felled safely prior to the work being undertaken. Clearly, the use of technology in the area with the resultant impact triggered these pre-existing residual risks into immediate danger.

The case highlights the importance of pre-commencement inspections of the work area. These assist the operator to identify and manage potential hazards. As shown earlier in the guidelines on working safety in wet weather for the construction industry, the presence of risk factors does not necessarily mean that work cannot be carried out. Instead, in order to work safely these risks have to be managed. As a prerequisite for safe work the operator must be aware of the hazards that are present and of their implications.

The risk management strategies for the physical environment/ technology interaction are three-fold. These are firstly to regularly monitor the condition of the physical environment and to anticipate what impact the use of technology will have on it. The purpose of monitoring is to trigger corrective action, which is the second step, and often involves reactive maintenance such as rectifying structural weaknesses. The final strategy to prevent degradation caused by the interaction of the physical environment and technology is to carry out proactive maintenance of natural and locational factors. Planned maintenance would have prevented the roof guttering in the earlier case, from becoming detached from the building during high winds. The focus of infrastructure maintenance is to prevent failures that cause hazards such as fallen objects and structural collapses. For example, when a dam wall fails, the water that is released has sufficient potential energy to cause damage and/or injury.

Similar strategies apply when managing the entropic risks associated with the interaction between the physical environment and processes. Any business activity that involves direct use of the physical environment such as forestry, mining, oil and gas extraction, and farming can cause hazardous changes to natural conditions. Blasting, for example, is a necessary part of the mining process. The shock wave from this activity can cause destabilization and also cracking in buildings and other infrastructure. Processes that emit fumes can cause corrosion of metal structures. An example is a case in which a chimney stack at a processing plant was emitting sulfur dioxide. This was being drawn into an air-conditioning unit in the administration building when wind conditions were unfavorable and as a result the unit and surrounding structures became corroded as the fumes combined with water to form sulfuric acid.[24]

Where processes cause the environment to deteriorate, the firm has to carry out regular monitoring to ensure that the risks do not have the potential to cause injury or damage. Safe work practices are required to minimize environmental damage and these should be adhered to for all standardized tasks. When unusual or nonstandard duties have to be carried out it is important to perform a JSA. In addition, a pre-commencement evaluation of risk factors should be undertaken whenever the task is carried out thereafter. The operator needs to determine how the physical environment will affect the process, for example, it is more dangerous to climb a ladder in wet and windy conditions than in dry conditions. In addition, the characteristics of the surface on which the ladder is placed, for instance its gradient and firmness, will affect the security of the ladder. These physical conditions influence the risks involved. If the work has to be done, the operator should adjust the method of work to account for these environmental factors. For example, if the ground is covered with rocks, these should be removed so that the ladder can be erected on a firm surface. Alternatively, workers should select an area in which the process is going to create the least harm. The objective of the pre-commencement evaluation is to create the best fit between the process and the physical environment to minimize the risk of injury/damage and to mitigate the likelihood of degradation.

In some circumstances such as in confined spaces the physical environment greatly increases the risks involved in a process. Enclosed areas allow the accumulation of by-products of the process such as heat and fumes, restrict the movement of the operator, and constrain access and egress. These factors have to be considered, before the work is undertaken, to anticipate the potential hazards that can be generated. For example, there was a case in which a welder was overcome by argon gas, when he entered a 750-mm pipe fabrication.[25] This gas was present because it had been used to create a shield behind an external weld zone. The pipe fabrication was several meters long and one section had been dammed and filled with the gas. When the welding had been completed the dams were vented to expel it, however, when the welder entered the pipe to remove the dams, he was unaware that the gas had accumulated in another section of the pipe. A second worker was also exposed when he went to assist the welder after the alarm was raised by an observer. The incident report identified the need for appropriate confined space procedures to be implemented. The accident would have been prevented if:

(a) a risk assessment, that would have identified the potential hazards and the appropriate control measures, would have been completed prior to work commencing;
(b) appropriate equipment would have been provided to test and monitor the internal atmosphere prior to any attempts to enter;
(c) employees would have been aware of the action to be taken in the event of an emergency, that would have eliminated exposure to risk of injury to the person who went to the assistance of the welder.[25]

Where processes cause the degradation of the physical environment, the entropy model indicates the need to take corrective action or reactive maintenance to eliminate, as far as practicable, the current level of danger. In the above case, this would involve repurging the pipeline and using monitoring equipment to ensure that conditions were safe. The second step would be to implement proactive maintenance strategies to prevent further degradation, with an appropriate remedy being to establish a safe work procedure for the welding of pipe fabrications. Set work practices reduce deviations that increase the level of risk. Processes are clearly defined to limit risks to an 'acceptable' level and to prevent additional risks from being introduced, such as the failure to use the required personal protective equipment in the presence of potential hazards.

The primary methods of managing both residual and entropic risks in the physical environment are sound planning and design, regular monitoring to detect changes in risk levels, the implementation of proactive maintenance strategies, and contingency planning including corrective action and emergency procedures when hazards reach an 'unacceptable' level. Each employee has some responsibility in carrying out these duties. Although the supervisor's role is to oversee risk management in the workplace, employees should also be accountable for monitoring the condition of their immediate work environment and technologies. This requires workers to be provided with basic training in hazard identification so that they are competent in monitoring known risks, aware of the impact behaviors have on these risks, and are able to identify changes in risk levels. These accountabilities can be detailed in position descriptions or work practice guidelines to formalize them into the workplace culture, as discussed later in this book.

## The physical environment audit

Under productive safety management, regular monitoring of the physical environment is a habit undertaken by the supervisor at a workplace level and also by each employee within their own work area. Annually, the firm also needs to undertake an audit of the physical environment that focuses entirely on hazards associated with natural and infrastructure conditions. Similar steps are taken to prepare for this audit as were taken to carry out the technology audit. Figure 5.3 illustrates the steps involved.

After preparing the guidance list, the auditor begins the questioning process. Areas of investigation include:

(1) Does the physical environment have any of the following residual risks?

Extreme temperatures
Precipitation and moisture
Solar radiation
Lightning
High winds

Elevation or heights
Instability
Poor surface conditions
Confined spaces
Drainage

Biohazards
Dangerous animals
Toxic vegetation
Noise (non-technological origin)
Submersion

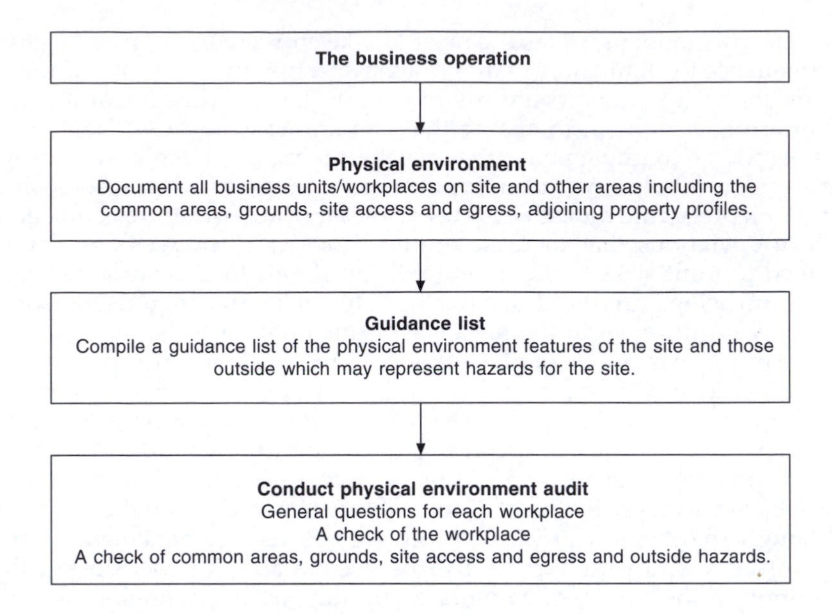


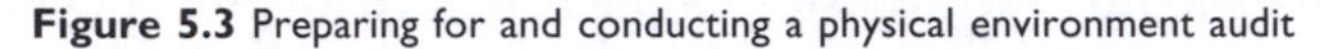

**Figure 5.3** Preparing for and conducting a physical environment audit

| | | |
|---|---|---|
| Atmospheric pressure | Vibration and movement | Insufficient light |
| Fire (non-technological origin) | Access/egress | Disturbing odors |

(2) How are residual risks being managed?
(3) Is there a better way to manage these residual risks?
(4) Are employees aware of these residual risks?
(5) Is the physical environment being maintained to eliminate entropic risk through sound housekeeping and other maintenance strategies?
(6) Have there been any incidents caused by physical environment risks since the last audit?
(7) What future actions are planned to compress residual risks?

To determine the risks associated with the physical environment, the seven process investigators (7 Ps) are used. During this audit, the 'place' is analyzed in detail. The 7 Ps are:

| | |
|---|---|
| Process: | What is being done? |
| Purpose: | Why do it? |
| Person: | Who is doing it? |
| Place: | Where is it being done? |
| Period: | When is it being done? |
| Procedure: | How is it being done? |
| Performance: | What are the outcomes – safety, productivity and quality? |

'What if' analysis is again applied to person, place, period and procedure to identify potential hazards associated with the interaction between the physical environment and other system factors. For example, the changing of a tire on a light vehicle can be analyzed as follows:

| | |
|---|---|
| Process: | Changing a tire on a light vehicle. |
| Purpose: | To replace tire that is in less than optimal condition (punctured/worn). |
| Person: | (1) Tradesperson assistant (planned maintenance)<br>(2) Vehicle driver (unplanned puncture) |
| Place: | (1) Light vehicle workshop<br>(2) Place chosen by operator following puncture |
| Period: | (1) Scheduled maintenance<br>(2) Following puncture |
| Procedure: | (1) Refer standardized procedure in light vehicle workshop<br>(2) Use jack provided to raise vehicle. Use wheel brace to remove tire. Replace with spare tire and tighten. Place punctured tire in vehicle |
| Performance: | (1) No incidents reported for this process in past year.<br>(2) 1 injury reported when operator injured back lifting the tire. |

'What if' analysis could result in the following questions being asked:

(1) What if there is oil on the floor of the workshop?
(2) What if the lighting in the workshop is poor?
(3) What if the vehicle driver is stranded on a major road? What additional risks are involved?
(4) What if the roadside is narrow and not level?

The purpose of the audit and analysis is to identify any areas of entropic and residual risk associated with the work environment. These risk areas are documented and recommendations developed using the four-fold strategy which is:

(1) To eliminate entropic risk in the short term;
(2) To establish maintenance strategies to prevent future degradation;
(3) To manage residual risk in the short term; and
(4) To compress residual risk in the longer term.

## Summary

The physical environment audit is one of the most critical audits undertaken by firms operating in high-risk environments. It is imperative that specialists, for example, geotechnical and environmental engineers, are involved in the auditing process. It is also important for the relevant workers to be given feedback regarding the findings of the audit of their work areas. Information allows employees to adjust their behavior to manage these risks. Feedback from the auditing process and other inspections provides workers with timely reminders of the risks present in the workplace and reinforces the need for vigilance, particularly in operations with high residual risks.

Environmental hazards are a characteristic of nature and design imperfections but they are also created when the company changes the operation's site. Consequently, firms have to develop strategies to manage these interactions to prevent degradation and the triggering of residual risks into imminent danger. Hazards may be identified and managed using walk-through inspections, targeted hazard inspections, audits, feedback from environmental monitoring systems, incident investigations and employee suggestions. These practices are particularly critical in high-risk environments such as construction sites, oil and gas operations, mining operations and manufacturing plants.

Of great importance, also, is the facilitation of employee input that raises the level of competence and safety consciousness of workers. The greater the level of environmental risk, the greater the need for vigilance and therefore, as part of its total risk management strategy, the firm needs to ensure that its HRM systems encourage the desired behaviors and attitudes that mitigate these risks. It is through the management of the people-related risks that OHS and HRM disciplines are integrated. In the following chapter this final system factor will be discussed. Part 2 will conclude with Chapter 7 in which a risk quantification method that assists the firm to prioritize its interventions will be presented. It allows the company to identify different activities or combinations involving the use of people, equipment and the work environment when undertaking a process. From this analysis, risk management strategies can be developed and included in the productive safety management plan discussed in Part 4 of this book.

## References

1. Kamp, A. and Le Bansch, K. (1998) *Integrating management of occupational health and safety and environment: participation, prevention and control*. Paper presented to the International Workshop: Policies for occupational health and safety management systems and workplace change, Amsterdam, 21–24 September, p. 6.
2. Byrom, N.T. (1999) *Health and Safety Management Systems, History, Rationale and Structures*. Paper presented to International Occupational Health and Safety Performance Management Systems Symposium, Toronto, Canada, 5–6 June, p. 6.
3. *Occupational Safety and Health Act 1984 (Western Australia)*, Western Australian Government Printers, Perth, Australia. Available on website: http://www.austlii.edu.au/au/legis/wa/consol_act/osaha1984273/
4. Department of Mineral and Petroleum Resources Western Australia (2002) *Guidelines for the Application of Environmental Condition for Exploration and Mining*. Available on website: http://www.dme.wa.gov.au/prodserv/pub/mining_info/11.html
5. Discussions with OHS Consultant, 2002.
6. Department of Mineral and Petroleum Resources Western Australia (2001) *Significant Incident Report No. 95*. Available on website: http://notesweb.dme.wa.gov.au/exis/SIR.NSF/6b390eea5649d21c48256097004aacb7/0dbcc37ef94a1cad4825668a0013ad5c?OpenDocument
7. Worksafe Western Australia (2001) *Significant Incident Summary Lightning strikes can kill*, dated 03/2001. Available on website: http://www.safetyline.wa.gov.au/PageBin/injrsign0209.htm
8. Department of Mineral and Petroleum Resources Western Australia (2001) *Safety Performance in the Western Australian Minerals Industry 1999/2000 – Accident and Injury Statistics*. Available on website: http://www.dme.wa.gov.au/prodserv/pub/pdfs/safeperform9900.pdf

9. Worksafe Western Australia (1993) 'Sick Building Syndrome', in *Safetyline Magazine*, issue No. 18. Also available on website: http://www.safetyline.wa.gov.au/PageBin/disegenl0003.htm
10. Worksafe Western Australia (2000) *Conditions for Legionella Growth – Evaporative Air Conditioning Unit*, dated 20/2000. Available on website: http://www.safetyline.wa.gov.au/PageBin/injrsign0186.htm
11. Halsey, W.D. and Friedman, E. (eds) (1984) *Colliers Encyclopedia*. Macmillan Educational Company, New York.
12. Department of Mineral and Petroleum Resources Western Australia (2000) *Fatigue Management for the Western Australian Mining Industry – Guideline*. Available on website: http://www.mpr.wa.gov.au/prodserv/pub/index.html#safety
13. Department of Mineral and Petroleum Resources Western Australia (2001) *Incident Report No. 447 – Incidents NOC on 21/3/2000*. Available on website: http://notesweb.mpr.wa.gov.au/exis/fyinew.nsf/9b4047000e180882c82563e70032a1bd/91aa21651de1c0ae482568e90030445f?OpenDocument
14. Department of Mineral and Petroleum Resources Western Australia (2001) *Incident Report No. 445 – Truck/Mobile Equipment Collision on 3/3/2000*. Available on website: http://notesweb.mpr.wa.gov.au/exis/fyinew.nsf/e7f8c6f8d521d0fec82563e70032a1c1/671c71dea367dbde482568dd00207a0b?OpenDocument
15. Mines Occupational Safety and Health Advisory Board (1997) *Report on the inquiry into fatalities in the Western Australian mining industry*. (Table 5.2) Available on website: http://www.mpr.wa.gov.au/news/fatal.pdf
16. Department of Mineral and Petroleum Resources Western Australia (2000) Potential hazards associated with mine fill, in *Safety Bulletin No 55*, dated 06/29/2000. Available on website: http.//www.mpr.wa.gov.au/news/june29_00.html
17. Butterly, N. (2000) Mechanised mining creates new safety hazards for industry, in *Gold Australia*, 16 October, pp. 1–2, Australia.
18. *Mines Safety and Inspection Act 1994 (Western Australia)*, Western Australian Government Printers, Perth, Australia. Available on website: http://www.slp.wa.gov.au/statutes/swans.nsf/be0189448e381736482567bd0008c67c/85cf936e5e10bbe8482568b600173550?OpenDocument
19. Worksafe Western Australia (2000) *Working Safely in Wet Weather – Construction Industry*. Available on website: http://www.safetyline.wa.gov.au/PageBin/workhazd0014.htm
20. Worksafe Western Australia (1994) A New Australian Standard gives Guidelines for improving safety with slippery floors in workplaces, *Safetyline Magazine No. 22*, May. Available on website: http://www.safetyline.wa.gov.au:81/
21. Department of Mineral and Petroleum Resources Western Australia (2000) The Aberfan Disaster . . . the village that lost its children. *Minesafe Vol. 11 No. 3*, September, pp. 6–7.
22. Department of Mineral and Petroleum Resources Western Australia (1999) *Incident No. 427 – Fixed Plant Incident on 12/10/1999*. Available on website: http://notesweb.mpr.wa.gov.au/exis/fyinew.nsf/9b4047000e180882c82563e70032a1bd/423d9765d1dde05d4825682c00192229?OpenDocument
23. Worksafe Western Australia (1995) *Supervisor killed by falling tree branch*, 13/1995. Available on website: http://www.safetyline.wa.gov.au/PageBin/injrsign0060.htm
24. Discussions with OHS Consultant, 2001.
25. Worksafe Western Australia (1998) *Work in confined spaces*, 10/1998. Available on website: http://www.safetyline.wa.gov.au/PageBin/injrsign0108.htm

Chapter 6

# Human resources

As explained in Chapter 1, a key advantage of the entropy model is that it shifts accident causation away from unsafe acts and unsafe conditions to a systems perspective. There is consequently less emphasis on worker error during incident investigation and greater focus on how the system has failed to provide workers with the competencies, safety consciousness and methods of work to operate in a safe manner.

Historically, the focus on human error has led to behavior-based safety (BBS) systems for OHS management which concentrate on the employee first and management second.[1] The danger with this has been that BBS can become an isolated program separate from the total management system. As a result, some key players in the behavioral safety field have shifted away from the premise that behavior management should be the central issue in OHS management.[2] There is an acceptance that it is one element in an overall safety initiative. This shift has involved a greater acceptance of multiple causal factors contributing to accidents.

Reason's 'resident pathogens' model, presented in Chapter 1, has contributed to this systems perspective. The old approach which centered on the person, viewed unsafe acts as arising from aberrant mental processes such as forgetfulness, inattention, poor motivation, negligence, recklessness and carelessness.[3] The systems approach acknowledges human fallibility and that errors will occur. The remedy, however, is not to blame the worker but to develop system defenses to mitigate these risks. The entropy model supports this approach by identifying human resources as one of four system factors. The analysis of residual and entropic risk sources in the human resources system factor, which is dealt with later in this chapter, supports the current points of view that:

- The situation in which the injury is actually caused by the worker, while existent, is extremely rare.
- The statement that 80 percent to 95 percent of accidents are caused by unsafe acts is wrong, because the primary causes come from management systems and the facility.
- The great majority of actual causes of injuries is an interaction between the worker and the facility.
- This point of interaction is the working interface, which should be the subject of attention.
- Figuring out how the worker interacts with the system changes the

focus of improvement from the worker to systems that enable safe behaviors.[2]

The identification of risks associated with the human resources system factor under productive safety management concentrates on this working interface. This is where the interaction between human resources and the other system factors in the entropy model occurs. According to the model, human resources need to be maintained at optimal performance and safety to reduce the probability of an accident, which presents a strong case for investment in training and development. In addition, the channel indicates that human resources are a capital factor along with physical and financial capital. Human resources and human capital are therefore one and the same. The model and channel are further linked because physical capital is made up of the physical environment and technology. Processes are also related to financial capital because work practices integrate the use of the other system factors. The level of performance and safety at which processes are undertaken has an effect on the firm's revenue potential and on cost control and, therefore, the level of financial capital.

## Human capital – employees are assets

Productive safety management reinforces the principle that employees are assets. The traditional approach to OHS management has portrayed them as a liability – the primary source of accidents in the workplace – and therefore, a cost. In addition, for accounting purposes, labor is recorded as an expense and technology as an asset. The channel suggests that the treatment of human resources as an asset is a necessary and important philosophical shift for firms intending to build sustainable businesses that are aligned internally and externally in the twenty-first century.

Figure 6.1 illustrates how the model and channel are integrated to reflect an economic and values-based definition of human resources. It shows that the system factors of the model are related to the capital resources in the channel. Processes provide the interface by which the physical environment, technology and human resources system factors interact, as shown by the arrows on the left-hand side of the figure. When system factors are maintained at optimal performance, safety and quality, the firm increases the value of its capital assets. This is not an increase in value in accounting terms but in practical terms that may be translated, for example, into extended physical capital life and reduced equipment downtime, as explained in Chapter 4. Financial capital is raised through increased efficiency, improved quality and reduction in rework and system failure costs. Optimization of human resources assets means that competencies are developed to a high level and this builds competitive advantage for the firm.

Figure 6.1 also shows that all capital factors are important to the firm. Some companies rhetorically use the catch phrase, 'Our people are our greatest asset'. In today's competitive, high-tech world, employees know that this is actually only a half-truth because, with technological innovation,

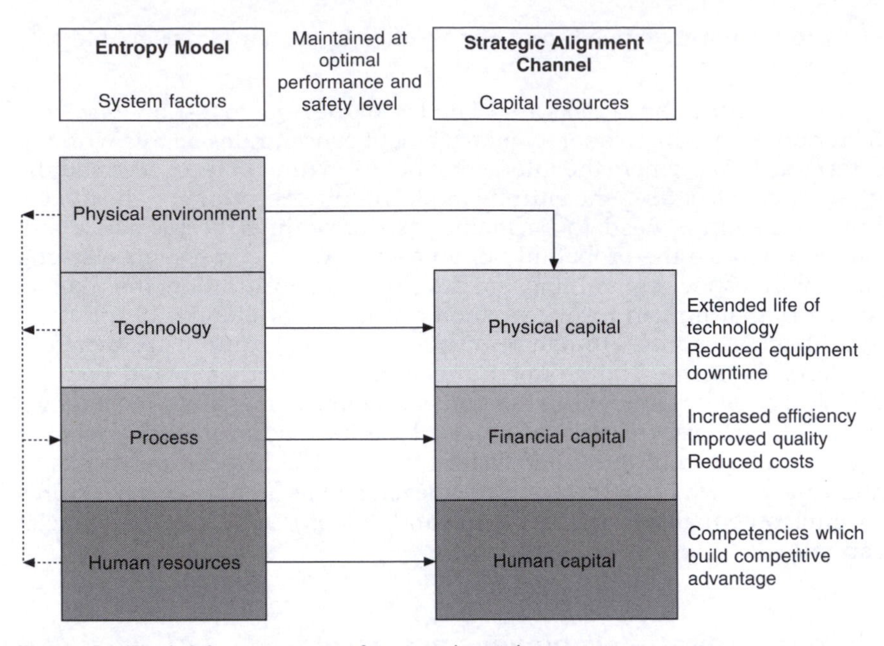


**Figure 6.1** The link between system factors and capital resources

the firm has increasing opportunities to substitute machinery and labor. In addition, business re-engineering that has resulted in downsizing and outsourcing has eroded the relationship between management and employees, sending a clear message to the workforce that firms no longer provide job security.

To illustrate this point, big businesses such as major oil and gas companies are exploring strategic alliances with suppliers of resources, including labor-hire companies, to gain a competitive and cost advantage. This outsourcing is an attempt by the industry to dilute capital risks.[4] Consequently, the labor market in this sector is becoming less permanent. The shift is towards single source alliances where one firm services all the resource needs for an outsourcing firm in a particular service sector.[4] In other words, the use of contracting companies in this industry is on the rise. The main benefits expected are cost savings and improved profitability, better safety performance, optimized service output and continuous improvement. Effectively, these firms are using their financial capital to source human capital rather than 'owning' it themselves. In the current climate, all assets are important for the company's success and the methods of obtaining them and substituting them allows firms greater flexibility than in the past.

As explained previously, the use of labor-hire companies and contractors does not reduce the company's duty of care to these employees nor does it reduce the dependence of the firm on the competencies of these individuals to achieve its goals. Regardless of whether the business acquires its human resources from the labor market or through outsourcing, its legal responsibilities and its reliance on the workforce are the same. In either case, the quality of human resources is important in terms of both productive capacity and risk management.

These current trends in labor usage are changing the nature of the employment contract. At the same time, social values are eroding the effectiveness of traditional management styles. These styles were characterized by power aligned with authority, linked to responsibility and systems that would detect errors immediately and point out the 'delinquent'.[5] The holistic approach to safety management which has emerged in recent years and which is the basis of productive safety management is driven by this shift in values. 'Blaming the worker' has always created a quandary because managers and supervisors have known that the worker does not choose to be injured. Likewise, managers do not intentionally set out to harm employees. This philosophical transition has opened the door for a new approach to safety management underpinned by shared values. These values have become increasingly important in the 'back to basics' approach to organizational management:

> Human behavior in the workplace is conditioned by a number of factors. It is not only governed by corporate imperatives, but is conditioned by the employee's values. Thus success of the enterprise depends in large measure on the extent to which these two value systems – the corporate and the personal – are in harmony . . . In other words, to achieve a safe workplace, we need leadership that supports safety, health and environmental excellence as a reflection of values.[6]

Behind such statements is a fundamental belief that the application of values in the workplace leads to sustained, desirable behavioral change, which in turn generates positive outcomes for both the firm and its employees. In Part 3: Behavioral change, the process of identifying these values and developing common goals based on them, that underscore the OHS management system, will be discussed. In the meantime, the key point in relation to risk management is that, as a result of these values, the employee is no longer seen as the primary cause of workplace incidents. This allows a shift away from the person approach to a systems perspective. The onus is therefore now on management to develop work systems which minimize risk, facilitate employee participation, and encourage leadership and achievement at all organizational levels. The starting point for the development of such systems, as explained by the channel, is for management to understand its responsibilities in relation to legal compliance and social responsibility – to align the firm with the external regulatory and social framework. In addition, employees need to be acknowledged as company assets.

## Legal provisions

Under 'duty of care' legislation, there are clear provisions related to the employer's responsibility to maintain human resources at a safe performance level. For instance, the Occupational Safety and Health Act 1984 (Western Australia) requires employers to:

provide such information, instruction, and training to, and supervision of, his employees as is necessary to enable them to perform their work in such a manner that they are not exposed to hazards;[7]

In addition, the Act encourages consultation and co-operation with health and safety representatives and/or other employees regarding OHS in the workplace. While it is extremely important for employers to understand their responsibilities for the safety of employees, it is equally important that workers understand their legal duty. This legislation includes 'duties of employees' as follows:

20. (1) An employee shall take reasonable care –
    (a) to ensure his own health and safety at work; and
    (b) to avoid adversely affecting the health or safety of any other person through any act or omission at work.
(2) Without limiting the generality of subsection (1), an employee contravenes that subsection if he –
    (a) fails to comply, so far as he is reasonably able, with instructions given by his employer for his own health or safety or for the health or safety of other persons;
    (b) fails to use such protective clothing and equipment as is provided, or provided for, by his employer as mentioned in section 19 (a) (d) in a manner in which he has been properly instructed to use it;
    (c) misuses or damages any equipment provided in the interests of health, safety or welfare; or
    (d) fails to report forthwith to his employer –
        (i) any situation at the workplace that he has reason to believe could constitute a hazard to any person and he can not himself correct; or
        (ii) any injury or harm to health of which he is aware that arises in the course of, or in connection with, his work.
(3) An employee shall co-operate with his employer in the carrying out by his employer of the obligations imposed on him under this Act.
(4) An employee who contravenes subsection (1) or (3) commits an offence.[7]

The duties of employers requires that employees 'are not exposed to hazards'. The entropy model indicates that exposure to hazards is unavoidable due to residual risks and the tendency of systems to degrade. The intent of the legislation therefore, is that employers do not expose employees to hazards unnecessarily, that such hazards are minimized to an 'acceptable' level of risk and that all practicable means are taken to prevent injury and damage.

The employer's duty is also to ensure that employees understand their obligations. Most firms start this education process during the induction of new recruits. At this stage, workers are made aware of the relevant provisions of the legislation. This can be referred to as 'know-what' – knowing what is required in broad terms. Does this enable them to perform their work safely? Clearly the answer is 'no.' Being aware of the legislation and understanding how one's own competencies and behavior affect compliance are two different levels of comprehension. Instruction in legislative requirements provides basic understanding of responsibilities. It does not ensure that the employee shares the same understanding of 'reasonable care' as his supervisor, colleagues or government regulators. In addition, the employee's capacity to manage his behavior in a manner that prevents unnecessary exposure to hazards needs to be imparted through on-going training and education. It cannot be assumed that induction provides the worker with all the required knowledge and skills. Induction is therefore only the beginning of safety awareness development.

Learning to work safely and the desire to work consciously requires the development of a sound understanding of the risks involved in the job. Employees need to be provided with the 'know-how' – the competencies – to do the job safely. In addition, for such learning to be permanent they also require 'know-why' – to appreciate the importance of this learning and behavioral modeling. As a result, employees should appreciate what risks are present in the workplace, the implications of these risks, be able to recognize changes in risk, and make conscientious choices to manage their responses to prevent injury. These choices include reporting changes and collaboratively modifying work practices to manage the risk. In some cases it will involve collectively deciding that the risk is too high and electing not to carry out the task until the risk has been reduced to an 'acceptable' level. The aim of knowledge-based training is therefore to attain an ability to assess risks objectively and to have a shared understanding of the nature and potential consequences of the risk.

Employees need to understand the risk factors associated with processes, technology and the physical environment to prevent incidents. They also need to appreciate the sources of entropic and residual risk associated with human resources – themselves. The reason for referring to workers as 'human resources' is to provide an arms-length objective term to shift the focus away from the individual and on to the system. The risks associated with this system factor are common to all employees and are generic. This provides a clearer understanding of 'human error' as a concept derived from the residual and entropic risks associated with this system factor, as explained in the following section.

Under productive safety management, employees are to be held responsible for their actions, however, it is the firm's duty to equip them with the capacity to make sound judgements in the presence of risks. Accountability for the quality of human resources, therefore, lies primarily with management who has control over strategy development, resource allocation and procedure implementation. It is management, for example, who determine recruitment, training, performance appraisal and supervisory practices, which in turn, affect the mix and level of competencies of the workforce.

An important caution should be asserted in relation to the shift from the worker perspective to the systems perspective. There is a danger that incident investigation may, having diverted away from 'blame the worker', migrate towards 'blame the manager'. This would erode the intent of developing a systems perspective. The entropy model suggests that human fallibility is derived from the inability of individuals, and likewise management systems, to have an ideal set of competencies to anticipate and control all organizational risks. The purpose of incident investigation is thus to enhance continuously the quality of these systems and to shift the firm further along the organizational learning curve. In other words, the emphasis is on continuous improvement and this involves, in some cases, learning from mistakes. Human resources, like the other system factors discussed in earlier chapters, have both inherent weaknesses and the tendency towards entropy, and as a consequence all human systems including management systems will be subject to these natural pre-conditions.

## Human resource hazards

As was the case with the other system factors, human resource risks fall into two categories – residual risks and entropic risks. The primary source of residual risk is incomplete KSAs. It is not possible for an individual to have a comprehensive set of competencies to deal with every situation that may arise in the workplace. As a result of this natural inadequacy, from time to time, workers will use technology inappropriately or make the physical environment degrade or deviate from standardized work practices. People, based on these limitations, will also make incorrect judgements including failing to appreciate the severity of risks. In some cases an error of judgement may result from having the required KSAs but failing to apply them. In other cases, lack of information may be the root cause of this error. It should be noted that this residual risk is not always about the quality of the choices that an individual makes in a given situation. In some cases, the employee's decision may be reasonable and rational yet still result in an incident. This is because the worker is constrained by cognitive limitations which mean that he is unable to assess all the possible consequences of an action prior to the event. It is only with hindsight that all relevant information may be revealed, which if known at the time, may have led to a different choice of response and prevented the incident.

Human resources also have a residual risk because of the limitations of the human body. When other system factors place excessive demands on the worker this greatly increases the probability of an accident. For example, in severe climatic conditions the body suffers to an extent that may be life-threatening. As explained in earlier chapters, the use of technology also places demands on the worker. Continuous use of computers, for instance, can lead to repetitive strain injuries, back or neck pain and eye discomfort. Processes which are physically demanding such as manual handling also stretch the limitations of the human body. This residual risk is therefore derived from the mismatch between the person's physical capacity and other system factors.

In addition to residual risk, the workforce is also subject to entropic risk which causes fluctuations in human resources quality. New employees are a significant source of this risk. They increase the level of residual risk by diluting the performance and safety level of the firm's human resources as a whole, and have a higher probability of introducing entropy in the form of deviations from safe practice when they interact with other system factors. This tendency stems from their lack of workplace-specific competencies. The extent of the degradation created will depend on the applicability of the new employee's previous experience and current competencies and also the speed at which he bridges the gap through learning. The length of this 'catch-up' period varies from employee to employee and depends on the learning curve, as illustrated in Fig. 6.2.

The figure shows that when the new employee commences, the human resources system factor drops in the level of safety and performance then gradually rises with his learning. Once the new recruit is familiar with the workplace and assimilated into the culture, the system factor is returned to the original level of quality. The implications of Fig. 6.2 for the firm are two-fold. The first is that it shows the importance of accelerating new incumbents' learning curves through training. The second is to treat new recruits, regardless of age, as potential hazards. When these employees start, for example, the supervisor should advise other workers that this group represents an increase in residual risk because of their incomplete KSAs, and that they have the potential to cause entropic risk to rise. Increased monitoring and support is required to prevent hazards from being introduced when these individuals interact with other system factors.

Young, inexperienced recruits present an even greater level of threat to organizational systems because they have a higher degree of residual risk. Their KSAs and level of safety consciousness are well below that of experienced workers, which explains the higher rate of accidents among young workers, particularly males, who tend to be more risk tolerant. The LTI rate from males aged 15 to 24 in Western Australia in 1996–1997 was 7.7 compared to other male workers who had a rate of 5.9.[8] This

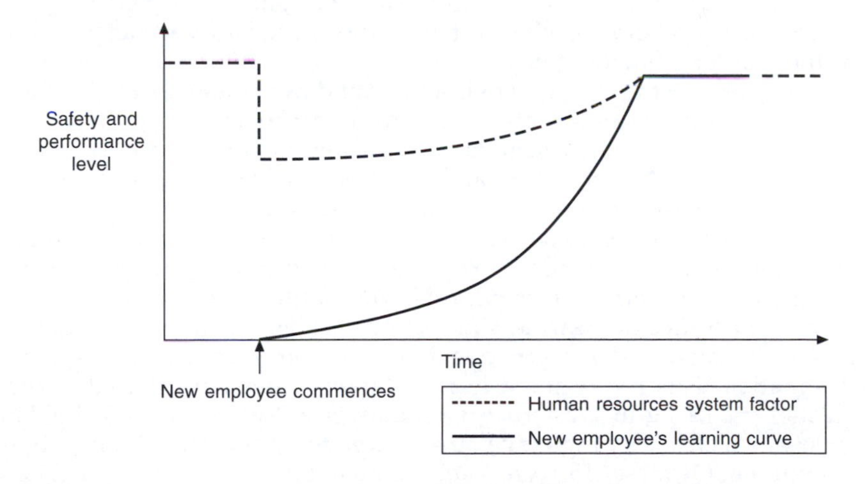


**Figure 6.2** The effect of a new employee on the human resources system factor

pattern of high incidence was not evident for young female workers, who had a rate of 2.2 compared to other female workers with a rate of 2.6. Interestingly, the duration and severity rates for young male workers were much lower than for other groups of workers. This appears to indicate that their level of residual risk in terms of their physical capacity to recover from injury is lower than for employees who are older.

Regardless of their level of competency or age, workers can have accidents as a result of degradation. A highly skilled worker, for instance, will be less capable of performing duties safely when suffering fatigue from working excessively long hours. Although the employee may have a low level of residual risk because of his well-developed competencies, the vulnerability to deterioration is similar to other less skilled workers. A major source of entropic risk in human resources, therefore, is fatigue. This condition can be defined as follows:

> Fatigue is a physical condition that can result when an individual's physical or mental limits are reached. This can happen following:
> - physical exertion;
> - mental exertion; or
> - inadequate or disturbed sleep.[9]

Workplace-related factors that contribute to fatigue include shifts and schedules, the type of work, commuting, and increased exposure to other hazards. In the earlier chapters on technology and the physical environment, it was explained that the greater the demand that these system factors impose on the worker, the greater the probability of an incident. Severe conditions require the worker to manage the risks involved. This taxes the individual's capacity to cope with these demands and to remain vigilant. These additional pressures, therefore, cause fatigue or a reduction in the operator's ability to work safely, which is also affected by the duration of activity. The longer the worker has to manage his behavior to prevent a hazard from translating into an injury, the more likely it is for an incident to occur. This was shown in Figure 3.11 in which the degradation of other system factors caused a severe deterioration in the worker's functionality.

The quality of the physical environment and technology therefore affects worker fatigue. Processes also have an impact, particularly the way in which work practices are structured, for example, the duration between breaks, the length of the shift, and the variety of tasks. Repetitive tasks can lead to boredom and induce loss of alertness. Some firms use job enrichment strategies to address this problem, which involves rotation and multiskilling to keep the employee stimulated and to prevent consistent exposure to the same set of hazards. The nature of the work can be a significant source of weariness in sectors such as the transport industry, in which workers drive for long hours, under pressure, to reach the destination. This has resulted in high fatality rates with 102 fatal truck crashes occurring in Western Australia over a 3-year period in the late 1990s,[10] which is very high given that this state's population is only about 1.5 million. Of these, 15 were likely to have been due to fatigue. There were a further 488 crashes which resulted in serious injury and 55 of these were tiredness-related. Other studies have found a higher level of significance with fatigue being a contributing factor in 25 to 30 per cent of fatal truck crashes. In response, the Western Australian State Government developed a 'Code of Practice for Commercial Drivers' to manage these risks, which covers issues such as operating standards for work and rest and measures for the management of fatigue.

Shift work is a further cause of fatigue that can reduce the worker's level of safety and performance. Approximately a quarter of Australian workers are on shift work.[11] Research has shown that this type of work system has negative health and social effects, including interference with family life, higher occurrences of health symptoms such as respiratory problems, high blood pressure and stomach ulcers, and more frequent use of medications and drugs. Shift work interferes with circadian rhythms or the body clock. The time of highest risk is in the middle of the night, around 3.00 a.m., when the basal metabolism is at its lowest. At this time, the body has a tendency to 'shut-down' and therefore alertness is greatly reduced and this leads to an increased probability of human error.

A further problem with shift work is that when extra hours are worked prior to or at the end of the session, this increases the worker's exposure to hazards at a time of least resistance. In addition, when the employee is required to recommence work at the standard time of the next shift, this reduces the recovery period. If, for example, a mechanic in a plant workshop is on nightshift for five nights from 11.00 p.m. to 7.00 a.m., the recovery time between shifts is 16 hours. If, however, he works overtime of 2 hours, this time is reduced to 14 hours. The more overtime encroaches into recuperation time, the greater the risk associated with fatigue.

Government regulators have been aware of the risk factors inherent in shift work and extended work hours. Legislation has been enacted to regulate work hours including the Mines Regulation Act 1946 (Western Australia). This Act includes provisions limiting the number of consecutive hours and shifts worked, for example, section 39 states:

> Hours of employment underground
> 39. (1) A person shall not be employed to work underground –
> (a) for more than $7^1/_2$ hours in any day unless he is a skipman or platman carrying out his duties as such on a normal working day;
> (b) for more than 6 shifts in any week; or
> (c) for a 6$^{th}$ shift without his express consent.[12]

A further concern with the length of shifts has been the severity and duration of exposure to airborne contaminants and other substances. A number of models has been developed to adjust the occupational exposure standards for longer work rosters. These models provide guidelines for the reduction of time exposed to these hazards, the main ones being the Brief and Scala, OSHA and the pharmacokinetic models. The use of these techniques is beyond the scope of this book, however, the key point is that most exposure standards have been developed for conventional workshifts, that is, five consecutive 8-hour work days followed by 2 days off.[13] When shifts are extended, exposure is also greater, and therefore, the worker receives a higher dose. The implication is that either the time of contact has to be reduced or the concentration of the hazard has to be reduced to keep the dose below the exposure limit.

Shift work and extended work hours induce both physical and mental fatigue. Some activities such as manual handling only cause physical tiredness. By definition, manual handling is any activity that requires the use of force to lift, lower, carry, move, hold or restrain a person, animal or thing.[14] This activity can cause a degradation of the worker's capacity to operate safely and efficiently and, in turn, increases the probability of harm. Manual handling injuries (MHIs) may result from:

- gradual wear and tear caused by frequent or prolonged periods of manual handling activity
- sudden damage caused by intense or strenuous manual handling or awkward lifts
- direct trauma caused by unexpected events.[15]

The first source of injury – gradual wear and tear – is an entropic risk of this activity. It causes the physical functionality of the worker to deteriorate with each subsequent lift or move. The second is an entropic risk which is a catalyst; it results from sudden damage. The final source occurs through the interaction with other system factors, for example, the physical environment, which causes the worker to slip while moving or carrying an object.

There are a number of factors which affect the risk of injury from manual handling and many of these relate to human resources risk factors including person-specific characteristics such as age, physical dimensions and any disabilities the worker may have. These characteristics determine the level of residual risk that a person has in relation to the task. A strong healthy male worker, for example, will have a lower level of residual risk in relation to this type of task than a female worker of small build. Behaviors can also influence whether risk translates into an actual injury. In particular, these behaviors include deviations from safe manual handling techniques such as sudden unexpected or jarring movements, awkward movements and holding static postures for a long time.[15]

The worker's familiarity with the task and the level of training received also affects the likelihood of MHIs. These are determinants of the worker's competence to perform such tasks safely. The nature of the interface between human resources and other system factors is a further variable. For example, how long and how often the task is required to be performed are process characteristics which affect the worker's rate and level of degradation. If the employee has the pace of work set by a production line then there are no opportunities to adjust the work rate to fit the condition of the employee at any given time. The way work is organized therefore influences the level of risk. For instance, if one person performs all the manual handling tasks instead of sharing these tasks among employees[15] the risk to that worker will be higher.

The design and layout of the physical environment and technology also contribute to the rate at which the worker experiences fatigue from these activities. It is, for example, more difficult to carry objects up and down stairs or in confined areas in which the object cannot be put down if the worker feels strained. The primary issue in the interfaces between human resources and other system factors is whether the employee has control to adjust organizational conditions to fit his person-specific capacity. If there is no avenue for minor modifications to such factors as the pace of the task and work conditions, then there is no way of compensating for the employee's state of degradation, particularly in relation to manual handling.

Physically demanding duties therefore increase the rate of entropic risk in human resources. The proportion of MHIs to all lost time injuries and diseases has risen. By way of example, 29.4 per cent of injuries were attributed to this activity in Western Australia in 1995–1996 compared to 27 per cent in 1988–1989.[14] Figure 6.3 illustrates the trend in MHIs and total LTIs during the period. In male workers 54.6 per cent of these injuries were caused by lifting and carrying, 43.3 per cent by handling which did not involve lifting, and 2.1 per cent by repetitive movement. MHIs have a very significant impact on productivity because of the high duration rates associated with these injuries. For male workers in 1994–1995, an average of 28.4 working days were lost for lifting and carrying, 31.6 for nonlifting handling, and up to 48.5 days for repetitive movement MHIs.[14] Of MHIs 89.6 per cent resulted in strain or sprains, 4.4 per cent with musculoskeletal disease and 3.7 per cent were hernias.

The importance of managing the risk associated with manual handling becomes particularly evident when the costs are considered. In this study, the overall average cost of MHIs was estimated to be $A9100 for male workers and $A10 in 125 for female workers.[14] Figure 6.4 illustrates the average costs incurred by sample business sectors. Not surprisingly, industries such as utilities (electricity, gas and water), construction and mining have high average costs. Interestingly, government administration was also in this upper group suggesting that manual handling is a significant risk factor in these office environments.

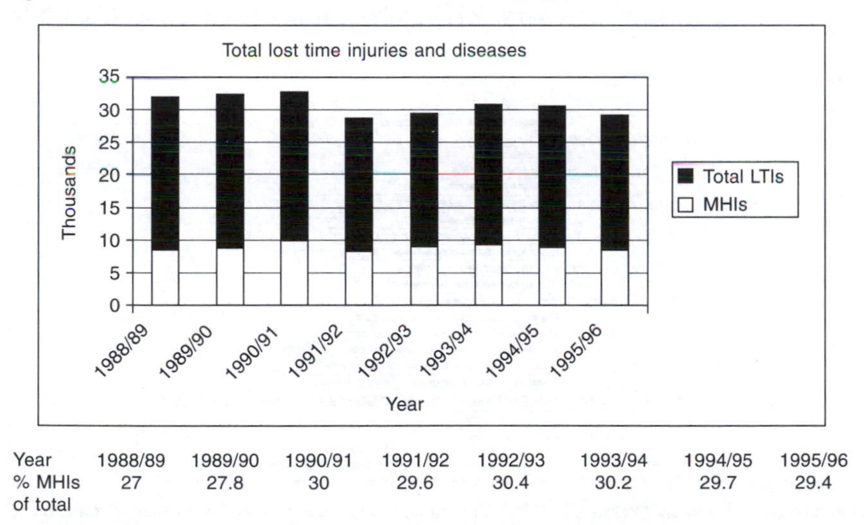


| Year | 1988/89 | 1989/90 | 1990/91 | 1991/92 | 1992/93 | 1993/94 | 1994/95 | 1995/96 |
|---|---|---|---|---|---|---|---|---|
| % MHIs of total | 27 | 27.8 | 30 | 29.6 | 30.4 | 30.2 | 29.7 | 29.4 |

Reproduced courtesy of Worksafe Division, Department of Consumer and Employment Protection, Western Australia (www.safetyline.wa.gov.au). Refer to Ref. 14.

**Figure 6.3** Manual handling and other injuries in Western Australia – 1988–1989 to 1995–1996

There are guidelines for manual handling issued by government regulators, which indicate that the risk of back injury increases for loads of more than 4.5 kg while in a seated position and more than 16 kg in other positions.[15] As the weight increases the percentage of healthy adults who can lift safely decreases. The guidelines also suggest that no one should lift objects over 55 kg. The physical abilities of the worker have to be considered and the risks minimized as far as practicable using the hierarchy of controls. Where possible the need for manual handling should be eliminated, however, if it cannot, then alternative methods of moving goods such as the substitution of manual handling for mechanized handling may be considered. The last resort is to implement work practices that reduce the frequency of lifting and loads involved, when alternatives to manual handling cannot be implemented.

Human resources have variable levels of residual risk as a result of their physical attributes and this affects the rate of fatigue or degradation when carrying out manual handling tasks. Some groups of individuals are therefore better able to cope with the demands of these duties than others. Similarly, individuals have varying abilities to cope with work-related stress. Stress is the psychological equivalent of physical fatigue. The difference is that the body recovers more quickly from muscular strain than it does mental strain.

In the broader sense, psychological stress can be defined as an emotional state that is experienced in situations where the person perceives an imbalance between the demands placed on him and the ability to meet these demands.[16] A more precise definition is given to work stress. It is:

> . . . the harmful physical and emotional responses that occur when the requirements of the job do not match the capabilities, resources or needs of the worker.[16]

Figure 6.5 illustrates two models of job strain/stress. Karasek's model

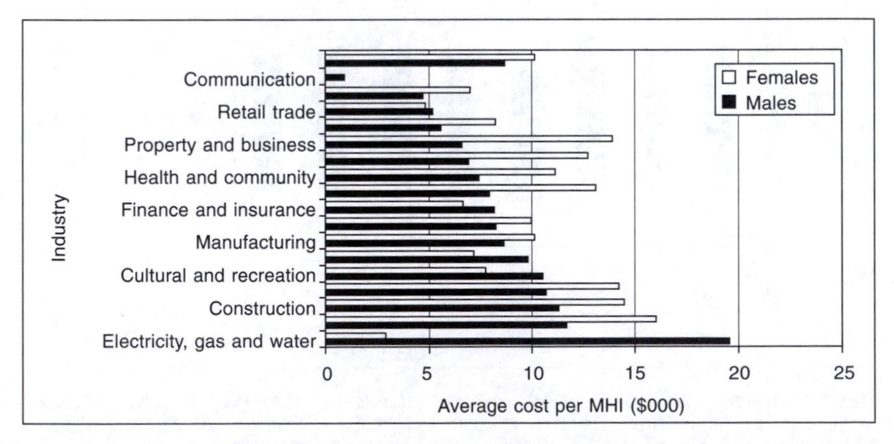


Reproduced courtesy of Worksafe Division, Department of Consumer and Employment Protection, Western Australia (www.safetyline.wa.gov.au). Refer to Ref. 14.

**Figure 6.4** Manual handling injuries in Western Australia – 1994–1995 – average cost by industry

looks at the combined impact of the level of job demand against the worker's job decision latitude or control. This control is expressed as the combination of job decision-making authority and the opportunity to use and develop skills.[17] The model indicates that high job demands and low latitude can cause high strain which, in turn, leads to the risk of psychological strain and physical illness. Karasek shows that high demands that are not threatening to safety, and high latitude can be stimulating because it encourages active learning and creates the motivation to develop new behavior patterns.

Frankenhauser's model uses effort and distress as the two variables that affect the release of stress-related hormones – adrenaline and cortisol. In the face of controllable and predictable stressors, adrenaline levels increase but cortisol decreases.[17] The result is effort without distress. Excessive or threatening demands, on the other hand, cause both hormones to increase and distress is experienced. High levels of effort and distress caused by organizational factors lead to the individual being at risk of hypertension and coronary heart disease.

The potential factors causing workplace stress are numerous. These include organizational change, inadequate communication, excessive workload, time pressure, lack of clarity, inadequate skills, poor work environment and technology, poor relationships and lack of support, and role conflict.[16] It is difficult to pinpoint exactly which factors lead to

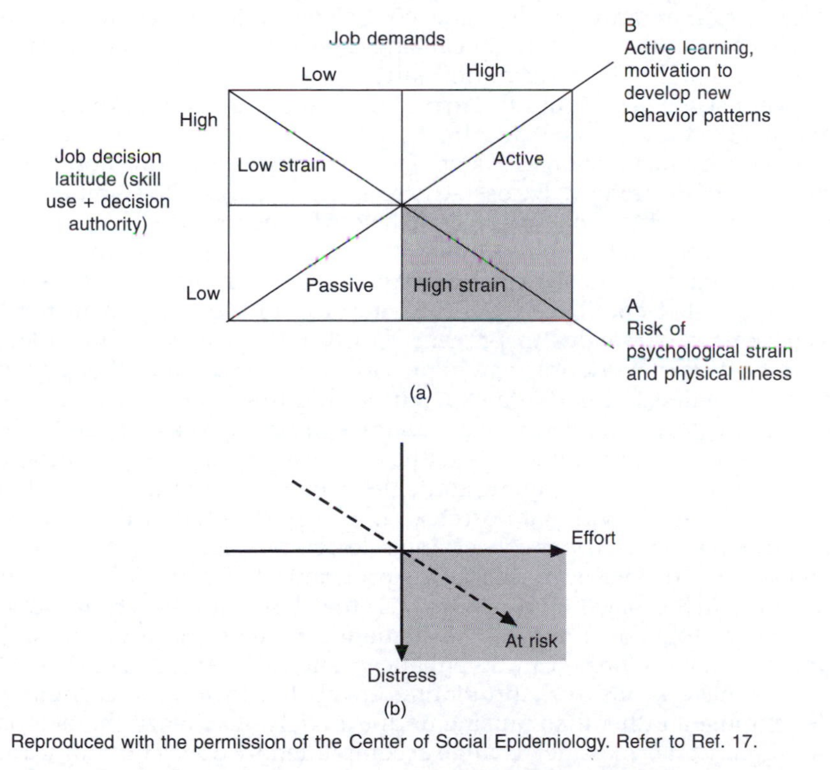


Reproduced with the permission of the Center of Social Epidemiology. Refer to Ref. 17.

**Figure 6.5** Job strain models: (a) Karasek's model, (b) Frankenhauser's model

stress and it is usually a combination of factors, some of which may be nonwork-related, that triggers symptoms of hypertension. In addition, the reactions of individuals to the same stimulants will vary so some workers may find conditions stressful while others do not. How then should the firm manage workplace stress? The entropy model suggests that system factors should be operated and maintained at optimal safety and performance levels indicating that, as an initial strategy, the demands on the worker imposed by the physical environment and technology should be minimized. Hazardous processes should be standardized where practicable. This helps to identify better the risks involved and reduces the stress induced by such risks. Developing reliable, high quality systems and a positive organization culture are the keys to reducing work-related stress.

The entropy model also indicates that despite the company's best efforts to manage organizational stress factors, it may have some employees who suffer hypertension or other symptoms. The reasons for this are two-fold. The first is the company's inability to control stressors outside the workplace including family problems, financial difficulties and dependency habits that may impinge on the worker's job performance. The second is the variability of individual perceptions and coping strategies which mean that some individuals are more at risk – they have a higher level of residual risk that makes them susceptible. In the presence of change and pressure, such employees experience a greater rate of degradation and are also less able to recover, with the result that stress factors accumulate potentially causing dysfunctional stress that inhibits a person's ability to perform normally.

What strategies should the firm develop to prevent workplace stress and to assist those who suffer from it? The entropy model indicates that when the human resources system factor degrades the firm should apply the four-fold strategy. Excessive stress, that reduces the ability of an individual to function safely, is a hazard. Supervisors, in particular, therefore, require the ability to identify this hazard, assess the risk and take corrective action. In extreme circumstances, this requires the removal of the affected employee from the interface with other system factors. Maintenance strategies to prevent harmful stressors may include the modification of work practices to provide sufficient breaks. Workers may also be rotated to distribute exposure to high-pressure tasks. The four-fold strategy also requires the residual risk of stress to be managed, which involves providing employees with the necessary training to recognize symptoms and develop coping strategies. Concurrently, the organizational climate should provide support including opportunities to air grievances, to participate in decisions concerning roles, responsibilities and workload, and access to employee assistance programs (confidential counseling services). In the longer term, residual risk may be compressed by improving system factors to minimize the demands of workplace hazards. The organizational culture can be enhanced such that the demands on the worker have a positive, stimulating effect that promotes learning and development rather than causing undue anxiety. A primary characteristic of this culture is managed change accompanied by open communication, as discussed later in Part 3 of this book.

In extreme circumstances workplace stress can result in violent behavior.[18] This is a profound rise in entropic risk to the point where the person can no longer work safely and becomes an imminent risk to other personnel and organizational systems. Only a small percentage of murders or assaults in the workplace are committed by colleagues.[19] Violent outbursts can result from a combination of personal, system or cultural factors. These include random hostility with no clear intent, intimidation used to achieve an end, expression of uncontrolled irritation such as dissatisfaction, displaced anger from past or nonwork-related situations, criminal activity, thrill-seeking and revenge.[20] Violence can also result from cultural, religious or political differences between sub-groups of employees. Figure 6.6 illustrates the sources of stress and aggression that can occur within the firm. (It excludes violence caused by members of the public or other persons outside the company.)

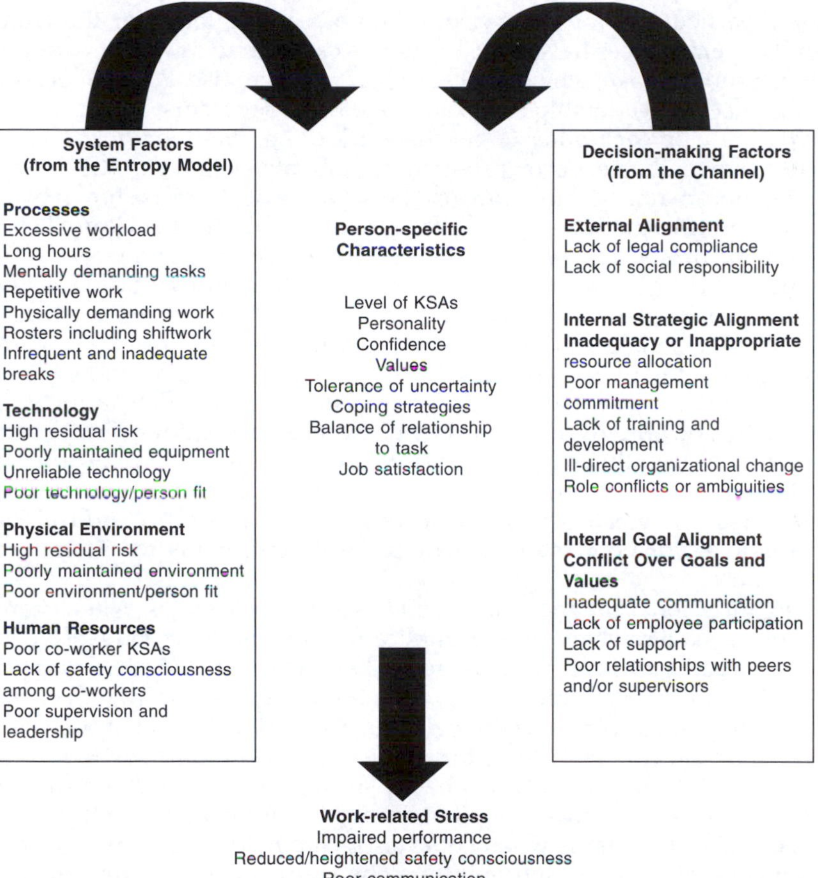


**Figure 6.6** Factors leading to stress and/or violence in the workplace

Figure 6.6 identifies the elements that can lead to workplace stress and, in some cases, to aggressive behavior, applying the entropy model and the strategic alignment channel. Stressors are divided into external sources which are system factors from the entropy model and decision-making factors from the channel, and internal sources related to person-specific characteristics. As shown on the left-hand side of the figure, each system factor places demands on the worker, as discussed in previous chapters. High levels of residual risk and rising levels of entropic risk force the worker to manage these hazards, particularly those from technology and the physical environment. Process issues include excessive workloads, long hours, demanding systems of work and insufficient recovery time. Human resource risks are also significant. When co-workers have poor KSAs and lack safety consciousness, this decreases overall efficiency and the level of safety. For those employees who are aware of the repercussions of an underskilled workforce, this can be a source of stress, particularly for supervisors who are accountable for the team's results. If employees have poor KSAs this can lead to low quality output, rework and wastage, which can cause frustration to rise. Poor supervision and ineffective leadership can exacerbate these problems. When workers have to rely on each other to get the work done, these inefficiencies can have a snowballing effect on productivity, job satisfaction and stress levels.

Decision-making factors can also be a source of stress as shown by the right-hand column in the figure. When, for example, the firm is not legally compliant or socially responsible, employees may experience internal conflict when the firm's practices do not fit with their own value system. In addition, workers may feel underappreciated by the company when it fails to manage risks and shifts the onus of safety on to employees. Internal strategic alignment decisions also have an impact on organizational stress. In particular, inadequate or inappropriate resource allocation places constraints on employees. They may be expected to achieve certain levels of output efficiently and safely, and yet, receive insufficient resources to achieve these targets. An imbalance between goals and the resources to reach them can create stress. Concurrently, lack of support and training, symptomatic of poor management commitment, tends to escalate this problem.

Organizational change is often cited as a source of stress. When change is ill managed it can have serious implications for employee job satisfaction. For this reason, the firm is encouraged to have a 'vision', 'mission', 'strategic intent' or 'directional intensity' to define its purpose and to reflect clear and challenging aspirations that will benefit all of its key constituencies.[21] Having a clear direction helps to reduce stress by eliminating the feeling that 'the goal posts are always being shifted'. Finally, lack of strategic alignment can cause role conflict or ambiguity. An example of this which was discussed earlier is when supervisors are held accountable for both production and safety, and yet, the primary push from management is increased output.

At the 'coalface', conflict over goals and values can create tension. When communication is inadequate and the management style nonparticipative, employees tend to feel disempowered. Karasek's model showed that lack of latitude that prevents workers from having input into decision-making and removes control from them, is a significant source of stress. Workplace relationships also have an impact on tension. On the one hand, positive relationships can reduce stress by compensating for negative task-related or environmental factors, whereas counter-productive relationships can escalate work-related stress, especially in workers with a high social need.[21] The level of support from peers and the supervisor is, therefore, important in managing the workplace climate and the stress levels of organizational members.

The center of the figure shows that the worker filters these external stressors through person-specific characteristics. These explain why people react in different ways when exposed to the same set of circumstances. The internal factors that affect this filtering process include the person's level of KSAs, personality and values. When, for instance, the decisions made by the firm do not fit with the employee's principles this can cause stress. Filtering is also affected by the level of confidence (sometimes referred to as the 'locus of control') in making judgements about the situation. Stress is increased by lack of confidence or uncertainty about one's ability to cope. In an ambiguous situation, the tolerance of uncertainty will determine the level of stress, as will the effectiveness of the worker's adaptation strategies. The extent to which the individual values relationships compared to valuing tasks will also have an impact. In stressful situations the worker may choose to 'back-off' from pushing his/her views on how the work should be done in order to maintain positive interaction with colleagues and the supervisor. Finally, the individual's overall level of satisfaction with the job will influence the extent to which they 'let-go' of stressful incidents.

Stress is problematic for the organization because it impairs a person's ability to function effectively, in terms of performance in the job and social skills. Symptoms of deterioration of the latter include poor communication or more serious behavioral problems such as aggression. Stress can have either a heightening or reducing impact on the level of safety consciousness. It depends whether the stressed individual becomes more 'fearful' of hazards and therefore is more aware, or whether stress leads to 'switching-off', and thus, lower alertness. On-going tension can lead to physical symptoms such as headaches, high blood pressure and heart disease. It can also cause psychological symptoms such as anxiety, depression and decreased job satisfaction.[22] The firm can monitor such factors by measuring observable behavioral changes such as reduced productivity, increased absenteeism, and high turnover (numbers of employees leaving the firm).

Stress causes degradation of the human resources system factor and should, therefore, be considered a workplace hazard. An International Labor Organization study of mental health policies and programs affecting workforces in Finland, Germany, Poland, the United Kingdom and United States showed that the incidence of mental health problems is increasing. As many as one in ten workers were found to suffer from depression, anxiety, stress or burnout. In some cases this led to unemployment and hospitalization.[23] In Western Australia, stress claims increased by 14.6 per cent from 1996–1997 to 1997–1998. The average cost per claim in 1997–1998 was $A27 519[24] and the number of claims continues to rise alarmingly. Stress is therefore becoming an increasingly concerning source of degradation of human resource capacity in organizations, and thus strategies need to be devised to manage it more effectively.

A further source of entropic risk in the human resources system factor is ill health. In some cases, minor sicknesses such as the common cold will not greatly affect the worker's capacity to perform his duties nor does the illness pose a major threat to other employees. Infectious diseases such as influenza are much more serious. Not only are they debilitating for the infected employee, they have the potential to cause severe degradation of the human resources system factor as the disease is spread. Many firms make influenza injections available to their workforce as a means of preventing large outbreaks.[25] Reports of the effectiveness of influenza vaccinations on absenteeism are mixed. In the 3M Company in the United States, when 2600 employees received shots in the 1996–1997 influenza season, sick time decreased by an average of 1.2 hours per employee compared with that of the previous season.[26] Working mothers had a reduction of 3.1 hours on average.

Human resources are therefore vulnerable to pathogens that cause disease. This susceptibility is a residual risk of this system factor. Infection can lead to varying degrees of deterioration that affect performance and the ability to remain alert, and hence, has safety ramifications in the workplace. The sources of disease fall into three categories. Firstly there are zoonoses – illnesses from micro-organisms that naturally live in animals and are transmitted from animals or their body fluids to humans.[11] Secondly, the environment, as discussed in Chapter 5, contains micro-organisms that can cause infection and disease, such as Legionnaire's disease. The final category is communicable diseases which are transmitted by micro-organisms whose natural habitat is people. These are transmitted from person to person or via contact with human body tissues and fluids, including hepatitis, AIDS, chicken pox, glandular fever and measles.

In some industries such as the medical sector, communicable diseases are prevalent as a workplace hazard. For all organizations in which first-aid treatment may have to be carried out the risk of blood-borne diseases such as hepatitis B and AIDS needs to be managed using the four-fold strategy. Appropriate interventions may include establishing procedures for taking corrective action when a worker is injured, enacting safe practices for the disposal of contaminated materials and for the maintenance of hygiene standards, and providing employees with training to manage the risk of contamination. Long-term residual risk compression strategies are also required. Currently, the residual risk of HIV AIDS is difficult to compress other than through safe behavioral practices, however, in the future, it may be reducible through inoculation against the disease.

Organizational systems make allowances for the risks associated with ill health through the provision of sick leave. Minimum entitlements are set out in industrial relations legislation and agreements. With the option of nonattendance available, the onus is on the employee to ensure that she is able to fulfill performance and duty of care requirements when at work. Employers have to deduce that when workers arrive on site they are fit for normal tasks. Should the health of a worker deteriorate during the course of a shift, then such conditions need to be reported, as the first step in taking corrective action. In addition to self-evaluation, the supervisor should monitor the health of workers. According to the entropy model, poor health is a hazard and should be managed using the four-fold strategy. For instance, corrective action should be taken if degradation from sickness becomes evident. Preventative measures such as vaccinations should be provided to address specific pathogenic hazards where possible. Reporting systems should be in place and flexible work systems available to manage the inherent risk of illness, and as a community, long-term solutions to such conditions should be sought through research and development of medical interventions.

While susceptibility to disease is a residual risk, the actual presence of morbidity is a source of entropic risk. When workers are ill they become tired and prone to deviations from safe work practice. Hence, entropic risk rises when employees lose concentration due to poor health. The greater the level of risk within other system factors, the more important it is for workers to be fit and vigilant. Employees' concentration levels may be affected by a number of external and internal factors including fatigue or sickness, however, there are also other variables that may contribute to this degraded state. Lack of knowledge of the need for attentiveness may be a factor, particularly among inexperienced workers. External pressures that place additional physical or mental demands on the operator such as distracting noise may further impair concentration. Environmental conditions, such as temperature and humidity, may also make it harder for the worker to remain focused on the task and to manage hazards. There are, therefore, a number of variables that determine how effectively the employee is able to mitigate hazards in the workplace.

The residual and entropic risks associated with this system factor thus far have been issues over which the individual employee has little control in the short term. The worker cannot, for example, reduce his physical limitations, bridge a shortfall in KSAs, master techniques to cope with stress, or eliminate the fatigue associated with shift work, without investing in long-term improvement or adaptation strategies. On the whole, therefore, these risk sources have to be managed by the organizational system. Workers can, however, take steps to minimize risk at the interface with other system factors by adhering to safe practices and by not engaging in risk-taking behavior.

The risk homeostatic model, provided earlier in Fig. 1.3, explains the process by which the individual makes behavioral choices in the presence of risks.[27] At point 1 in the model, it showed that people have an expected utility of action alternatives. This simply means that they expect some benefits from the various choices available to them which affect their target level of risk – the level that they consider to be 'acceptable'. For some individuals the 'acceptable' level may be higher than the firm's 'acceptable' level. They then use perceptual skills to measure the level of risk involved in a particular activity. The target and perceived levels are compared using decisional and psychomotor (processing) skills, which leads to an adjusted action or choice of behavior.

What can the firm do to affect the worker's choice of behavior? It has to consider ways of influencing the worker's motivation for taking the risk. According to the risk homeostatic model, motivation can be modified. The strategies are to:

Reduce the expected benefits of risky behavior alternatives.
Increase the expected costs of risky behavior alternatives.
Increase the expected benefits of safe behavior alternatives.
Decrease the expected costs of safe behavior alternatives.[27]

This model puts forward a strong case for the use of reward and punishment strategies. Certainly, there is a place for such responses in the firm; specifically, incentives for desired behaviors and disciplinary procedures for intentional violations. These, however, are end-points. They treat the outcomes or the symptoms of behavioral choices and what they fail to address are the underlying factors that determine the target and perceived levels of risk. These are the factors which cause one individual to consider a situation to be risky while another considers it to be 'acceptably' safe. These variations and the subsequent risk-taking behavior are indicative of a higher level of residual risk in some individuals.

Poor KSAs are the primary source of this type of human resources risk. These include inadequate task-related skills which can lead, for example, to the inappropriate use of technology, the failure to follow procedures or poor housekeeping. In the underground mining sector, factors contributing to risk-taking behavior were found to include lack of skills, turnover and inexperience, and lack of awareness of the consequences.[28] Individual attributes affecting 'unsafe acts' therefore center on the worker's knowledge of and experience with risk and its consequences. These are also affected by perceptive skills that involve the ability to evaluate accurately and therefore manage risks, together with workplace-specific competencies. Provided that the worker has the KSAs to operate safely and provided that she does not deviate towards risk-taking behavior by mentally overriding these KSAs in a given situation, then the employee should be adequately prepared to interface with controllable workplace hazards.

One of the most common examples whereby employees deviate from their normal behavioral patterns is when shortcuts are taken. When established behavioral practices are ignored this can be indicative of either individual or systemic factors. The former was explained by the risk homeostatic model. The expected utility or benefit that the person derives from taking the risk determines the target or 'acceptable' level of risk. Saving time or effort is often the perceived benefit of risk-taking. The other causal factor in these circumstances can be overconfidence on the part of the individual. In this case the worker has an unrealistic perception of his own abilities that leads to the false belief that the risk is within his capacity to manage it. The final individual factor that can lead to risk-taking behavior is indifference to the consequences although this is uncommon.

An individual can therefore choose to take risks independently of the organizational system. He can elect to act in violation of the rules, procedures and unspoken codes of behavior. These deviations from safe practices, which include horseplay and vandalism, attract punitive responses from the firm. It is thus uncommon for the individual to make a conscious choice to behave contrary to broadly accepted and understood norms of co-workers and management. More often than not, risk-taking behavior is induced by systemic factors:

> When risk taking behavior was reported by employees or supervisors it was invariably multifactorial although production pressure was clearly the major contributing factor.[28]

Other issues that have reportedly led to 'unsafe acts' include inadequate resources for initial and on-going training. Too rapid promotion of relatively inexperienced trades/professional personnel has also been cited as a problem. The lack of management commitment to safety, and retention by some managers and workers of an outdated risk-taking culture that is resistant to change, can cause this status quo to be maintained. It has also been reported that some companies are failing to recruit new employees using value-based selection[28] which identifies those applicants who are more likely to fit a safety culture.

These factors, however, are secondary to production pressure in triggering risky behavior. HRM practices have been, in part, guilty of perpetuating this production focus with the implementation of solely production-based incentive schemes. As a result, collectively, workers have been encouraged to pursue output to the detriment of safety. In these circumstances, individuals who are risk-averse may experience peer-pressure to ignore risks in order to achieve output requirements. Collectively, therefore, the 'target level of risk' may be increased, because, as shown by the risk homeostatic model, the benefits of risk-taking are perceived to be greater than the potential consequences. These output-based incentive schemes are generally of two types – production bonuses and piece-rate schemes. A report by the New South Wales Minerals Council, conducted in 1997, found a very high level of adoption of production bonus/incentive schemes in the mining industry.[29] Most of the schemes were based on production, for example, tonnage mined or kilometers developed, rather than productivity. The report concluded that such systems could encourage risk-taking behavior. In contrast, in Western Australian underground mines it was found that there was a significant trend away from incentive-based remuneration towards salary-based systems.[29] The effect this has had on safety results is yet to be evaluated.

Current trends in OHS management have to a significant extent laid the 'myth of the careless worker' to rest, in philosophical terms at least. There is a much stronger emphasis on maintaining organizational systems to manage risks. The barriers to safe behavior have been found to be derived from management systems and from facilities/equipment in 73 per cent of incidents.[2] This suggests that:

> the greatest risk reduction will come from attention to those two subjects and that the emphasis following the identification of at risk behaviors should be to determine the reality of their causal factors – the antecedents.[2]

These 'antecedents' are systemic factors such as the quality of the physical environment and technology, the level of control in processes, and the firm's commitment to competency training. Worker behavior is thus driven primarily by contextual factors, however, a balance of responsibility for safety needs to be achieved. This means that total ownership should not be shifted away from the worker to the system.

The entropy model indicates that the quality of human resources – their level of residual risk and tendency towards entropic risk – is an important issue in risk management. There is within the human resources system factor an element of 'personal choice' over which the system has limited control. Hence productive safety management does not advocate the disposal of behavioral-based programs within the OHS management system. There is still a significant need for such matters to be addressed through both procedural and cultural interventions. While overemphasis on the worker can fail to pay attention to management leadership and result in dysfunction or less than optimum labor–management relations,[1] a culture in which employees forfeit responsibility for their actions and 'blame' the organization can be equally unhealthy. Both employee and management actions thus need to be integrated into a systems perspective of OHS management in which the accountabilities of respective parties are clearly understood.

The need for behavior-based elements is particularly evident in relation to the use and abuse of drugs and alcohol. Although the use of substances can be prevented at work, the organizational system has no direct control over consumption outside the workplace. The best that the firm can achieve in this regard is that the worker is not affected by substances when he is at work. Studies in the United States have found that 70 per cent of illicit drug users aged 18 to 49 work full-time, and more than 60 per cent of adults know someone who has gone to work under the influence of alcohol or drugs. In addition, drug-using employees are 3.6 times more likely to be involved in a workplace accident and five times more likely to lodge a workers' compensation claim.[30]

Some may argue that random drug testing is an abuse of civil liberties. In the presence of contrasting social opinion about the use of such methods, the key to having the system accepted and supported by employees is providing them with the 'know-why'. This requires the firm to explain to the workforce why it considers testing to be a valid part of its risk management strategy and how this relates to the impact substances have on human resources' residual and entropic risks. Dependence and habitual use of narcotics, for example, can be considered to be a residual risk because the firm cannot assist the affected person to eliminate this risk in the short term. Ordinarily, with behavioral-related issues, companies pursue win–win outcomes to create a fit between the individual and the organization, however, in the case of drug use, the risks to the affected employee, other employees and systems must be considered 'unacceptable'. For this reason, most firms adopt a 'no tolerance' policy and terminate the employment of known drug users.

Other substances, such as alcohol and some prescription medications, even in low doses, can increase the rate of degradation. In Western Australia in the late 1990s, it was found that alcohol contributed in part to 33 per cent of all road deaths, with almost 25 per cent having a blood alcohol concentration (BAC) of 0.15 per cent or more.[31] Employees who operate machinery, drive a vehicle or rely on motor coordination, face an increased risk of injury if affected by alcohol and/or drugs.[32]

The risks of substance use cannot be considered in isolation. As explained in previous chapters, the greater the level of risk posed by the physical environment, technology and processes, the greater the reliance on the operator to manage these risks. When the worker is under the influence of drugs their ability to manage these risks is also impaired. Substance use has also been found to have a profound negative effect on productivity, absenteeism, turnover and medical costs. One accident costs a business an average of $US12 000 to $US16 000 according to recent studies.[30] Notably, accidents caused by substance use also represent an area of high liability with the injured party not being the abuser in 80 per cent of serious accidents.

The use of alcohol or drugs in the workplace and their recreational use to the extent that it affects workplace safety and performance, are not moral issues. From a system's point of view, employees who are dependent have a higher level of residual risk. This residual risk should be compressed and this may be done through testing as a means of screening job applicants, particularly in high-risk environments. Recreational use which results in residual concentrations in the blood while at work increases the likelihood of entropic risk. Employees who are identified as being under the influence should be removed from the workplace to prevent their interaction with other system factors. This response fits the four-fold strategy which requires corrective action to be taken when systems are degraded. As a follow-up, maintenance strategies that include education programs, random testing and hazard monitoring should be implemented.

> The aim of the strategy should be to eliminate or reduce alcohol and other drug related harm as far as practicable. This objective should be achieved through a three-tiered approach:
> - preventing harm through such steps as providing information and education;
> - management of hazards through introducing procedures for dealing with affected persons at the workplace; and
> - provision in the strategy for the return to usual work duties of affected employees.[32]

With substances such as alcohol there is a threshold level usually set in legislation or guidelines. What happens if the worker is below this level? Is this acceptable? The entropy model indicates that the firm has to consider whether BACs below the legal limit are likely to have an unacceptable impact on safety and performance. This becomes an issue to be addressed within the organizational culture. In a low-risk environment the firm may tolerate some level of use, for example, if the job entails negotiations with clients over lunch. In contrast, in a high-risk environment, the company should operate a 'no tolerance' policy whereby even levels below the legal threshold are considered a performance issue. The firm should also consider whether system factors are contributing to the use of drugs or alcohol among its workforce. Issues to be considered include the availability of such substances, for example, does the company's business involve selling alcohol? Low job satisfaction, isolation from family and friends and stress can also increase usage.[33] Some workplaces provide 'Friday afternoon drinks' or social events which involve alcohol consumption. Again, the issue is not a moral one, but concerned with the appropriateness of such behavior within a context of other risk factors.

Policy and practices therefore need to be tailored to the operational situation and applied fairly and consistently. In addition, all employees need to be provided with the 'know-why' and the 'know-how' to comply with the system. The steps are:

(1) Write a substance abuse policy;
(2) Train supervisors;
(3) Educate employees;
(4) Provide an employee assistance program;
(5) Implement drug and alcohol testing.[30]

There is usually little resistance to the introduction of such systems if explained well. Employees understand that working next to someone who is high on drugs or under the influence of alcohol is unsafe for that person and potentially, for themselves.[30] In principle, therefore, few employees would argue against it. The greatest difficulty with such systems is for the individual to break the social cycle that encourages such behavior, particularly for example, in isolated mining towns which have a culture of 'work hard, play hard.' Education in this regard is the key; that is, providing employees with the 'know-how' to manage their recreational practices so that they do not impact negatively on work performance and safety.

The discussions about human resources residual and entropic risks have highlighted that the firm, through its OHS system, can proactively manage many of these hazards. Some risks, however, are caused by the worker's personal behavioral choices. There is therefore, a case for the inclusion of behavioral-based elements in the OHS management system. The risk homeostatic model explained that punishment and reward approaches are appropriate to modify undesirable behavior such as intentional violations of safety. Deterrent and detection systems are valid methods of encouraging appropriate behavior in some circumstances. These situations include those where the firm has no direct control over the person's actions outside work hours and yet this behavior has a direct and significant impact on safety and performance at work. This includes the consumption of drugs and alcohol during leisure hours and makes random testing a necessary risk management strategy in hazardous industries.

## Managing human resource risks

Table 6.1 summarizes the residual and entropic risks associated with the human resources system factor. Residual risks stem primarily from incomplete KSAs and physical limitations. Degradation occurs particularly when new employees are introduced into the workplace. Human resources 'symptoms' that lead to entropic risk are fatigue, ill health, stress and loss of concentration. The higher the level of demand placed on the worker

**Table 6.1** Strategies to manage the risks associated with human resources

| Source of hazard | Risk management strategy |
|---|---|
| *Residual risk* | |
| Incomplete knowledge, skills and abilities (KSAs) | • Induction<br>• Task-specific training<br>• Risk management training<br>• Training prior to introduction of changes in processes, technology and/or physical environment<br>• Effective communication systems |
| Physical limitations and fragility | • Modification of the physical environment, technology and processes to fit the worker<br>• Matching the worker to the demands of the job<br>• Provision of sick leave entitlements<br>• Job rotation to share physically demanding tasks<br>• Job enrichment<br>• Training in manual handling techniques<br>• Procedures for the management of biohazards<br>• Preventative health measures |
| *Entropic risk* | |
| The new employee | • Value-based recruitment and selection procedures<br>• Induction<br>• Mentoring<br>• Pre-operational training<br>• Hazard monitoring of interaction of new employees with other system factors |
| Fatigue | • Reducing the demands imposed by the physical environment, technology and processes through modification and maintenance<br>• Regular rest and meal breaks<br>• Job rotation and/or enrichment strategies<br>• Management of manual handling hazards<br>• Monitoring of operator alertness by supervisor and co-workers<br>• Training in self-monitoring and management strategies<br>• Effective management of shift work rosters and practices<br>• Flexible systems of work<br>• Education to minimize fatigue |
| Ill health | • Multiskilling and flexible systems of work<br>• Monitoring of operator alertness by supervisor and co-workers<br>• Training employees in self-evaluation of physical capacity<br>• Healthy lifestyle programs |
| Stress | • Legally compliant and socially responsible operations – hazard minimization<br>• Planned and well-managed organizational change<br>• Appropriate and adequate resourcing strategies<br>• Development of common goals and values<br>• Stress management programs<br>• Workplace violence policy and contingency plans<br>• Dispute resolution procedures<br>• Employee assistance programs |
| Loss of concentration | • Regular rest and meal breaks |

**Table 6.1** *(Contd)*

| Source of hazard | Risk management strategy |
|---|---|
| | • Monitoring of operator alertness by supervisor and co-workers<br>• Multiskilling (minimization of repetitive work)<br>• Effective communication systems<br>• Reinforcement of vigilance through training<br>• Modification of workplace to improve attentiveness where practical |
| Safety violations | • Organizational culture and practices which discourage violations and risk-taking<br>• Communication of safety policy and duties of employees<br>• Risk identification and management training<br>• Scorecard based incentive programs<br>• Disciplinary procedures<br>• Procedures for dismissal where appropriate |
| Drug and alcohol abuse | • Education<br>• Substance abuse policy<br>• Monitoring of worker capacity by supervisor and co-workers<br>• Pre-employment drug/alcohol testing<br>• Random testing<br>• 'Moderation' culture<br>• Healthy lifestyle programs<br>• Employee assistance programs (EAP) |

by other system factors, the more these 'symptoms' are likely to occur and the greater the probability of an incident. Management-controlled factors, such as the condition of the physical environment and technologies, the degree of certainty in processes, the structure of rosters and work systems, and the organizational culture have a significant impact on the quality of this system factor.

As explained earlier, worker behavior also affects the level of entropic risk, particularly individual risk-taking choices such as intentional acts to violate safe practices and expected behavioral norms. These 'unsafe acts' greatly increase the probability of a workplace incident. Workers may also have behavioral problems outside the workplace that affect their ability to operate safely and productively, such as financial difficulties, family problems, a gambling habit or drug/alcohol dependence. As a general rule, the private life of the worker is exactly that – private. In relation to most of these personal matters, it is up to the employee to seek help through the company's employee assistance program (EAP) or other avenue. The information provided to the EAP should be treated with strictest confidence and not used for any other purpose. In the case of drug and alcohol use, the firm is within its legal rights to intervene with the worker's personal habits by randomly testing all members of the workforce for substances in the interests of safety and to fulfill the firm's duty of care.

Table 6.1 identifies training as the primary means of addressing the residual and entropic risks associated with human resources. It is an extremely important component of productive safety management and is discussed in detail in Chapter 9. In broad terms, the purpose of training is to develop human resource competencies and safety consciousness to the level of optimal safety and performance. To achieve this each employee needs three sets of KSAs as a minimum standard of proficiency. The first are task-specific skills, for example, the ability to operate machinery correctly and safely. Secondly, risk management competencies are needed which include the ability to identify hazards, appreciate their severity and potential consequences, and choose an appropriate course of action. This usually involves reporting followed by participation in the development of risk management solutions. These safety competencies can be acquired through instruction and 'on the job' training by providing a broad understanding of residual and entropic risk, the potential sources of these risks, and how the operator's behavior may affect them. The final set of KSAs is workplace-specific entailing the appreciation of site hazards and the required methods of risk mitigation. As a result of such training, workers should know defined operating procedures to control hazards within their workplace, such as the appropriate use of personal protective equipment.

The process of addressing the gap between an employee's incomplete KSAs and those competencies required for safe and efficient performance is a gradual process that begins with induction. During the course of the worker's employment with the firm, KSAs do not necessarily increase at a steady rate proportionally to experience. Whenever modifications to processes, technology and/or the physical environment occur, KSAs decline because change requires relearning. As a result, an important risk management strategy is to provide training prior to and during the introduction of alterations to system factors. The impact that training and communication strategies have on the level of risk and probability of an accident following the implementation of change can be illustrated. Figure 6.7 shows conceptually how such strategies affect the level of residual risk.

As explained earlier, people have limited cognitive capacity which means that they are not always able to appreciate all the variables and the implications of these variables in a given context. Current competencies are more easily applied if the conditions are familiar, however, when change is introduced, the residual risk rises. The greater the significance of the change, the longer it will take for employees to adapt and acquire the level of competence held prior to the change. If a business undertakes a total upgrade of its technologies, for example, then its human resources residual risk will rise significantly and only return to previous levels when employees have attained the KSAs to operate as safely as they did with the old technologies. This process takes time because it requires both instruction and familiarization. In both Company A and B in Fig. 6.7, therefore, there is a significant increase in residual risk. Although not shown, there would also be a greater tendency towards entropic risk because of errors and deviations made during the learning process.

In the figure, Company A is proactive in its human resources risk management strategies. It provides employees with the information they need to work with the new technology and begins training prior to its introduction. Company B does not provide any training and the decline

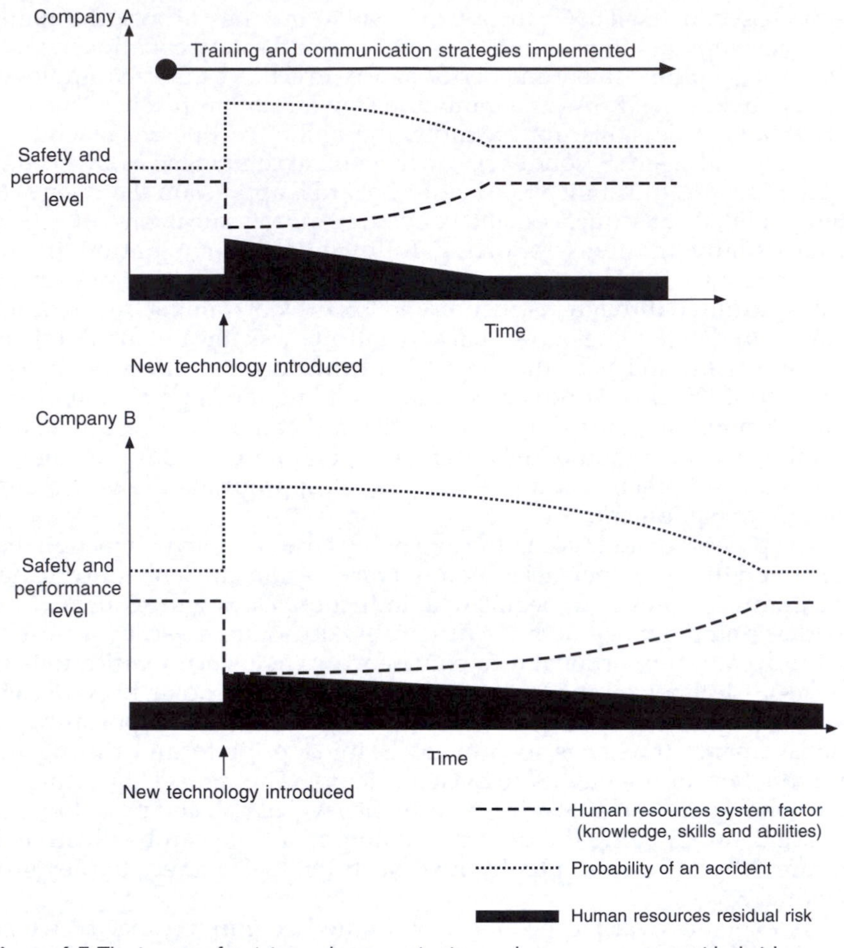


**Figure 6.7** The impact of training and communication on human resources residual risk following the introduction of new technology

in human resources quality is more severe because in Company A employees have been prepared with pre-operational training. Company A's training and communication strategies accelerate the employees' learning curve (shown by the dashed line) and with time the system factor returns to a high level of safety and performance. In Company B, the employees have to learn by doing. Eventually they return to the original level of safety and performance, but in the meantime, they have a higher likelihood of an accident. Their residual risk takes longer to compress. The figure shows that training is required to counter any rise in residual risk caused by workplace change. On-going, effective communication systems are necessary to reinforce this training.

Reiterating, human resources are subject to fluctuations in residual risk whenever significant change is introduced. These risks relate to their KSAs. They also have physical limitations that have to be managed. The primary means of managing such risks are summarized in Table 6.1.

Firstly, the firm has to modify the physical environment, technology and processes to fit the worker, as far as practicable. In addition, where physical hazards such as manual handling are required as part of the job, it is important to match the worker to the task so that the job is within the physical capabilities of the employee.

Human beings are also susceptible to illness. As a result, they need time off from work occasionally, without impacting negatively on their income. For this reason, in countries such as Australia, firms are legally required to provide sick leave entitlements to permanent employees. This is set out in the Minimum Conditions of Employment Act 1993[34] and also in relevant workplace agreements or awards.

This human fragility includes susceptibility to biohazards. At-risk businesses are required to take preventative maintenance measures including sterilization and disposal of contaminated wastes. When such hazards are detected corrective action is required to limit the risk. For example, containment practices have been used in recent years to quarantine farms affected by foot and mouth disease in the United Kingdom. There are biohazards that are relevant in most industries. All workplaces, for instance, require strategies to manage the risk of communicable diseases such as hepatitis B and HIV AIDS. For all practical purposes, a worker with a blood-borne disease like HIV is not a hazard as such. The biohazard is safely contained within the person's body and does not pose a threat to other workers under normal circumstances. If however, the employee requires medical treatment then it is the firm's responsibility to ensure that first-aid personnel have had the relevant training to prevent infection.

Other communicable diseases are more readily spread, particularly respiratory infections, which are transmitted via air-borne droplets from coughs and sneezes. In addition to the health implications, the firm should consider the impact of infections such as influenza on employee attendance and organizational productivity, and determine whether preventative health measures may be appropriate. Some jobs, for example, horticulturalists and sanitation officers have a higher risk of environmental biohazards such as tetanus. The employer should provide relevant employees with free vaccinations for the prevention of these diseases.

The discussions thus far have indicated that workers have a level of residual risk that cannot be fully compressed because of incomplete KSAs and physical limitations. This risk has to be managed in the short term by modifying other system factors to fit human capacity in a generic sense and by providing training. Partial compression is possible in the longer term through on-going development of competencies. In addition to this residual risk, human resources do not consistently maintain a high level of safety and performance because, like other system factors, they have a tendency to degrade. Many of the management issues to correct the entropic risk caused by new employees were discussed previously. In the first instance, the firm can use value-based recruitment and selection procedures to identify those job applicants who are safety conscious and less likely to commit safety violations. (A word of caution should, however, be given at this point. The firm must ensure that its interview questions and other selection methods do not ask whether the applicant has ever had an accident. Nonselection on the basis of employment safety history is open to claims of discrimination, and is another means that 'blaming the victim' may be perpetuated.)

Once the preferred candidate has been selected for a vacant position and the offer of employment accepted, the process of integrating the new employee begins. Within the first days in the workplace, new recruits should be put through induction. Mentoring may also be used to accelerate the employee's learning curve and to monitor the employee's interaction with other system factors. Mentoring, in addition to its social benefits, can help to minimize entropic risk caused by the new starter's incomplete KSAs.

As explained previously, regardless of the worker's level of experience, fatigue can be a significant source of human resources degradation. It is brought on by and exacerbated by the demands imposed by other system factors. To reduce fatigue, it is therefore necessary to reduce these demands by designing or modifying the physical environment, technologies and processes as far as practicable to fit the needs of employees. Once the onset of fatigue begins, there is no other way to rectify it than to take a break, therefore, it is important for firms to provide workers with regular rest and meal intervals to allow this recovery. Respite from the task is necessary to prevent physical fatigue and to remedy mental tiredness that leads to loss of concentration. Entropic risk caused by muscular exhaustion may be managed by providing rest and using job rotation/enrichment systems of work. Techniques to control manual handling hazards include task modification such as reducing the weight and physical size of the load, the number of lifts, applying the correct handling technique, and removing any other risks such as tripping hazards, which make the process more dangerous.

Fatigue is a human resources entropic risk and is particularly relevant to shift workers. Supervisors should be given some responsibility for monitoring operator alertness to ensure that team members are capable of continuing to work safely. Once the firm has reduced the demands of other system factors to an 'acceptable' level of risk, the primary responsibility for managing fatigue rests with the employee. The worker is the only one who knows how he feels and therefore, in hazardous industries, employees should be taught to monitor their own condition for symptoms such as loss of concentration and a feeling of physical weakness. Ideally, this training should also cover lifestyle modifications to minimize the likelihood of tiredness such as regular sufficient sleep, exercise and a balanced diet. Shift workers particularly need this information to manage the disruptions to their body clock.

What should the supervisor do if a worker is fatigued to the extent that he represents a hazard to himself, to co-workers and organizational systems? The first response is to take corrective action by removing the employee from high-risk interfaces involving the operation of hazardous equipment or dangerous procedures. Assignment to low-risk duties is appropriate and this requires that the firm have some flexibility in its systems of work. Is this approach to fatigue management open to abuse? Will some employees use this as a means of avoiding hazardous duties? That depends on the firm's strategies. Counseling may be provided to those employees who are fatigued on a number of occasions within a given period of time. This would need to reinforce the use of appropriate coping strategies such as getting sufficient sleep, help the employee identify lifestyle issues which may disrupt recovery, and point out the need to implement these strategies to maintain sound work performance.

The main factors the firm should consider when developing its fatigue management strategies are:

- The structure of the work schedules and rosters;
- Irregular and unplanned work schedules;
- Potential for call-out of shift work employees for breakdown or absences to result in sleep deprivation and fatigue;
- Shift length in relation to the physical and mental demands of the work and commuting arrangements;
- Proximity of residence or accommodation;
- Method of travel to and from work available to employees and the risk of commuting accidents;
- Environmental factors – eg heat, humidity, noise levels, vibration etc; and
- Ability to access a balanced diet and adequate rest.[9]

The risk management strategies for fatigue and ill health are similar. Multiskilling and flexible systems of work allow those workers who are in suboptimal condition from time to time, to temporarily reduce their exposure to risk. Clear guidelines are required, however, to ensure that healthy workers do not end up 'carrying' unhealthy workers. The firm may develop a 'fit for work' policy to address this issue.

Employees who elect to attend work with a minor illness should ensure that if their health deteriorates during the course of the shift that they are still capable of working safely. Poor health that hinders the worker's ability to manage hazards is itself a hazard and should be monitored and controlled. Human resource degradation in combination with demanding or unfamiliar processes, in high-risk environments or while in charge of hazardous equipment can cause the worker's capacity to cope to be exceeded. For this reason, supervisors while carrying out workplace inspections should evaluate the condition of all four system factors including the health and alertness of human resources. In the longer term, 'wellness' programs can be used to help workers develop good habits that reduce their susceptibility to illness and enhance both their productivity and quality of life.

The management of these human resources risks is inter-related as described in Table 6.1. Fatigue and stress are, for example, treated in similar ways. The company has three avenues by which it can address workplace stress. The first is to reduce the stressors. Karasek's model indicated that the combination of high job demands with high decision latitude generates beneficial stress or stimulation. When Frankenhauser's model is considered concurrently, however, it is clear that these 'high job demands' should not involve conditions which are threatening. Hazardous systems that put the worker at risk of injury cannot, therefore, be considered stimulating. The first step in controlling job stress is to achieve legal compliance and to minimize hazards as far as practicable.

The quality of change management is a further stress variable. The key issues here are whether the change is perceived by the workforce to be challenging or threatening, alienating or participative, and whether the process causes employees to be empowered or disempowered. These stressors can lead to lower productivity and an increased possibility of accidents, injuries and environmental incidents.[35] In addition to planning and managing change well, the firm should monitor the impact of the change process. Specifically, it should look for early indicators of attitudinal and behavioral change that can lead to low morale, poor motivation and potentially to accidents. Indicators include declining job performance, sloppy work habits, uncooperative and quarrelsome communication and general negativity.[35] Participation and communication are the most important means of remedying change-induced stress, and these are issues discussed in Part 3.

The second way in which the firm can manage stressors is to ensure adequate and appropriate resourcing. For middle managers, in particular, the lack of alignment between the company's objectives and the resources needed to achieve those objectives can be a significant source of anxiety. In other words, the failure to attain internal strategic alignment creates imbalance in the firm and unreasonable pressures on supervisors and employees.

A further issue that relates to resourcing is the need for HRM practices which ensure that people are employed in accordance with their capabilities and skills. A mismatch between the person's KSAs and the demands of the job can be a source of stress. Selection systems should identify the person who is the best fit for the job in terms of competencies and attitudes, which allow the incumbent to work safely and efficiently following a reasonable period of adjustment. Sometimes, in the recruitment process, the 'best' person for the job still has a shortfall of KSAs which needs to be addressed through training. Workplace or task-specific instruction may be necessary before the person is exposed to the hazards in the workplace depending on the nature of these hazards and the applicability of KSAs currently possessed by the worker.

The final group of stress management strategies is contingencies that deal with its consequences. This includes developing a workplace violence policy and an action plan to deal with aggressive behavior. There are usually two methods of addressing these occurrences. The first is to exercise disciplinary procedures for serious offences. The second is to provide counseling services to assist the aggressor and the persons to whom the aggression has been directed, to reconcile their differences. The firm should also establish a dispute resolution process to arbitrate issues of concern. A further option that may be made available to employees is the provision of help via an employee assistance program. This involves access to counselors who can assist the worker to deal with stress from both work-related and nonwork-related sources. The purpose of such programs is to help employees overcome symptoms, manage stress better in future, and return to a safe productive level of functionality.

Stress and most human resource risks have a competence, capacity and/or attitudinal dimension. Loss of concentration falls into these categories. Its primary causes are fatigue, ill health, unstimulating or repetitive work, and lack of understanding of the need for vigilance in the presence of risks. The first issue, as discussed earlier, can be addressed by providing workers with regular rest and meal breaks. Proactive firms use safety records to identify those times at which the highest number of incidents has occurred historically as a result of fatigue, and roster in an additional break or remedial strategy prior to this peak time. Workers who are most at risk are those that work alone. Communication assists employees to remain alert and therefore, team-based work systems are an effective means of preventing loss of vigilance.

Multiskilled work teams can also be used to minimize repetitive work. The danger with incessantly performing the same activity is that it allows the worker's mind to deviate from the task. If risk conditions suddenly change for the worse, there is a response delay because the worker has to 'switch back' to the present, identify the hazard, assess the risk and then act. In relation to safety, the purpose of job design, therefore, is to provide sufficiently demanding and stimulating activities to keep the worker's mind in the present – to maintain alertness – without exceeding his capacity to the extent that it creates dysfunctional stress.

The tendency of the mind to deviate from the task is a common human trait and is a significant source of entropic risk. From the individual perspective, the main method of preventing this risk is self-correction. The firm's role is to provide training so workers are aware of this tendency and of the need for vigilance. The education process can be reinforced through informal workplace communication and reminders on stickers, t-shirts, posters and other triggers. Attentiveness may also be improved by modifying the workplace to minimize conditions that inhibit the ability to focus on the task, such as distracting noise. Extremes of heat or cold can also hinder concentration. The ideal temperature for safe, comfortable work is around 20°C (68°F). Researchers have found that above this, accidents increase by 4 per cent per degree centigrade.[11] Accident rates also increase below this temperature.

Many of the variables that cause workers to become degraded through loss of physical capacity or mental concentration are induced by the work environment, technologies and processes. Seldom do individuals consciously introduce entropy by defying safe practices while knowing the risks involved, without having felt compelled to do so by deadlines and other external pressures.[28] Most safety deviations are caused by organizational factors, but there is also a behavioral dimension that is entirely within the control of the individual. These attitudinal traits determine whether or not the individual will commit violations or will take other risks such as attending work under the influence of drugs or alcohol. Although these inappropriate behaviors are rare, the firm still has to have contingencies to address them in place.

Overt violations of safety have to be remedied within the OHS management system. The primary means of preventing undesirable behavior is to develop an organizational culture that discourages risk-taking. Reckless acts often result from the failure to appreciate the hazards involved. Hence, it is important to put all employees through training so that they understand the two categories of risk, the sources of these risks in their workplace, and how these can be controlled. Employees should only be held accountable for safety when the company has made reasonable efforts to provide them with the KSAs to manage effectively their behavior in the presence of these risks.

Concurrently, the firm has to ensure that its HRM practices are aligned with the OHS management system. For example, incentive schemes should not encourage unsafe behaviors. Bonus systems that are determined by output tend to compromise safety because workers are rewarded according to production quantities without concurrently encouraging safe work practices. (As explained later in Part 4, productive safety management uses a scorecard-based incentive program which contains a number of measures including production targets, safety results and system factor quality measures.)

The firm also has to have in place a disciplinary procedure, including an avenue for dismissal in serious cases, to deal with intentional violations of safety such as horseplay. In this regard the current definition of 'unsafe acts' is problematic:

> An unsafe act is any act that deviates from a generally recognized safe way of doing a job and increases the likelihood of an accident.[36]

Examples under this definition include operating without authority, use of defective equipment and taking an unsafe position, for example, working on a fragile roof without crawl boards.[36] If deadline pressures cause an employee to take an unsafe position, then where is the fault, with the worker or with the management system? Does this constitute an 'unsafe act'? Where an 'unsafe act' is a symptom of an underlying organizational problem then it cannot be thought of as unsafe in terms of placing accountability for it on the worker. To make disciplinary procedures precise and effective, a more appropriate definition is required to establish the boundaries of responsibility between the employee and the organizational system. An exact definition is:

> An unsafe act is a behavior consciously taken by an employee in defiance of the safety rules, procedures, training and expected norms of behavior, and independently of external pressures stemming from the management system. This act must occur after the firm has provided sufficient training and supervision for the employee to have a reasonable understanding of the risks involved and potential consequences, for it to be considered unsafe.

There is clearly delineation between where the accountability of the employee ends and where the firm's responsibilities to provide adequate training and supervision begin using the above definition. This becomes particularly useful when dealing with the issue of drugs and alcohol in the workplace as both enforcement and educational strategies are appropriate to deal with this issue. Most firms that operate in a high-risk environment or that involve the use of heavy or dangerous technologies implement a substance abuse/drug and alcohol policy. These statements usually cover two key issues. The first is to prohibit the possession, solicitation, secretion or consumption of any nonprescription or prohibited drug on the company site or when driving company vehicles. The second covers alcohol consumption prohibiting the intake of alcohol on the job and usually states the requirement that it is the employee's responsibility to ensure his fitness to perform normal duties.

Company policy is part of the enforcement approach that, for medium- and high-risk operations, should also include pre-employment and random testing. Urine sampling is an intermittent process and cannot ensure that on a daily basis employees are in a fit condition to work. It is, therefore, also necessary for supervisors to monitor whether workers are abiding by the policy.

Substance abuse is a very serious matter affecting the safety of the user and his co-workers. In Australia there is a social stigma associated with 'dobbing in a mate' (telling on a friend). Unfortunately, in the case in which a mineworker died from crush injuries when she lost control of the haul truck she was driving, the reluctance to report such matters resulted in a fatality. The incident report raised the problem of drugs and alcohol in the workplace[37] after a quantity of prohibited substance was found in the cab of the truck. Three children lost their mother because of the failure to report this hazard. Clearly, in a very practical sense, employees have a duty of care towards each other. In addition to the enforcement role, the firm has a responsibility to provide education and to develop a 'moderation' culture by supporting healthy lifestyle programs and by providing an avenue for those employees who have a problem to seek help through an employee assistance program.

Education and training are very important management strategies to address human resources risks. They provide employees with the 'know-what', 'know-how' and 'know-why' to create behavioral modification that leads to safe and efficient work practices. Even if the firm invests extensive resources to develop its workforce, however, tangible behavioral change will not occur if there is lack of management commitment. In particular, if resources are not dedicated to the reduction of hazards, if the firm lacks leadership, and if HRM systems do not reinforce production and safety as compatible goals, then there will be a limited return on this investment. Weaknesses in the management system will dilute the effectiveness of these education and training initiatives. It is essential that behavioral change begins in the management ranks, as discussed further in Part 3.

## The human resources audit

To develop sound strategies for the management of human resources risks, the firm has to identify the nature of these risks. It does so by conducting an audit of this system factor. The audit helps to identify how its workforce affects the level of risk through its interaction with other system factors and allows the development of workplace-specific strategies. For example, a firm with a stable, experienced workforce will have a different set of risk factors than a workshop for trainees. The audit process follows similar steps to the others discussed in previous chapters. The steps to be taken are shown in Fig. 6.8.

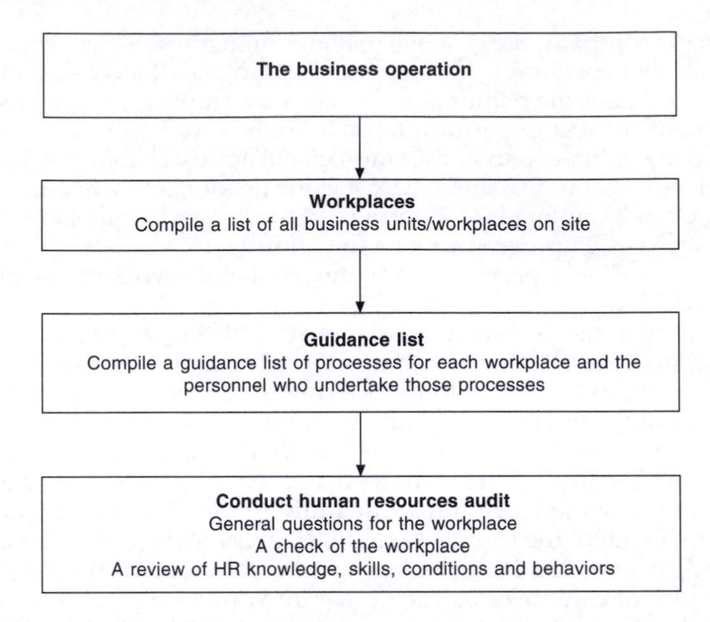


**Figure 6.8** Preparing for and conducting a human resources system factor audit

After preparing the guidance list, the auditor begins the questioning process. Issues to be investigated include:

(1) What measures are taken during recruitment and selection to ensure that new employees are suited to the demands of the job?
(2) Does the company use value-based selection techniques?
(3) Have all new employees received induction prior to starting work?
(4) What issues are covered during induction and are these sufficient?
(5) What feedback has been received concerning the quality of induction?
(6) What has the incident rate been for new employees compared to experienced employees?
(7) What is the company's safety record – LTIs, near misses, duration rates etc?
(8) Was a shortage of KSAs or vigilance a contributing factor in any of these incidents?
(9) Has training been provided whenever major system factor changes have been introduced?
(10) Has management been trained in risk identification and management, appropriate decision-making and leadership?
(11) Have employees been given training in risk identification?
(12) Has knowledge retention been measured following training and what were the results?
(13) Do any employees have a condition that may impede their ability to operate in a safe manner, and how may system factors be modified to accommodate the condition?
(14) Does the company use surveys to monitor employee perceptions and attitudes? If so, what have the results been?

(15) What formal communication systems does the company use to promote safe work practices and to deal with disputes?
(16) What are the rates for absenteeism and turnover? Are they too high?
(17) Have there been any safety violations in the past year and what actions were taken?

These questions help the auditor to get the 'big picture' about the condition of the human resources system factor. In particular, they identify the level of KSAs and safety consciousness of the workforce, company commitment to training, opportunities for participation, and the extent of risk-taking behavior in the company.

The risks associated with human resources are then analyzed further. The seven process investigators (7 Ps) are used and 'Person' is evaluated in detail. The 7 Ps are:

| | |
|---|---|
| Process: | What is being done? |
| Purpose: | Why do it? |
| Person: | Who is doing it? |
| Place: | Where is it being done? |
| Period: | When is it being done? |
| Procedure: | How is it being done? |
| Performance: | What are the outcomes – safety, productivity and quality? |

The example of changing a tire on a light vehicle can be used again. The review is as follows:

| | |
|---|---|
| Process: | Changing a tire on a light vehicle. |
| Purpose: | To replace tire that is in less than optimal condition (punctured/worn). |
| Person: | (1) Tradesperson assistant (planned maintenance)<br>(2) Vehicle driver (unplanned puncture) |
| Place: | (1) Light vehicle workshop<br>(2) Place chosen by driver following puncture |
| Period: | (1) Scheduled maintenance<br>(2) Following puncture (reactive maintenance) |
| Procedure: | (1) Refer standardized procedure in light vehicle workshop<br>(2) Use jack provided to raise vehicle. Use wheel brace to remove tire. Replace with spare tire and tighten. Place punctured tire in vehicle. |
| Performance: | (1) No incidents reported for this process in past year.<br>(2) One injury reported when driver injured his back lifting the tire. |

'What if' analysis is used to consider the impact of the worker's competencies and level of vigilance on the risks involved in the task. The results of this analysis could include:

(1) What if the driver does not know how to change a tire safely?
(2) What if the driver has a back condition? How will this affect the level of risk?

(3) What if the driver parks too close to the road when breakdown maintenance is required?
(4) What if the operator does not know the correct tire pressure and over-inflates the tire?

When auditing human resources it is incorrect to assume that employees have the required competencies to undertake tasks, regardless of whether the task appears to be simple or 'common knowledge' to the auditor. The auditor has to bear in mind that all human resources are characterized by incomplete KSAs and have a tendency to degrade through fatigue and loss of concentration. Unfamiliar tasks and/or demanding conditions exacerbate these limitations.

Once the audit has been completed and risks have been documented, recommendations can be developed using the four-fold strategy. Corrective action should be used to address shortfalls in the immediate term and may include using increased supervision to prevent new recruits from introducing degradation. On-going refresher courses, annual medical check-ups and regular monitoring practices may be used to maintain the human resources system factor. The residual risk from incomplete KSAs and physical limitations also has to be managed in the short term and compressed in the longer term. One of the key considerations in addressing residual risk within this system factor as a whole is the reduction of turnover. When well-trained, experienced employees leave the firm this leads to a rise in residual risk. Employee turnover is a very significant issue in risk management and can be addressed using HRM practices and strategies that develop a positive safety culture. Once strategies have been developed to control the human resource system factor risks, they can be included, together with other strategies, in the productive safety management plan described in Part 4.

## Summary

The following chapter is the final one of this part. It presents a method to quantify the risks inherent in different activities, which involves combining the hazards identified during the four system audits. Specifically, these are the risks involved when human resources, the physical environment, technology and process risks interface. The results of this technique allow managers to more confidently identify high priority risks and thereby allocate resources to manage the hazards associated with business activities more effectively.

## References

1. Fulwiler, R.D. (2000) Behavior-based safety and the missing links. *Occupational Hazards* Jan 2000 issue, Penton Media, Inc. in association with The Gale Group and Looksmart. Available on website: http://www.findarticles.com/cf_dls/m4333/1_62/59557571/p1/article.jhtml?term=fulwiler
2. Manuele, F.A. (2000) Behavioral safety: looking beyond the worker (changing theories of facility and worker at risk interaction). *Occupational Hazards* Oct 2000, Penton Media,

Inc. in association with The Gale Group and Looksmart. Available on website: http://www.occupationalhazards.com/full_story.php?WID=2644
3. Reason, J. (2000) Human error: models and management. *British Medical Journal* issue March 18, 2000, British Medical Association and The Gale Group. Available on website: http://www.occupationalhazards.com/full_story.php?WID=1874
4. Boudreaux, W.G. (1999) Supply chain management: why pursue single-source supplier relationships. *Offshore* issue May 1999, PennWell Publishing Co. and The Gale Group, USA.
5. Pierce, F.D. (2000) Safety in the emerging leadership paradigm. *Occupational Hazards* June 2000, Penton Media, Inc. in association with The Gale Group and LookSmart. Available on website: http://www.findarticles.com/cf_dls/m4333/6_62/63825876/p1/article.jhtml?term=pierce
6. Topf, M.D. (2000) Including leadership in the safety process. *Occupational Hazards* March 2000, Penton Media, Inc. in association with The Gale Group and LookSmart. Available on website: http://www.occupationalhazards.com/full_story.php?WID=1779
7. *Occupational Safety and Health Act 1984 (Western Australia)*. Western Australian Government Printers, Perth, Australia. Available on website: http://www.austlii.edu.au/au/legis/wa/consol_act/osaha1984273/
8. Worksafe Western Australia (1994/5) *State of the Work Environment No. 26: Injuries and Diseases in Young Workers, Western Australia 1994/95*. Available on website: http://www.safetyline.wa.gov.au/PageBin/injrstat0052.htm
9. Mines Occupational Safety and Health Advisory Board (2000) *Fatigue Management for the Western Australian Mining Industry, State of Western Australia*. Available on website: http://www.mpr.wa.gov.au/safety/mining/pdfs/fatiguemanagement.PDF
10. Worksafe Western Australia (2000) Driver fatigue sleeping giants. *Safetyline Magazine No. 47*, July. Available on website: http://www.safetyline.wa.gov.au/PageBin/pg008221.htm
11. Mathews, J. (1993) *Health and Safety at Work – Australian Trade Union Safety Representatives Handbook*. Pluto Press, Australia.
12. *Mines Safety and Inspection Act 1996 (Western Australia)*. Western Australian Government Printers, Perth, Australia. Available on website: http://www.slp.wa.gov.au/statutes/swans.nsf/be0189448e381736482567bd0008c67c/85cf936e5e10bbe8482568b600173550?OpenDocument
13. Department of Mineral and Petroleum Resources Western Australia (1999) *Adjustment of Exposure Standards for Extended Workshifts Guideline*. Available on website: http://www.dme.wa.gov.au/safety/mining/pdfs/adjustmentofexposurestandards.PDF
14. Worksafe Western Australia (1995) *State of the Work Environment: Manual Handling Injuries and Disease 1995/96*. Available on website: http://www.safetyline.wa.gov.au/PageBin/injrstat0113.htm
15. Worksafe Western Australia (2001) *Code of Practice Manual Handling*. Available on website: http://www.safetyline.wa.gov.au/pagebin/codewswa0072.htm
16. Worksafe Western Australia (2001) *Work-Related Stress – Different meanings to different people*. Available on website: http://www.safetyline.wa.gov.au/PageBin/disegenl0005.htm
17. Schnall, P.L., Landsbergis, P.A. and Baker, D. (1994) Job strain and cardiovascular disease, in *Annual Review Public Health 15*: pages 381–411. Summaries called: *Brief Introduction to Job Strain* and *Definitions and Formulations of Job Strain*. Available on website: http://workhealth.org/strain/hpjs1.html
18. Job Stress Network (2001) Reducing occupational stress – an introductory guide for managers, supervisors and union member. Available on website: http://workhealth.org/prevention/prred.html
19. National Institute of Occupational Safety and Health (United States) (1996) NIOSH report addresses problem of workplace violence, suggests strategies for preventing risks, *HHS Press Release Monday July 8 1996*. Available on website: http://www.cdc.gov/niosh/violpr.html
20. Worksafe Western Australia (2001) *Violence in the Workplace*. Available on website: http://www.safetyline.wa.gov.au/PageBin/workhazd0013.htm
21. Katzenback, J. and Smith, D. (1993) Top management's role: leading to the high-performance organization, *The Wisdom of Teams*. Harvard Business School Press, Boston, Massachusetts, USA, pp. 239–258.
22. Robbins, S., Waters-Marsh, T., Cacioppe, R. and Millett, B. (1994) Work Stress, *Organizational Behavior Concepts, Controversies and Applications*. Prentice Hall, Sydney, Australia, pp. 337–345.

23. International Labor Organization (2000) Costs of workplace stress are rising with depression increasingly common, *Press Release*. Available on website: www.ilo.org/public/english/bureau/inf/pr/2000/37.html
24. Center for Human Services of Griffith University (2000) *Occupational Stress: Factors that contribute to its Occurrence and Effective Management*. Available on website: http://www.workcover.wa.gov.au/pdf/occupational%20stress.pdf
25. University of Michigan (2000) Flu vaccine might not be cost effective in working adults, but it's still needed, *News Release Tuesday October 3*. Available on website: http://www.umich.edu/~newsinfo/Releases/2000/Oct00/r100200a.html
26. American College of Occupational and Environmental Medicine (2001) Employers Reduce Sick Day Costs with Free Flu Shots, *New Release 30 August*. Available on website: http://www.acoem.org/news/news/default.asp?NEWS_ID=83
27. Stellman, J.M. (ed.) (1998) *Encyclopedia of Occupational Health and Safety* 4th edn. International Labor Office, Geneva.
28. Mines Occupational Safety and Health Advisory Board (1998) *Risk Taking Behavior in the Western Australian Underground Mining Sector – Risk Taking Behavior Working Party Report and Recommendations*. Available on website: http://www.dme.wa.gov.au/prodserv/pub/pdfs/riskreport.PDF
29. Mines Occupational Safety and Health Advisory Board (1999) *A Review of Incentive-Based Remuneration Schemes in the Western Australian Mining Industry – Report and Recommendations from MOSHAB Incentive-Based Remuneration Working Party*. Available on website: http://www.dme.wa.gov.au/prodserv/pub/pdfs/remun.PDF
30. Nighswonger (2000) Just Say Yes to Preventing Substance Abuse (in the Workplace), in *Occupational Hazards, April 2000*. Available on website: http://www.findarticles.com/cf_dls/PI/search.jhtml?type=all&magR=m4333&key=nighswonger
31. Injury Control Council of Western Australia (1999) *Position Paper – Road Crash Trauma*. Available on website: http://www.iccwa.org.au/road.html
32. Worksafe Western Australia (2001) *Guidance Note – Alcohol and other Drugs at the Workplace – 2 Strategies*. Available on website: http://www.safetyline.wa.gov.au/PageBin/guidwswa0055.htm
33. Worksafe Western Australia (2001) *Guidance Note – Alcohol and other Drugs at the Workplace – 4 Risks*. Available on website: http://www.safetyline.wa.gov.au/PageBin/guidwswa0057.htm
34. *Minimum Conditions of Employment Act* (1993) State Law Publishers, Perth, Australia. Available on website: http://www.slp.wa.gov.au/statutes/swans.nsf/html/agency+DOPLAR+acts?opendocument
35. Topf, M.D. (2000) Managing change (organizational change and employee health and safety), in *Occupational Hazards July 2000*. Available on website: http://www.occupationalhazards.com/full_story.php?WID=2224
36. Stranks, J. (1991) *The Handbook of Health and Safety Practice* 2nd *edn*. Pitman Publishing, London.
37. Department of Mineral and Petroleum Resources Western Australia (2001) *Significant Incident No. 12*. Available on website: http://notesweb.mpr.wa.gov.au/exis/SIR.NSF/6b390eea5649d21c48256097004aacb7/48edaea26b1156ac4825609d004bdaee?OpenDocument

Chapter 7

# Risk quantification and strategy development

There are a number of steps involved in risk management. The first is to identify workplace hazards. This may be undertaken using the system factor auditing processes described in Chapters 3 to 6, or as part of a routine identification program of workplace inspections. The second is to determine how resources are best allocated to minimize these risks. To do this effectively, the firm needs the ability to quantify the hazards and to evaluate their relative significance so that the most severe risks are given priority over those that are much more easily controlled. After this, strategies may be implemented and the processes of monitoring and maintenance established. Generic models of the risk management process, together with qualitative measures of consequence or impact, likelihood and a risk analysis matrix, are available in documentation such as Australian/New Zealand Standards AS/NZS 4360:1999 Risk Management[1] or the national equivalents.

## Risk assessment tools

The tools currently used for the purpose of OHS risk evaluation include the risk assessment code and the risk scorecard as explained earlier in Chapter 3 (refer to Fig. 3.3). These methods are particularly useful for short-term remedial action. When a workshop supervisor identifies an oil spill as a slipping hazard, as an example, the code allows him to assess the likelihood and consequences of the risk. From the code, the threat is 'almost certain' to result in an accident with 'moderate' repercussions and therefore the spill has to be cleaned up immediately. The scorecard can be used in the same way. It helps the supervisor evaluate the probability of the risk as 'almost certain', exposure as 'frequent' in a high use area and the consequences as a potential 'casualty'. The spill is therefore a 'very high risk' requiring immediate rectification. At the operational level, such tools provide a method that prompts corrective action or short-term intervention.

From a strategic perspective, however, these risk management tools are limited as they do not provide a system-wide method of hazard reduction. These constraints can be shown if different activities are used as examples and are compared. For instance, assume that activity 1 involves an experienced truck driver reversing to the edge of a mine waste dump over which there is a substantial drop which is protected by an earthen barrier. From the scorecard, the probability of an accident is 'conceivable', exposure is 'frequent', and the consequences are 'very serious', resulting in a score of 50. This can be compared to activity 2 which involves an apprentice using a fixed circular saw in a workshop. The probability of an injury to a finger or to the hand is 'quite possible', exposure is 'frequent' and the potential consequences range from 'first aid treatment' to 'serious injury'. The score depending on the nature of the injury may be somewhere between 30 and 200. From these results, a strategy to minimize the risk of injury to apprentices in the workshop would take priority over the other activity.

The scorecard does not consider the effect of risk modifiers that either increase or decrease the risk, such as the level of training received by the worker or the impact of work systems, such as shift work. The failure to consider systemic weaknesses that exacerbate risks can lead to resources being allocated to areas, which in reality have a lower level of risk. When using the code or scorecard, therefore, assumptions have to be made about the condition of system factors. The score of 50 is accurate, for example, if the truck driver remains vigilant at all times and is highly competent. His experience and safety consciousness allows him to manage the risks effectively. If, however, the driver works overtime for a number of consecutive days, he may not be able to operate safely at all times because of fatigue. This method therefore can fail to identify underlying factors that increase the probability of an accident.

The effectiveness of the scorecard relies on the risk assessor's awareness of all risk elements and the integration of these into the evaluation, particularly in determining the 'probability' of an accident. In the apprentice workshop example, the firm may have already implemented a high level of on-the-job training and regular inspections that reduce the likelihood of an accident. The auditor may easily overlook the significance of these interventions and therefore, evaluate workshop safety as a higher priority than activity 1. When the two activities are compared thoroughly, work rosters that cause driver fatigue may be a much more significant risk source than the use of tools in the apprentice workshop. If the firm develops safety interventions based on the code or scorecard in this example, it may fail to address covert organizational problems in high-risk areas. In addition, management cannot be certain that it is making best use of limited resources. A more effective risk evaluation method which allows managers to manipulate the quality of each system factor to pinpoint these weaknesses would increase certainty and improve resource allocation.

The entropy model indicates that the level of risk in an organization is not static because of the tendency of systems to degrade. How then can the firm evaluate its level of risk given that only residual risk is constant in the short term and entropic risk is variable? How can it identify the systemic risks that require correction in the short term and preventative measures in the long term? How can it plan for continuous risk reduction to ensure legal compliance and social responsibility? Clearly the company requires a method to quantify its risks in relative terms so that primary causal factors are identified and addressed.

In the following section a method is proposed that is intended to assist management to better understand these underlying factors, to improve the evaluation of relative risk levels, and therefore, enhance OHS decision-making and resource allocation. The tools presented are referred to as 'scorers'. These form a risk ranking system which allows high-risk factors to be differentiated from lower risk factors. The values derived from applying the scorers are therefore relative rather than absolute.

As stated earlier the first step in risk management is to identify workplace risks using system factor audits. These provide a risk profile for the workplace. When the audit results are compiled, a table of system factors can be developed, as shown in Fig. 7.1. Different activities are generated by identifying the process, the people who undertake the process, the technology used, and the physical environment in which it is carried out. The questions to be asked are:

(1) What process is undertaken?
(2) Who does it?
(3) What technologies are used?
(4) Where is the process undertaken?

An activity therefore is the combination of system factors, so that:

$$\text{Activity} = \text{Process} + \text{Human resources} + \text{Technology} + \text{Physical environment}$$

Figure 7.1 shows two examples. The first is the process of changing a vehicle tire. The technology involved includes a tire, jack, wheel brace and an air-pressure pump. A mechanic undertakes this task in the light vehicle workshop. The second example is the use of a nail-gun for building maintenance. This work is undertaken at various building sites by maintenance personnel. The examples show that when system factors are combined, activities are developed. The risks within the system factors of these activities can then be identified and quantified using the scorers presented later in this chapter. To do this, residual and entropic risks

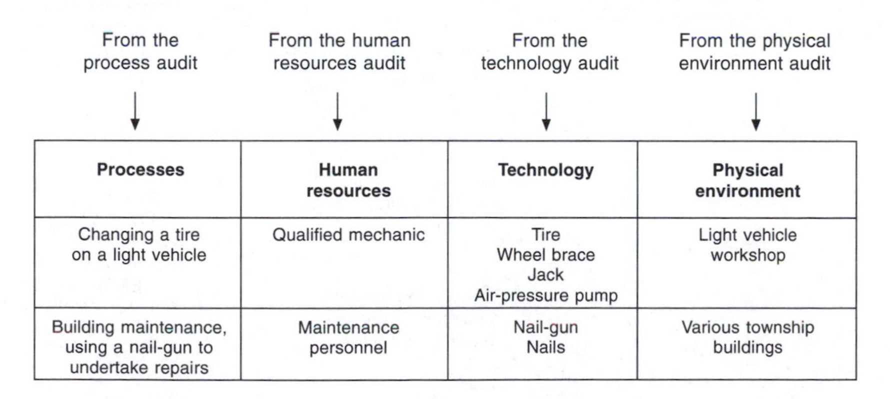


| Processes | Human resources | Technology | Physical environment |
|---|---|---|---|
| Changing a tire on a light vehicle | Qualified mechanic | Tire<br>Wheel brace<br>Jack<br>Air-pressure pump | Light vehicle workshop |
| Building maintenance, using a nail-gun to undertake repairs | Maintenance personnel | Nail-gun<br>Nails | Various township buildings |

**Figure 7.1** Developing activities from the system factor audits

have to be dealt with separately because the two categories of risks require different strategies to manage them, as explained earlier using the four-fold strategy. The identification and quantification of residual and entropic risks leads to residual risk management strategies (RMS) and entropy prevention strategies (EPS), respectively. There are, therefore, two groupings of risk control interventions under the productive safety management approach. Figure 7.2 illustrates the relationship between these risks, the four-fold strategy and the interventions that will be later included in the productive safety management plan.

The figure shows that residual risk from the entropy model, according to the four-fold strategy, has to be managed in the short term and compressed in the longer term, leading to the development of RMS. Entropic risk has to be addressed using corrective action in the short term and through maintenance strategies to prevent future entropic risk as both a short-term and on-going strategy. These risk control interventions are EPS. Each strategy type is discussed below.

## Residual risk management strategies (RMS)

The development of RMS begins by quantifying the residual risks within each of the system factors in a given business activity, in relative rather than absolute terms. Effectively, the method that will be described identifies areas of higher risk against those of lower risk in a particular system factor. For example, an electric saw has a higher residual risk than a hand saw because it uses electricity and has moving parts. An underground mining environment has a higher inherent danger than an office building because it is potentially unstable, confined and has limited illumination and other risk factors. The process of climbing a ladder is more hazardous than carrying out administrative duties. Using this method, system factors

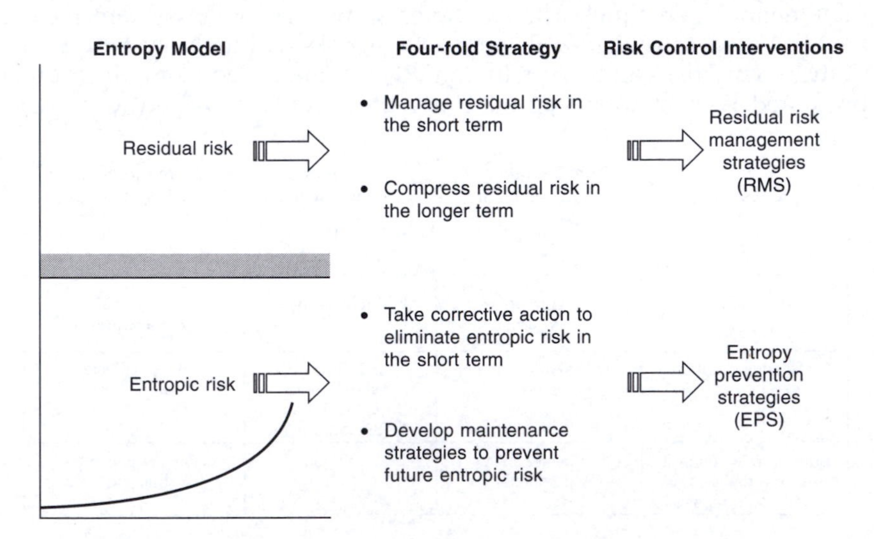


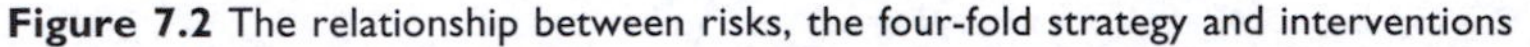

**Figure 7.2** The relationship between risks, the four-fold strategy and interventions

are considered in turn to quantify the level of residual risk in each, before combining them again into an activity.

## Processes

As discussed in Chapter 3, the firm begins its risk analysis by identifying the processes that it undertakes using the process audit. How can the firm determine the relative level of residual risk in these work practices? Every process involves the transfer of energy within a system. Electrical energy, for instance, is used to generate mechanical energy that in turn can be used to project a nail into a piece of wood or crush rocks in a processing plant. The level of energy inherent in a process has the potential to cause negative consequences for organizational systems. Energy can, if released in an unintended manner, lead to injury or damage. This concept was originally proposed in the energy transfer theory of accident causation. It suggests that a worker incurs injury or equipment suffers damage through a change of energy and that for every change there is a source, a path and a receiver.[2] The level of energy determines the potential severity of consequences, for example, a spanner falling from a height of 10 meters will have a higher level of energy than one that falls from 2 meters.

The severity of residual risk thus depends on the level of energy involved in a process. There are different types of energy that may be present or generated. An activity that increases height has potential energy, for instance when workers or equipment are raised above the ground they gain this energy. Climbing a ladder has an increasing residual risk the higher up the worker goes, to the point where and above which a fall is 'certain' to result in a fatality. Likewise, hoisting a pallet of bricks has a higher risk as the height increases. These processes are hazardous because they can lead to either people or objects falling. Figure 7.3 illustrates the residual risks associated with potential energy and heights. The arrow shows that when the process is undertaken close to the ground, the object and the worker have a low residual risk. At heights, they have a high residual risk attributable to the danger of falling or impact from falling objects, which are particularly relevant hazards in the high-rise construction industry.

Process residual risks also arise from the energy associated with speed, referred to as kinetic energy. The higher the kinetic energy involved in a process, the higher the level of residual risk. For the dangers associated with a particular system factor to be analyzed, other system factors have to be assumed to be constant and ideal. This is referred to as the principle of separation of system factors. For instance, under fixed human resources, technological and environmental conditions, the process risk associated with travelling at 100 km per hour is greater than travelling at 60 km per hour because of the higher kinetic energy involved.

Vibration is also a type of kinetic energy that involves oscillations of movement. The main factors affecting the level of inherent threat are the frequency or rate at which the vibration occurs and the amplitude measured in terms of displacement, velocity or acceleration.[3] The processes

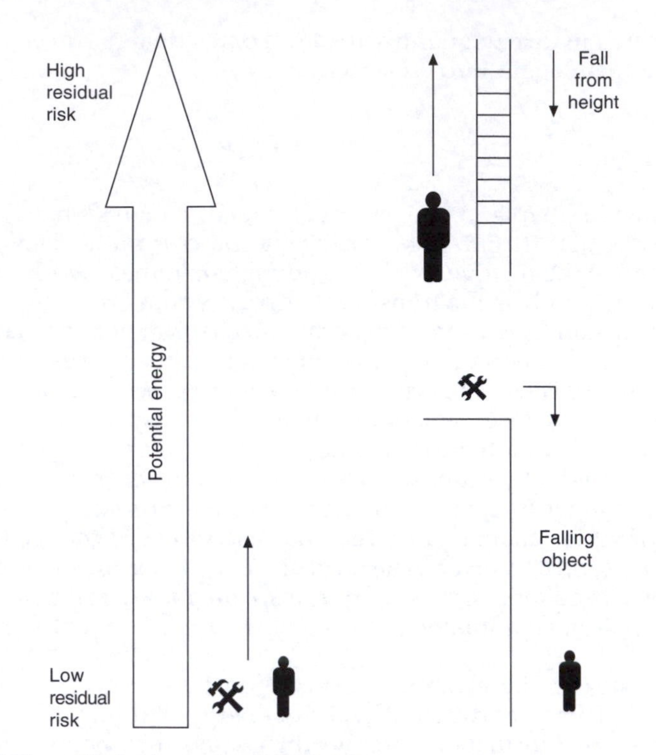


**Figure 7.3** The residual risk associated with heights

undertaken by astronauts to prepare them for space travel, for example, have a higher residual risk than the vibration hazards associated with car travel. When quantifying residual risk, the auditor must consider the risk within a static time frame and apply this approach to all the processes undertaken to determine relative residual risk levels. The level of exposure is not considered in this method because it complicates the determination of relative residual risk. Further, because residual risk is fixed in the short term, it is assumed that regardless of when the worker undertakes the process the level of danger is the same. Exposure is more significant when considering entropic risk because it determines the firm's susceptibility to degradation, as explained later.

In addition to potential and kinetic energy, business processes often involve energy transfer in the form of mechanical force, which can be considered in two ways. The first is when the human body is used mechanically, for example, when pushing or lifting equipment such as in manual handling that requires the body to apply effort to carry out the task. The greater the exertion required the higher the level of residual risk. In keeping with the principle of separation of system factors, the auditor needs to consider the process independently of who undertakes it. This is so that the level of residual risk associated with the process can be determined relative to other processes. To quantify the risk involved in manual handling, the process has to be defined clearly, for instance, lifting 30 kg and carrying it 20 meters or pushing a 30 kg wheelbarrow for 20 meters. The former has a higher residual risk because of the additional force required to do the work.

The second way in which mechanical energy applies is using some form of technology. A wheelbarrow or other equipment that reduces the force required to shift materials decreases the residual risk associated with handling. Technology can therefore be used to compress residual risks within processes. In contrast, however, work practices that expose employees to mechanical energy can have a high level of residual risk. Incidents that involve workers being struck by or caught between mechanical devices are examples of this. The greater the forces involved the greater the danger, as illustrated by the case described earlier in which a mechanical fitter died when he was trapped in an electric shovel's components.

Processes can also generate thermal energy or involve gases or liquids under pressure. For instance, the manufacturing of canned foods that involves pressure-cooking under high temperatures has a greater residual risk than basic catering processes. Workers can receive fatal burns when system failures release this energy. Work practices that use electrical energy to undertake a task also have a high level of inherent danger. Using a nail-gun, for example, has a higher residual threat than using a hammer, because the operation involves the use of electrical energy to give a nail kinetic energy causing it to become a projectile. Processes that generate radiation also result in such risks proportional to the level of emission.

The final energy type associated with this system factor is sonar energy. Noise is usually generated by the use of technology such as operating a jackhammer. The process that involves noise emissions can be easily defined for example – 'using a jackhammer to break a concrete path' – 'doing X using Y'. The different types of energy that may be transferred or generated whilst undertaking a process are summarized in Table 7.1. Symbols have been developed so that the risk assessor can code the energy types for a particular process. Examples of the potential consequences associated with each energy type have been included. Electrical energy, for instance, can cause a worker to be electrocuted, a technology to short, or start a fire. Mechanical energy is associated with manual handling and with moving machine parts. The former can result in strains, whilst the latter can cause crush or impact injuries and damage.

As mentioned earlier, there is an important rule applied when quantifying these process residual risks – the principle of separation of system factors. When the principle is enforced, processes are thought of independently of other system factors. They are analyzed as if in an 'ideal' system where other system factors are perfect – free of residual and entropic risks. This means considering the activity of traveling at 100 km per hour, for example, in an ideal physical environment, by an 'ideal' person on a 'perfect' technology. The level of energy, therefore, depends on the speed at which the person or object is going, which in turn, determines the level of residual risk relative to other processes. Using the method of isolating processes from other system factors allows different processes to be compared. In later stages of analysis, other factors are considered, for instance whether the process is undertaken in a car or on a motorbike. The residual risk associated with a particular technology is

**Table 7.1** Types of energy transferred or generated by processes that result in residual risk

| Energy type | Symbol | Characteristics | Examples of potential consequences |
|---|---|---|---|
| Kinetic | ⇒ | Speed<br>Acceleration<br>Vibration | Collision<br>Disorientation<br>Nausea |
| Potential | ⇓ | Heights<br>Pressure | Fall from height<br>Falling objects<br>Explosion |
| Mechanical | ⇒ ⇐ | Manual handling<br>Moving machine parts | Strains<br>Crush – caught between moving parts<br>Impact – struck by moving parts<br>Damage |
| Thermal | ′′′ | Hot objects<br>Hot gases or liquids | Burns<br>Fire |
| Electrical | ϟ | Electricity | Electrocution<br>Shorting of technology<br>Fire |
| Radiation | ☼ | Handling of radioactive materials<br>Radiation used in process | Radiation poisoning<br>Site contamination |
| Sonar | ◀ | Noise | Industrial deafness |

assessed, before determining the inherent weaknesses of the rider and the dangerous characteristics of the physical environment.

A key concept in the quantification of relative residual risk is that this risk affects the probability of an incident independently of time because, according to the definition provided by the entropy model, residual risk cannot be compressed in the short term. The auditor assesses the process in static time to eliminate the concept of exposure. This is important because residual risk does not usually change dramatically in organizational systems unless there are unforeseen circumstances or catalysts that act as a trigger.

There are a number of steps involved in quantifying process residual risks as illustrated in Fig. 7.4. The first step is to assume that other system factors – technology, physical environment and human resources – are ideal, applying the principle of separation of system factors. The second step is to define the process and this requires quantifying it in terms that indicate the level and type of energy, for example traveling at 60 km per hour or climbing 60 meters, or including a generic specification of the technology used – 'doing X using Y'. Step 3 involves considering the process in static time followed by the identification of the types of energy involved. These are evaluated for potential consequences assuming the 'most likely' outcome of failure, for example, using a hammer and nail may result in bruising or a minor fracture of the hand or finger. The final

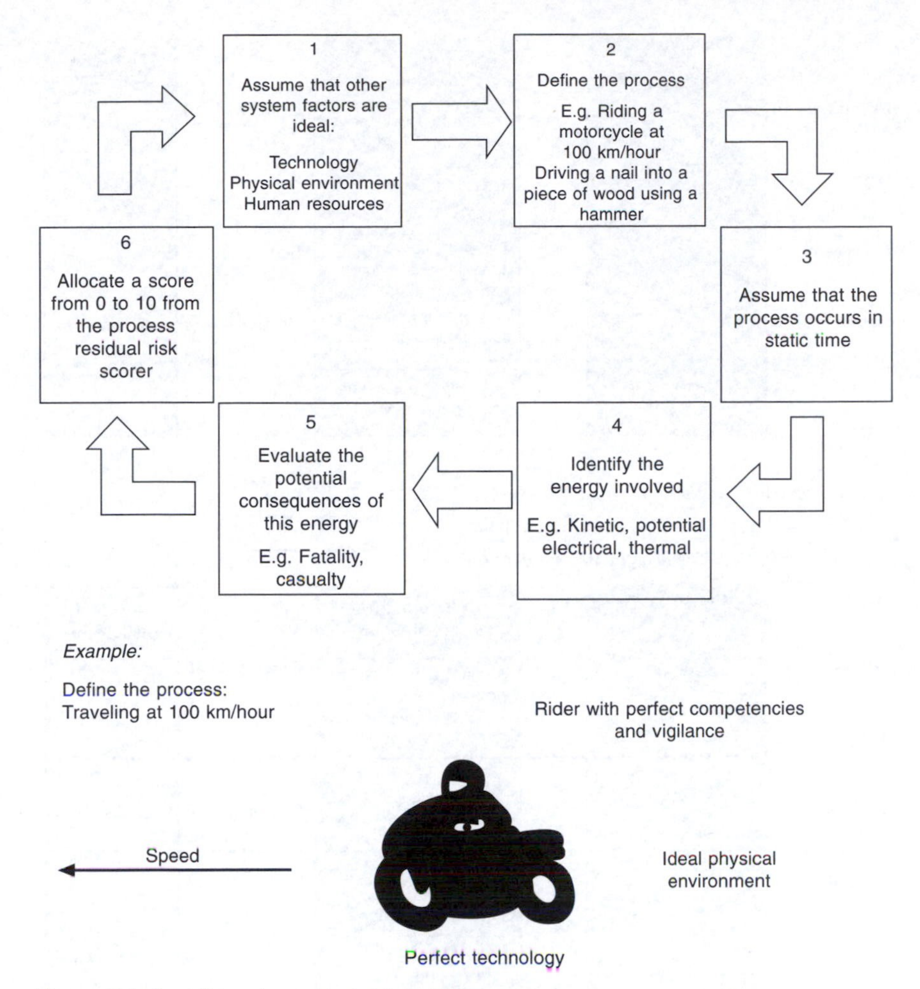


**Figure 7.4** Quantifying the residual risks within a process

step is to allocate a relative score based on these consequences using the process residual risk scorer, given in Fig. 7.5.

The scorer shows that for processes, the level of energy is the primary risk parameter. This is defined as:

> The primary risk parameter is the key variable that determines the level of risk in a system factor.

The scale runs from one to 10 being low to extreme energy, respectively. These correlate with the potential consequences of this energy, which range from first aid treatment and/or some damage to numerous fatalities and/or catastrophic damage. The scale does not begin at zero, because it is impossible to completely eliminate residual risk. In addition inherent danger is not dependent on exposure because it can result in injury or damage at any time, which is a particularly important concept for high-risk environments, such as the mining and construction industries in

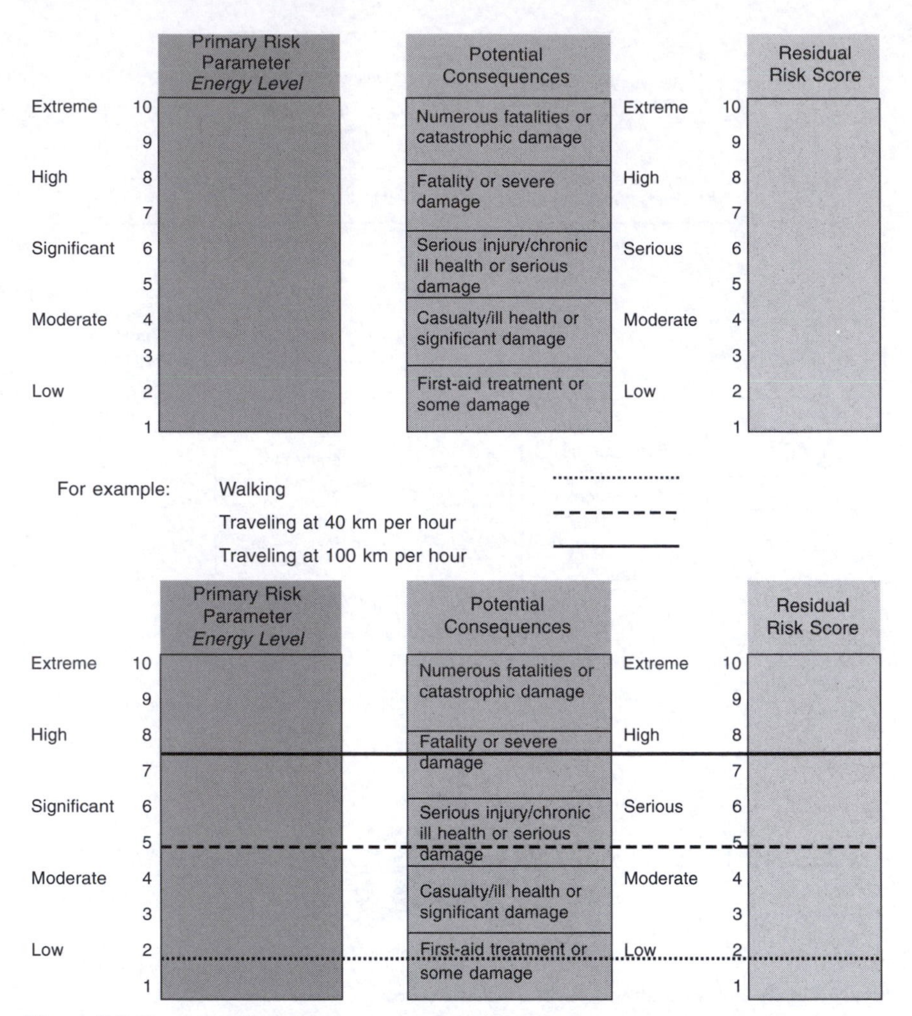


**Figure 7.5** Process residual risk scorer

which residual hazards are ever present. The exposure rate or time variables are therefore not relevant parameters in the allocation of a residual risk score. The constant presence of residual risks makes it possible, for example, for a person who operates an air-leg drill in an underground mine to be injured in the first 5 minutes of the shift. It is also possible for a person to work for years as a driller and not have an injury that results from the inherent danger of this process. It depends on whether inherent dangers translate into imminent threats and the combination of other risk factors, including the effects of degradation.

Figure 7.5 illustrates that in this exercise the auditor anticipates the 'most likely' consequences of the process based on the level of energy involved. This does not, however, mean that should this energy be released that the consequences predicted by the auditor will become evident. There have been instances where employees have walked away unhurt or the number of injuries sustained has been well below the systems damage incurred by the firm. In particular there have been cases involving underground mine disasters where significant destruction has resulted from blasting induced structural failures yet the number of fatalities incurred have been far less than the numbers which could have been lost. For example, in the Northparkes E26 gold and copper mine at Goonumbla in Australia, four people (two drillers, a manager and an engineer) were killed and a further 57 others trapped for 4 hours when the mine collapsed on 24 November 1999.[4]

> At around 3 p.m. the orebody gave way, dropping into the cave and pushing a massive rush of air down the exploration tunnel and through the decline and ventilation shafts. The wind, estimated at more than twice the force of a hurricane, caused major damage and killed Stuart, Colin, Ross and Michael.[5]

The case illustrates the impact of suboptimal fit between the process of underground mining and the physical environment, which results in these operations having a higher inherent danger than open-cut mining practices. There is also variation in the degree of residual risk dependent on the mining method. A significant contributing factor to the danger in this incident was that this technique, known as block caving, can result in air blasts from rock falls. The case also highlights that exposure, as thought of in the current sense as the amount of time the worker is in contact with a hazard, is irrelevant in relation to residual risk because the organization is constantly vulnerable, as shown by the entropy model. The firm is subject to these residual risks 24 hours a day, therefore, a mine tunnel can collapse during shutdown time. A construction site can destabilize when operations have ceased for the day. Business systems are constantly susceptible to residual risk and the level of this risk depends on the overall compatibility between the processes being undertaken, the environment and other system factors. This lack of fit affects the likelihood that the energy introduced or stored will be released.

Figure 7.5 illustrates the method of allocating a score to process residual risks. The first step is to determine this energy level. The result is then transferred to the 'potential consequences' column. Higher energy levels lead to greater potential injury or damage. The final step is to transfer across and allocate a score from one to 10 to determine the level of residual risk involved in the process.

Figure 7.5 also provides examples to illustrate how the score can be determined. The score itself is less important than its value relative to other process scores. The method proposes that provided that the analysis is applied consistently, the firm can identify those processes which require more stringent management in the short term and compression in the longer term. In the example, walking, traveling at 40 km per hour and at 100 km per hour are compared. Walking involves the lowest level of residual risk because the human body experiences a low level of kinetic energy. Traveling at 40 km per hour, on the other hand, leads to a higher level of kinetic energy that can result in serious injury. A speed of 100 km per hour can result in a fatality or severe damage and therefore has the highest level of residual risk, scoring approximately 7.5.

## Technology

The residual risks associated with technology, which include plant, equipment, tools and chemicals, are determined in a similar manner. Chemicals are evaluated for the level of residual risk on the basis of toxicity and/or reactivity. The higher the level of toxicity, the greater the residual risk. Plant, equipment and tools are evaluated using the level of energy involved in the same way as processes were analyzed with energy being the primary risk parameter. Technologies can also have safety features that reduce the level of risk. These are referred to as positive risk modifiers. The definition for these variables is:

> A risk modifier is a parameter or characteristic of a system factor that either increases or decreases the level of risk. Negative risk modifiers increase the level of risk whereas positive risk modifiers lower this level.

Traveling at 100 km per hour on a motorbike, for example, has a higher residual risk than traveling at the same speed in a car because cars have more safety features, the most obvious being a structure that surrounds and protects the driver.

There are a number of rules applied to determine the relative residual risk of a technology that relate to the level of energy involved. Firstly this energy depends on the movement of the equipment, for example, a hand saw has a lower residual risk than a bench saw which has a moving blade driven by an electric current. A chainsaw has an even higher score because it places additional demands on the operator who has to both hold the saw and operate it correctly. The bench saw, on the other hand, is fixed and only requires the worker to bring the material for cutting against the blade. As a general rule of thumb, fixed equipment has a lower residual risk than mobile equipment.

The other parameter affecting the level of energy is the size of the plant or equipment. For example, large chemical storage facilities have a higher risk than smaller ones because the former has a larger volume of stored chemical energy and in the event of systems failure, the potential severity of injury or damage is greater. These parameters were relevant in the Bhopal disaster where a significant quantity of highly toxic product was released into a densely populated area. In terms of equipment risks, an impact involving a small aircraft is considerably less destructive than an airbus traveling at the same speed and under the same conditions and therefore the airbus has a higher residual risk when the threats to passenger and environmental safety are combined.

The level of energy is partly mitigated when the technology has inbuilt safety features. These are positive risk modifiers. Plant that has automatic shutdown capabilities in the event of failure has a lower level of risk than plant that requires manual shutdown. Cars equipped with seatbelts, roll bars and airbags are safer than those without. Where the design of the equipment prevents the possibility of contact between its hazardous components and other system factors or reduces consequential damages, the equipment has a lower residual risk.

The final consideration when determining inherent danger is the fit between the technology and other system factors. As discussed in Chapter 4, residual threats are higher when plant is installed incorrectly or does not fit the physical environment. An example is a case described by an OHS practitioner of an underground ore crushing plant that was installed with insufficient space between the top of the crusher and the back of the crusher excavation/cavity. As a result, large rocks occasionally became wedged between the machinery and the back wall. When this occurred a worker was sent up to lever the boulder down putting the employee at considerable risk of injury. There are other examples where poor fit creates additional hazards. Badly designed equipment and tools which do not account for the physical needs of the operator have a higher residual risk than ergonomically designed technology. The questions the auditor needs to ask to determine the level of inherent danger in the technology system factor are shown in a flowchart in Fig. 7.6.

The first question is to determine whether the technology is in the plant/equipment/tool category or chemical category. The latter have a toxicity or reactivity that relates directly to the level of residual risk. If plant/equipment/tools are being assessed for residual risk the next step is to determine whether they are fixed or mobile/moving. The size or scale of the equipment is then considered and the types of energy involved identified, which provides an analysis of the total energy associated with the technology. If consequential damages are partially mitigated by the safety features of the equipment this reduces the residual risk. The technology residual risk scorer is then used to determine a relative score and this is illustrated in Fig. 7.7.

This scorer shows that the primary risk parameter for technology is energy or toxicity. The level of risk may be modified by in-built safety features which reduce the potential consequences of systems failure. An example of this is the on-board fire suppression systems in some large earthmoving mobile machines. Features such as these reduce the potential damage or injury likely from the release of inherent energy threats. The next step is to transfer the potential consequences of technology failure to determine the residual risk score. When considering these possible outcomes, the assessor, particularly at the time of incident investigation, needs to consider what could have happened rather than what actually happened. For instance, if a large prefabricated concrete slab falls over at a construction site, the possible consequences are 'fatality or severe damage' or worse. If an incident of this type happens and no injuries or damage are incurred, the event cannot be scored downward. The scorer is valuable in reminding investigators that the energy involved is the primary parameter of the risk, which in turn relates closely to the potential consequences. An event of this type therefore represents a major warning of the dangers inherent in construction work using this type of building material.

To explain how the technology residual risk scorer may be used, Fig. 7.7 includes examples. The process, 'cutting wood' is applied to each of

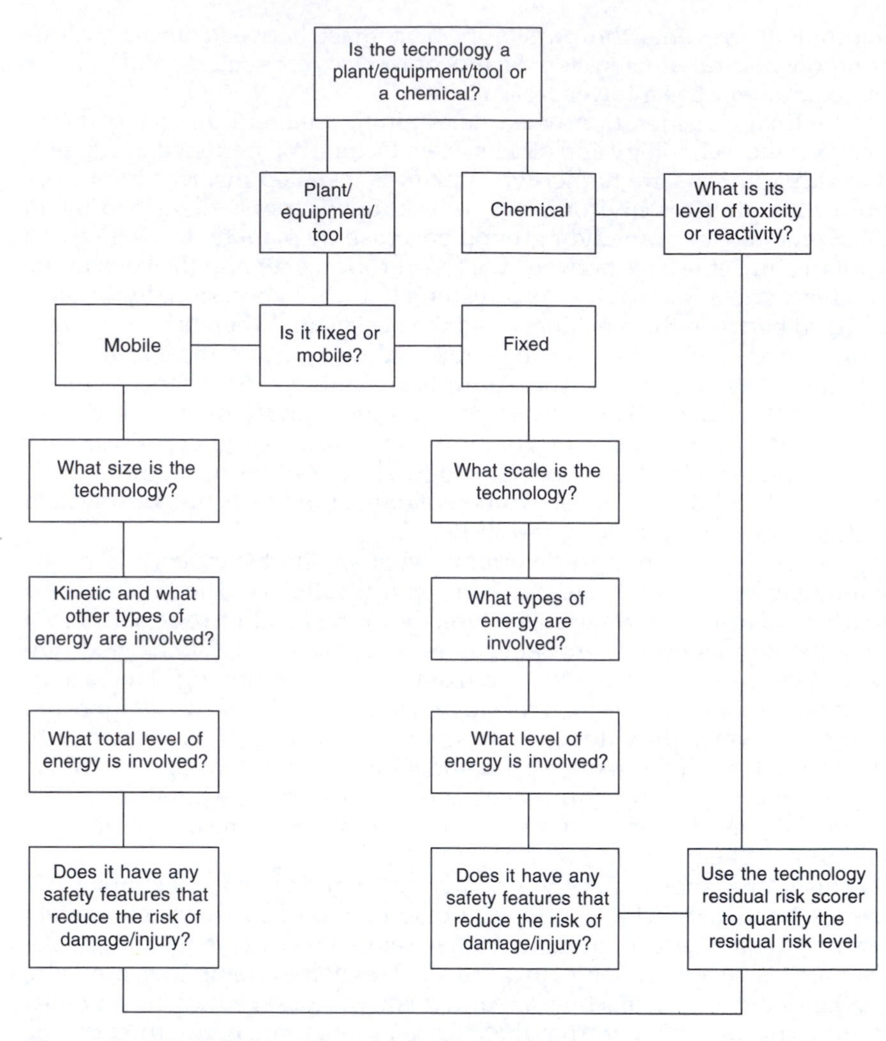


**Figure 7.6** Determining the level of residual risk in technology

the technologies. Under the principle of separation of system factors, the environment, process and worker are assumed to be 'perfect' so that relative values for different equipment can be determined. As shown, a hand saw has a lower residual risk than a fixed bench saw, which is in turn, safer than a chainsaw. None of this equipment has in-built safety features capable of reducing the risk, specifically, the equipment cannot differentiate between wood and human flesh and cutout to prevent injury. The relative scores given are 2, 4 and 6, respectively. The scores indicate, for example, that training resources are best directed at instruction for workers who operate chainsaws, followed by bench saws then hand saws. In fact, certification is usually required for the former because of the high level of inherent danger associated with such equipment and the processes undertaken using them.

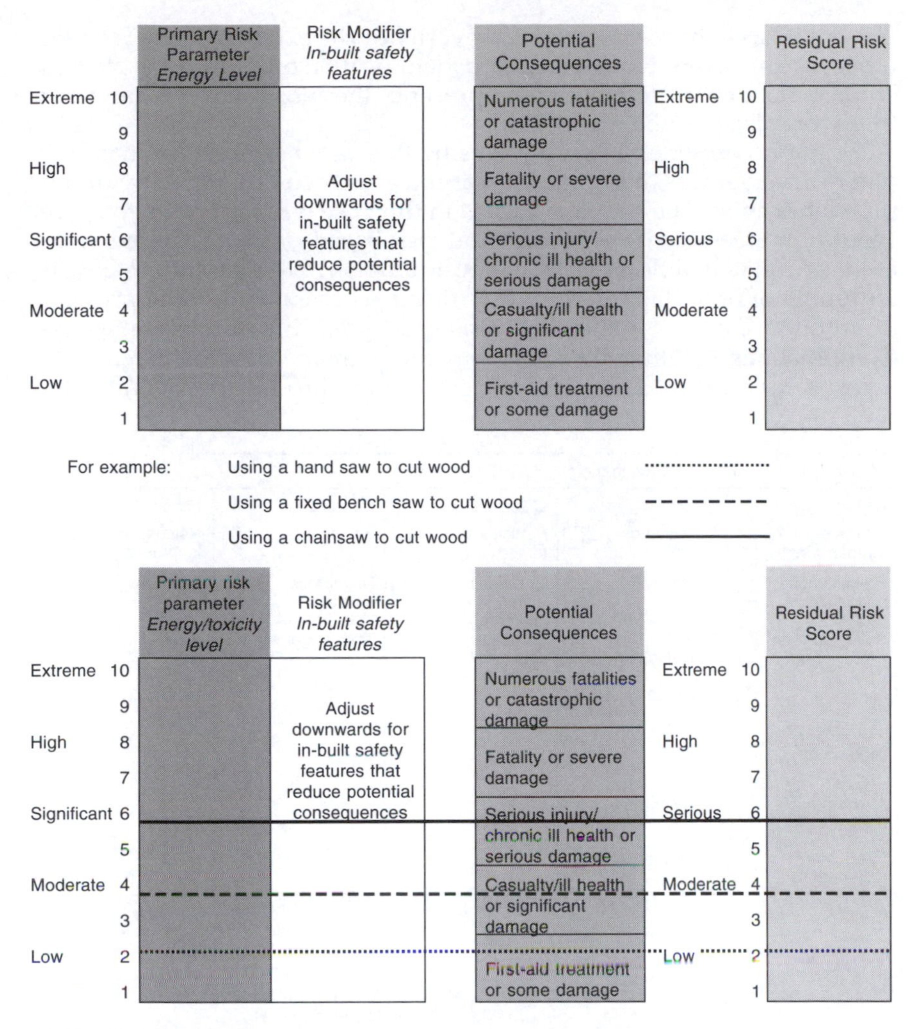


**Figure 7.7** Technology residual risk scorer

## Physical environment

As explained in Chapter 5, the physical environment includes locational and infrastructure characteristics and can comprise a complex set of hazards. From Table 5.4 it was shown that the main sources of residual risk in this system factor are site design shortcomings and a poor fit at the interface between the environment and other system factors. Table 5.1 listed the categories of hazards in this system factor as climatic, structural, chemical, biological and other.

A profile of physical environment inherent dangers for each site is obtained from the physical environment audit. How can the firm in the face of this complexity ascribe a single score to the residual risks in a workplace? The method proposes that the first step is to identify the hazards that are relevant from those listed in Fig. 7.8. Where a hazard is present it has to be assessed for its severity or the extent of lack of fit with other system factors. From this hazard identification procedure, the physical environment characteristic which presents the most extreme danger in that workplace can be identified.

Figure 7.8 provides an example using the case described in Chapter 5, where a welder was overcome with argon gas when he entered a 750-mm pipe fabrication. The primary hazard in this case was the confined space. Would an assessment of the residual risk based on the confined space alone adequately reflect the level of residual risk in this situation? Clearly, a number of other factors were significant. These included the presence of argon gas, poor ventilation in this space and difficult access/egress. Combinations of other hazards with the primary hazard increase the level of residual risk. The case where a mine production operator fell

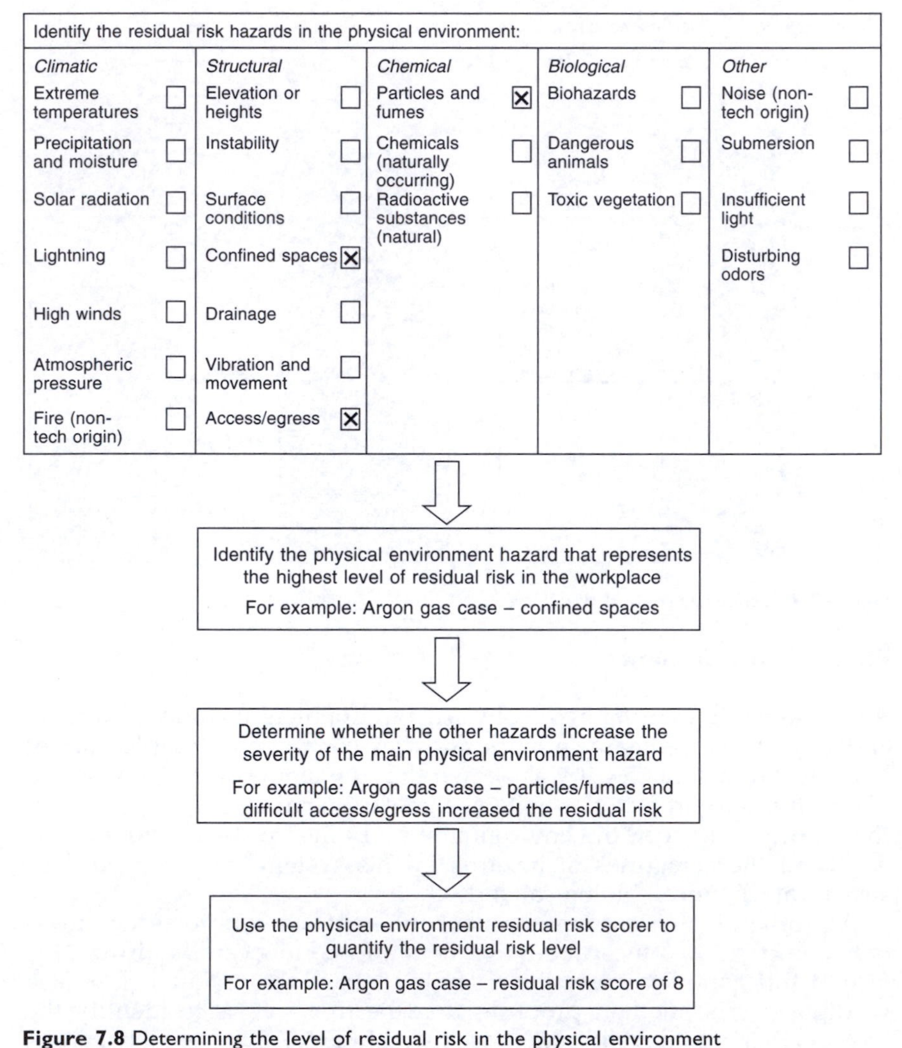


**Figure 7.8** Determining the level of residual risk in the physical environment

when he stepped into a 3-meter cavity in the berm because the area was insufficiently lit is a further example of physical environment residual risk factors combining to increase the overall level of inherent danger. In this case, the primary hazard was the cavity that resulted in a fall injury, whilst insufficient lighting was a secondary hazard. As a further example, when slippery conditions accompany heights the probability of injury and damage also increases due to the presence of another secondary hazard.

The combination of these physical environment conditions is particularly evident in relation to traffic conditions. When roads are gravel rather than sealed with bitumen, they have a higher level of residual risk. In combination with unfavorable weather conditions, such as heavy rain, these dangers are greatly exacerbated. This requires the driver to take remedial action by reducing the residual risk in other system factors, for example, decreasing process risks by reducing speed and where possible, selecting a vehicle type that is better suited to the situation.

To determine the residual risk in a workplace, therefore, the auditor has to identify the most significant risk – the primary risk parameter – and assess whether other hazards cause an even higher level of inherent danger. These additional hazards are negative risk modifiers that increase the overall level of risk with the combination affecting potential consequences.

Physical environment residual risks can lead to negative outcomes for organizational systems in three ways as shown in Fig. 7.9. The first is by introducing undesirable energy into the business that puts it at risk, for example, the kinetic energy of high winds or the potential energy of a mine fill as it saturates with water and starts to slip. The second is that

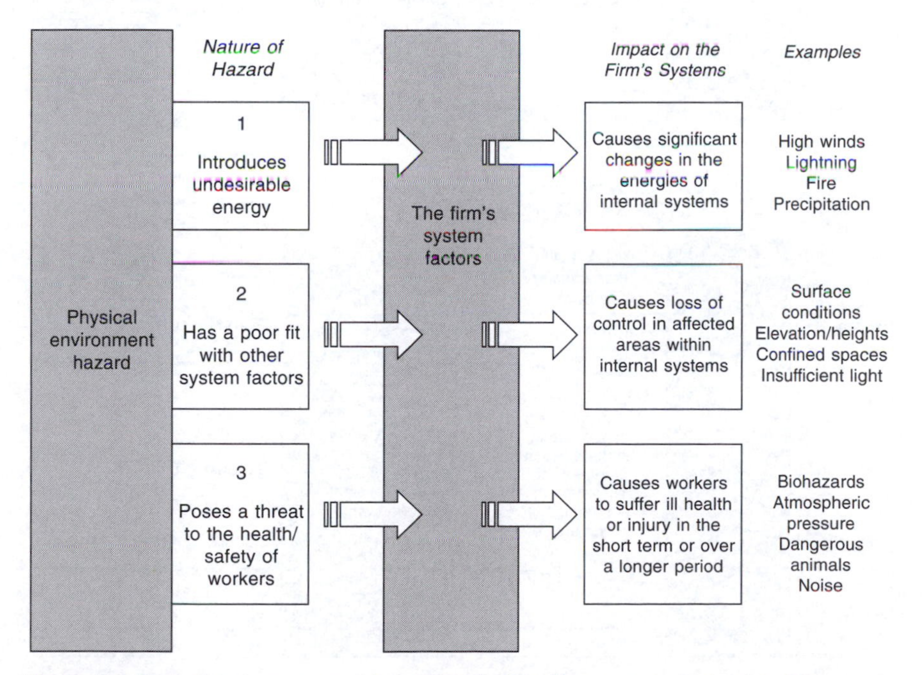


**Figure 7.9** Physical environment residual risks and their effect on organizational systems

the physical environment causes loss of control in other system factors, for example, when a slippery surface causes a truck to skid or an employee to fall. The third is that the residual risk is a direct threat to the health and safety of human beings, such as biohazards, fumes and dangerous animals. The extent to which physical environment factors introduce destructive energy or cause an unwanted change in energy of other system factors or directly threatens human health, is the level of residual risk.

The simplification of physical environment residual risks into one of these three categories allows the relative severity of the hazards in a particular workplace against another workplace to be determined. In effect, the assessor sets out to compare one work area, such as a mobile plant workshop, against another, such as the front-gate security office, to identify which is a higher residual risk area. The physical environment residual risk scorer is used to allocate a score for each work site or section of the work site as shown in Fig. 7.10.

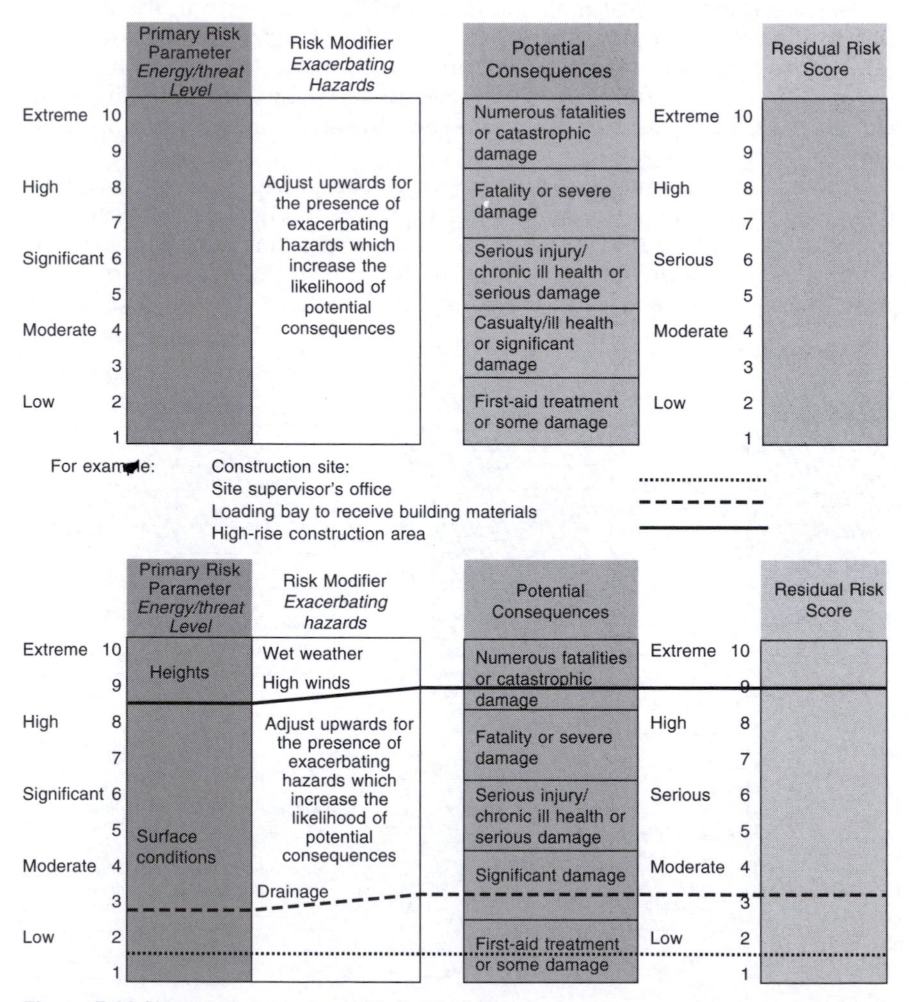


**Figure 7.10** Physical environment residual risk scorer

The first step in applying the scorer is to quantify the level of energy or threat imposed by the most severe hazard in the workplace – to give a level to the primary risk parameter. The level is adjusted upwards for the presence of exacerbating hazards or negative risk modifiers, for example, the inherent risk in slippery surface conditions is worsened by precipitation and poor drainage. This adjustment for the presence of risk modifiers may be incremental or substantial, for example, when the confined space in the earlier incident was combined with argon gas the risks rose significantly.

The result after taking exacerbating conditions into consideration correlates with the potential consequences, as shown in the second part of the scorer. Again, when the risk becomes an imminent danger and translates into an incident, there is no guarantee that the potential consequences predicted would eventuate, as this is also dependent on other contributing variables, such as the timing of the incident. The level is then transferred across to obtain a residual risk score. The lower part of the figure contains an example. It shows that the supervisor's office at a construction site has the lowest level of residual risk. The loading bay where building materials are received has a higher risk due to poor surface conditions and inadequate drainage. The highest risk work area on site is the construction area where elevations are a significant hazard exacerbated by wet weather and strong winds. The scores are 1.5, 3 and 9, respectively.

Although the final score relates to potential consequences it is important not to by-pass the primary risk parameter and risk modifier assessments because these are the underlying systemic conditions which determine the level of risk. Identifying these is the first step in developing effective risk management strategies. In this example, the score highlights the need for safe work procedures to mitigate the risks associated with heights in the construction area. In addition, guidelines are required to determine whether work should proceed in wet weather and high wind conditions. Other remedial strategies needed are structural changes to improve surface conditions and drainage in the loading area and communications to employees about the presence of slipping hazards in this workplace. The residual risk scorers therefore provide an indication of the types of RMS that should be implemented.

## Human resources

In Chapter 6, the residual risks associated with human resources were identified. These were incomplete KSAs and the physical limitations/fragility of human beings. The age and experience of workers has a significant effect on the former. Research has shown that workers below the age of 24 years are at least 30 per cent more likely to be involved in an accident at work.[6] Age and experience therefore have a substantial impact on human resources residual risk and are the primary risk parameter. In addition, the physical limitations of the human body suggest that the greater the demands of manual tasks, the greater the residual risk. For instance, in an office role where tasks are sedentary, the residual risk associated with physical limitations is not much of a concern, whereas on the factory floor where goods are lifted and carried by workers it is a significant risk issue. The physical demands of the job are therefore a negative risk modifier for this system factor.

A further issue affecting human resources residual risk is the level of workplace-specific competence and familiarity. It was shown in Fig. 6.2 that all new recruits, regardless of previous experience, cause a rise in residual risk. This is because they have to undergo a learning curve to adjust to the organization and to develop awareness of workplace-specific hazards. The rate of workforce turnover, involving the loss of experienced workers from the firm and their replacement with new employees, is also a serious issue in risk management. There is a direct relationship between turnover and the extent of risk-taking behavior in companies.[7] To quantify human resources residual risk, therefore, the length of tenure of workers has to be taken into consideration. Any employee group with duration of incumbency below a reasonably expected learning time has a higher level of residual risk than workers who are longer-serving and who are integrated into the organization's safety system.

The final factor affecting the level of human resources residual risk is training. Formal instruction should address inherent cognitive limitations in three stages: induction, competency training and risk awareness training. The purpose of training therefore is to compress the shortfall in KSAs so that the worker is able to operate safely and productively. Job demands, length of tenure and the level of training are modifiers affecting human resources residual risk.

The human resources residual risk scorer contains the primary risk parameter and these risk modifiers, as shown in Fig. 7.11. It begins on the left-hand side by finding the average age of workers in a specific job classification. A classification is a group of employees in a given workplace who share the same or similar job description, for example, all electricians on the construction site, all haul truck drivers at a mine, or all flight attendants working for an airline. It is important to treat employees on a group basis, rather than an individual basis, to avoid discrimination and a negative impact on management–worker relations. The age groups are given according to their relative levels of residual risk. For the sake of simplicity the underlying assumption is that collectively younger workers tend to have a higher residual risk because of limited KSAs compared to older workers. The scorer presumes that workers aged 35 and over have had sufficient opportunity in the workforce to develop their KSAs to a satisfactory level of competence and that they are conscious of the need for safety as a result. (There will be individual differences within these categories, however, for the purposes of ascribing relative risk levels, the method proposes that these approximate classifications are sufficient for the firm to identify higher risk groups.)

In the second part of the scorer, adjustments are made upwards for jobs that are physically demanding. This takes into consideration the potential for the demands of the job to exceed the capacity of the worker to cope in the short term and also the longer term impact on the health of the employee. For manual handling jobs, for example, the latter may include the development of calcification on the vertebrae of the spine that can limit the employee's ability to do this type of work in the future.

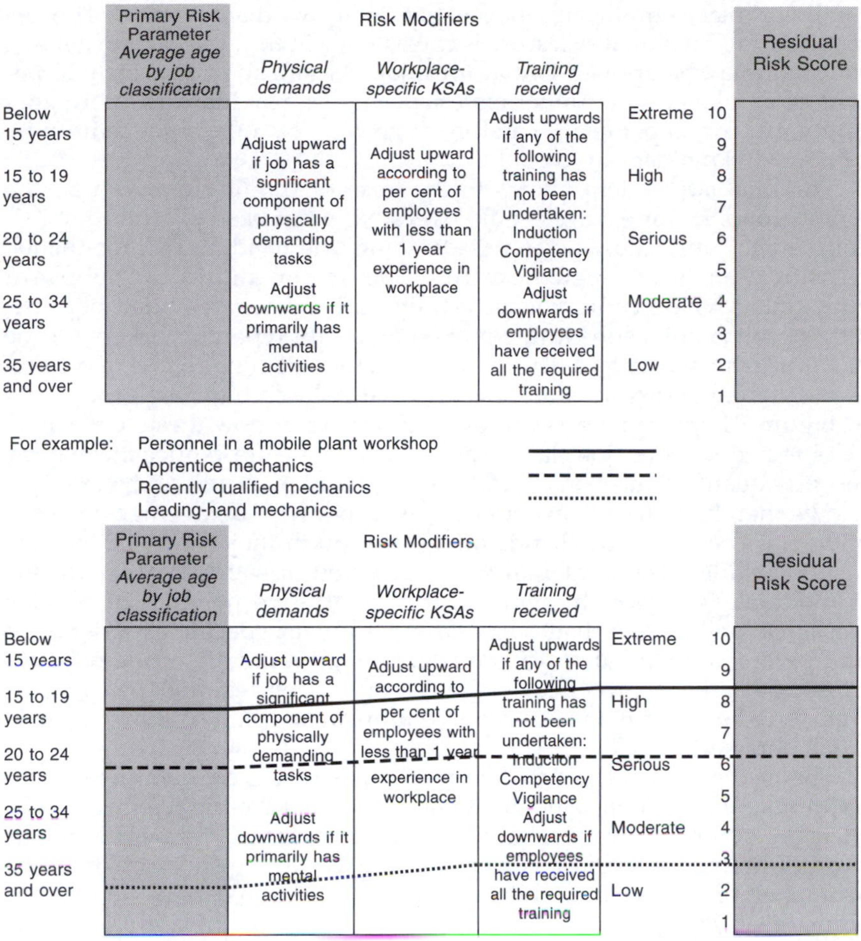

**Figure 7.11** Human resources residual risk scorer

If the job does not have a significant physical component, the residual risk level is modified downwards then transferred across to the next column, 'Workplace-specific KSAs'.

As stated earlier, workforce turnover has a significant impact on safety. The scorer makes a general assumption that it takes the average worker at least one year to gain sufficient workplace-specific KSAs to operate safely and efficiently. In reality, the time taken will depend on a number of factors including previous experience, the relevance of that experience to the new work context, the complexity of the job, and the extent of hazards in the workplace. Most of these issues, except for the latter, are factored into the primary risk parameter. The assumption of a one-year learning curve, though arbitrary, allows the auditor to make relative comparisons of different employee groups in a particular workplace and across different business units. The greater the percentage of employees with less than a year of experience in the workplace as a proportion of total workplace employees, the steeper the upward adjustment. The key to allocating a residual risk score is consistency in the approach. A general rule that may be applied is that for a job classification in which 25 per cent of employees are 'under-experienced', the residual risk score goes up by 0.5. For 50 per cent, therefore, it goes up by 1.0, 75 per cent up by 1.5, and 100 per cent up by 2.0.

The final adjustment relates to the level of training received by the work group. If, for example, employees have not been given induction, competency and safety awareness training, the residual risk for the job classification is adjusted upwards. The scorer allows a downward adjustment where comprehensive training has been provided because, although it is not a substitute for experience, it prepares workers for the hazards in the workplace. The final level in the 'Training received' column is transferred across to obtain the residual risk score for the group.

Figure 7.11 provides an example that illustrates how these scores may be obtained for three job classifications. These are apprentice mechanics, recently qualified mechanics, and leading-hand mechanics. The example shows that the latter are experienced workers with an average age over 35 years. Their jobs are 'hands-on' even though they have supervisory responsibilities. Their residual risk is adjusted upwards because the job is moderately physically demanding. Twenty-five per cent of this job classification has less than one year's workplace-specific experience so an upward adjustment of 0.5 is made. Most, but not all, of the required training has been provided to this group so the residual risk level is transferred directly to the risk score column and a result of approximately 3.0 is obtained.

The next group is the qualified mechanics who have completed their apprenticeships. Their average age is 21 years. Like the leading-hands, their job is moderately demanding so an upward adjustment is made. Twenty-five per cent have less than one year workplace-specific experience as a result of increased recruitment in this job classification. An upward adjustment is made for this. They have undertaken comprehensive training during their apprenticeships and all new recruits have been appropriately inducted and trained since joining the firm, though their knowledge of risk management is somewhat lacking. The residual risk score of this group is 6.0.

The final group is the new apprentices who have an average age of 17 years. The job type has a consistent level of physical demand and is adjusted upwards as was done for the other classifications. The profile of this group comprises first, second, third and final year apprentices. Some of them therefore have less than one year's experience and have not received the required competency and vigilance training. The residual risk score is adjusted upwards to approximately 8.5. The example shows that in the mechanical workshop, the highest risk group is the apprentice mechanics and the lowest group is the leading-hands.

The scorer highlights the importance of workforce maturity and experience as residual risk minimizers. It also identifies risk modifiers, such as training. Through its training and development strategies the firm can compress its level of human resources residual risk. Other appropriate strategies include the reduction of turnover and limiting the physical demands of the job. RMS specifically for the apprentice group could include close supervision, on-the-job training, formal classroom instruction, limits on the types of tasks undertaken to lower-risk duties, restriction from high-risk work areas, and standard day shift work rosters. The RMS control the risks at the interface between these workers and other system factors, as well as addressing competence shortfalls and lower levels of risk consciousness.

## Total residual risk scores

Once each of the system factors has been scored for the level of residual risk they can be combined into activities. Activities are defined as:

> An activity involves human resources undertaking a process or procedure using technology within a designated workplace.

This can be represented as the equation given earlier:

$$\text{Activity} = \text{Process} + \text{Human resources} + \text{Technology} + \text{Physical environment}$$

The table used to develop these activities is given in Figure 7.12(a). The first step in calculating the total residual risk scores (residual risk level for each activity undertaken in a workplace) is to list the processes that are carried out and enter the respective score from the process residual risk scorer. The next step is to identify the people who undertake the process, list the job classifications and include their respective scores from the human resources residual risk scorer. The technology used to carry out the operation is then identified and assessed followed by the physical environment. As stated earlier, the four steps are:

(1) What process is undertaken?
(2) Who does it?
(3) What technologies are used?
(4) Where is the process undertaken?

These steps lead to the development of different business activities for which a total residual risk score can be calculated. The formula is as follows:

$$\text{Total residual risk score} = P \times HR \times T \times PE$$

where: P, process residual risk score; HR, human resources residual risk score; T, technology residual risk score; PE, physical environment residual risk score.

The system factors are multiplied to generate this score for a given activity because the residual risk of one system factor exacerbates the residual risk in other system factors. As an example, the probability of an accident is extreme if apprentice mechanics, who have a relatively high residual risk compared to other workers, undertake complex dangerous tasks such as removing large components from heavy machinery, using manual winches in a poorly lit, hazardous physical environment. On the

(a) Blank table

| Systems residual risk profile for workplace: ............................. | | | | | | | | |
|---|---|---|---|---|---|---|---|---|
| *Processes* | *Score* | *Human resources* | *Score* | *Technology* | *Score* | *Physical environment* | *Score* | *Total residual risk score* |
| | | | | | | | | |
| | | | | | | | | |

(b) Examples

| Systems residual risk profile for workplace: Road haulage route Chicago to New York | | | | | | | | |
|---|---|---|---|---|---|---|---|---|
| *Processes* | *Score* | *Human resources* | *Score* | *Technology* | *Score* | *Physical environment* | *Score* | *Total residual risk score* |
| Driving a road train at 100 km/hour | 7.0 | Truck drivers – permanent workforce | 4.0 | Brand X model Y double trailer road train | 6.5 | Main highway between Chicago and New York (good weather) | 4.5 | 819 |
| Driving a road train at 100 km/hour | 7.0 | Truck drivers – contract as required | 5.0 | Variable – owner/operated on contract basis | 7.5 | Main highway between Chicago and New York (good weather) | 4.5 | 1181 |

| Systems residual risk profile for workplace: Light vehicle workshop | | | | | | | | |
|---|---|---|---|---|---|---|---|---|
| *Processes* | *Score* | *Human resources* | *Score* | *Technology* | *Score* | *Physical environment* | *Score* | *Total residual risk score* |
| Changing a tire on a light vehicle | 4.0 | Recently qualified mechanic | 6.0 | Tire<br>Jack<br>Wheel brace<br>Tire pressure pump | 3.0<br>3.0<br>2.0<br>3.5 | Light vehicle workshop | 3.0 | 252 |
| Changing a tire on a light vehicle | 4.0 | Apprentice mechanic | 8.0 | Tire<br>Jack<br>Wheel brace<br>Tire pressure pump | 3.0<br>3.0<br>2.0<br>3.5 | Light vehicle workshop | 3.0 | 336 |

**Figure 7.12** The calculation of total residual risk score for different activities

other hand, if the same group undertakes a simple standardized task, such as changing a tire on a light vehicle, using simple manual tools and equipment, in a well-designed and illuminated workshop, the residual risks are much lower. This multiplication effect was also explained earlier using the activity involving a driver operating a poorly designed truck in wet, slippery road conditions. The lower the competencies of the driver, the greater the speed, the fewer the safety features of the truck, and the worse the road conditions, the greater the residual risk.

Figure 7.12 contains two examples in part (b). In the first, a company compares the use of casual contract drivers to permanent employees for its road haulage services between Chicago and New York. It is shown that the contract drivers have a higher residual risk because they have variable competency levels and the firm does not have direct input into driver training. In addition, the vehicles used do not meet the standards of the firm's own fleet. When the scores are calculated, the use of these operators is shown to have a significantly higher level of risk than when the firm uses its permanent workforce and own trucks.

In the second example, the risk analysis shows that the apprentice mechanics in the company's light vehicle workshop have a higher level of residual risk than the qualified mechanics when undertaking a tire change under the same conditions. Each of the technologies used in the process is identified and the equipment with the highest residual risk is selected to calculate the total residual risk score for this activity. When the firm compiles a list of all activities undertaken in the workshop it can determine those with higher scores and focus on reducing the risks involved in these. It can also ensure that high residual risk factors do not combine, for instance, apprentices are not involved in complex processes unless they take a secondary role and are under strict supervision.

This method of activity-based risk analysis is more comprehensive than approaches using the risk scorecard and 'what if analysis'. It may be used to develop a system-wide residual risk profile for a given workplace and allows high-risk areas to be readily identified. By highlighting some of the underlying systemic factors that lead to residual risk, managers may prioritize risk management strategies and allocate scarce resources more confidently.

When the total residual risk score for an activity is calculated, the result can range from 10 000 ($10 \times 10 \times 10 \times 10$) to 1 ($1 \times 1 \times 1 \times 1$). Once all the scores for different activities in a workplace are calculated they can be graphed to identify peak hazards. It is proposed that high-risk activities should be given greatest priority. According to the four-fold strategy, resources should be allocated to ensure that such risks are effectively managed in the short term and compressed in the longer term. In some cases, the firm may find that its processes are of such high residual risk and are pervasive to the extent that the best method of risk reduction is to change the way a process is carried out so that workers are removed from danger. Examples of where this has occurred include the use of robots or remote control technologies to carry out hazardous work in factories and underground mines.

A firm should consider wholesale systemic changes when it identifies that one or more of its system factors has an unacceptable level of residual risk, such that the risk cannot be managed sufficiently by substantial compression of inherent risks in other system factors. For instance, in operations with an extreme physical environment residual risk, extensive investment in technology, human resources and work practice improvement will not necessarily compensate for these hazards, unless the interface between the worker and environment is changed dramatically. The process of carrying out under-sea maintenance on oil rigs can be used as an example. In the North Sea, the residual risk associated with the physical environment is very high. If divers undertake this maintenance work, investment in safer equipment, training and procedural controls cannot mitigate the risks because the hazards associated with the environment are severe. If, however, the worker is isolated from immediate danger and the task carried out remotely then residual risk is reduced substantially.

The process residual risk score decreases significantly because the employee is no longer exposed to high energy levels. The technology score drops because high-tech maintenance systems often have internal monitors that cause the system to cease operating when there is a failure. The human resources score declines because the work no longer involves physical demands. Only the physical environment score remains the same. Changing the way in which a process is carried out can, therefore, greatly reduce the total level of residual risk. However, concurrently such changes can be a drain on the firm's financial resources. The company therefore needs to undertake cost–benefit analysis to evaluate the feasibility of such changes.

Recapping, the purpose of developing total residual risk scores for different workplace activities is to provide a profile of the residual risks within a workplace and across the organization. From this profile the firm can rank the scores, prioritize its risk areas, and develop RMS to make better use of limited resources by having an enhanced understanding of systemic weaknesses. The steps involved in developing RMS are illustrated in Fig. 7.13.

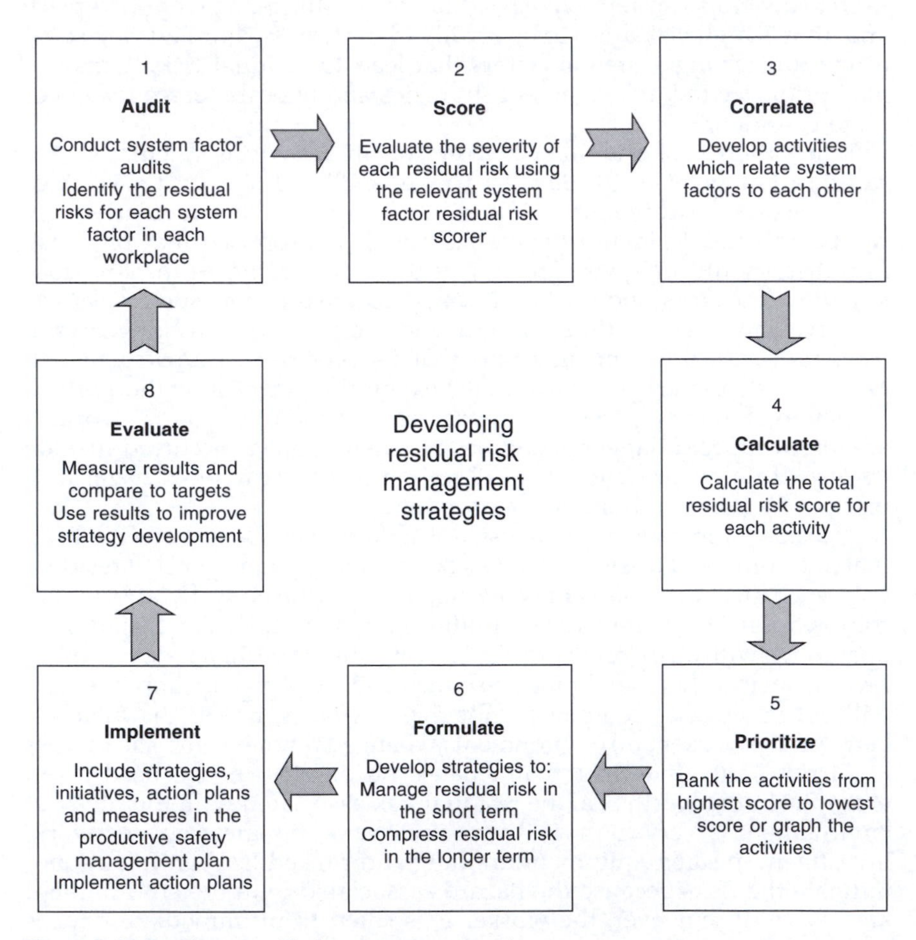


**Figure 7.13** Developing residual management strategies

The first step is to conduct system factor audits to identify the risks for each system factor in a work area. These are then scored using the relevant residual risk scorer. In the workplace residual risks combine or interface when activities are carried out. Once these interfaces have been correlated and all the different activities identified, the total residual risk score for each activity can be calculated. The listing or graphing of these scores allows the activities to be ranked from highest to lowest risk and prioritized. In the sixth step, strategies are formulated based on the severity of the danger. It may be found, for instance, that there are pervasive residual risks within one system factor that affect a number of activities. For example, there may be a significant level of residual risk amongst human resources attributable to high turnover and lack of training. Human resources residual risks therefore should be a priority area in the OHS budget. Concurrently, the HRM budget should also include resources for the development of strategies to reduce turnover.

The following step involves expanding strategies into initiatives, action plans and measures so that they can be implemented. The details of such strategies should be included in the productive safety management plan. The final step is to evaluate the effectiveness of these interventions, to improve strategy development and to pursue continuous improvement. The strategy development process is a cycle and thus system factors are again audited at the end of the cycle to measure the change in residual risk.

Figure 7.14 provides an example of a total residual risk profile for a light vehicle maintenance workshop. The scores for the various activities

| 840 | 1200 | 990 | 440 | 5600 | 3400 |
|---|---|---|---|---|---|
| 770 | 680 | 500 | 280 | 7800 | 2540 |
| 3780 | 350 | 550 | 1200 | 660 | 570 |
| 190 | 140 | 3800 | 380 | 990 | 540 |
| 440 | 180 | 870 | 2120 | 1980 | 410 |

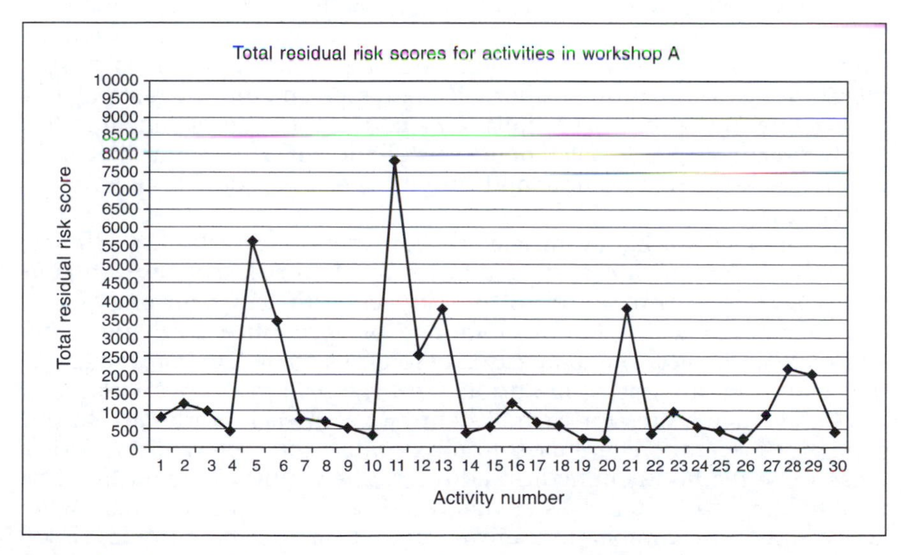


**Figure 7.14** Residual risk profile for light vehicle maintenance workshop

are listed in the top of the figure, then graphed below. It shows that, in descending order, activities 11, 5, 21, 13, 6 and 12 have the highest residual risks. The results should be used to trigger an investigation into the sources of residual risk in each system factor in these activities so that options for short-term management and long-term compression can be determined. It may be found, for example, that personnel who do not have the required certification are carrying out high-risk duties. A remedial strategy would be to immediately change work practices so that only qualified employees undertake the task. If the firm requires labor flexibility, then in the longer term, the unqualified workers may be given training to do this work in the future.

The total residual risk profile data may be manipulated in a number of ways to make the information useful for management decision-making. Each company should determine a method that best suits its requirements and resource availability. The following example is presented as one method that may be used. The graph can be divided into quartiles with ranges 7500 to 10 000, 5000 to 7500, 2500 to 5000, and 1 to 2500. The activities that score in the upper quartile are of highest priority and resources allocated to compress these as the first round of residual risk reduction. After this, the second quartile may be addressed and so forth. A further method of manipulating the data is to plot the separate system factor scores for the workplace's activities to identify the system factor that is the greatest source of residual risk. A workplace may, for example, have 20 different activities. The total score for each system factor can be obtained by summing the scores down the systems residual risk profile, as shown in Table 7.2. By averaging these total scores (total score divided by the number of activities) the system factor with the greatest level of risk is readily identified.

In example A, the road haulage firm, the system factor with the highest score is technology followed by processes. This indicates that the firm has to pay particular attention to the quality of its fleet in order to manage this risk. The large process score highlights the need to establish procedural standards. For example, with increased speed the residual risk rises, therefore the firm should consider self-regulation and introduce standards that define the upper-speed limit. This could, for example, include the development of an anti-speed policy, installation of cruise-control devices on the trucks, driver training and provision of additional time to make deliveries.

In example B, the light vehicle workshop, there are 30 activities. Human resources have the highest system factor residual risk score attributable to the significant proportion of employees with less than one year's experience and to the low average age of apprentice mechanics and qualified mechanics. The comparison of total system factor scores suggests that worker training is a high priority for the firm in this workplace.

The average scores for each system factor in each workplace in the company's operations are useful data because they allow priority risk areas to be readily identified. In addition, the results obtained the first time the firm's residual risks are analyzed can be used to compare with late results. The comparators are referred to as the baseline levels of residual risk, defined as the average of the residual scores for each system

**Table 7.2** Using the systems residual risk profile to compare systemic residual risk levels

| Processes | Score | Human resources | Score | Technology | Score | Physical environment | Score | TOTAL RESIDUAL-RISK SCORE |
|---|---|---|---|---|---|---|---|---|
| *Example A: Systems residual risk profile for workplace: road haulage route Chicago and New York* | | | | | | | | |
| Driving a road train at 100 km/hour | 7.0 | Truck drivers (permanent workforce) | 4.0 | Brand X model Y double trailer road train | 6.5 | Main highway Chicago and New York (good weather) | 4.5 | 819 |
| Driving a road train at 100 km/hour | 7.0 | Truck drivers (contract as required) | 5.0 | Variable – owner/operated on contract basis | 7.5 | Main highway Chicago and New York (good weather) | 4.5 | 1181 |
| Activity 3 to Activity 20 | Sum ↓ | | Sum ↓ | | Sum ↓ | | Sum ↓ | |
| TOTAL SCORE for system factor | 119 | | 89 | | 128 | | 94 | |
| AVERAGE SCORE (divide by 20) | 5.95 | | 4.45 | | 6.4 | | 4.7 | |
| *Example B: Systems residual risk profile for workplace: light vehicle workshop* | | | | | | | | |
| Changing a tire on a light vehicle | 4.0 | Recently qualified mechanic | 6.0 | Tire<br>Jack<br>Wheel brace<br>Tire pressure pump | 3.0<br>3.0<br>2.0<br>3.5 | Light vehicle workshop | 3.0 | 252 |
| Changing a tire on a light vehicle | 4.0 | Apprentice mechanic | 8.0 | Tire<br>Jack<br>Wheel brace<br>Tire pressure pump | 3.0<br>3.0<br>2.0<br>3.5 | Light vehicle workshop | 3.0 | 336 |
| Activity 3 to Activity 30 | Sum ↓ | | Sum ↓ | Sum highest score of the technologies used | Sum ↓ | | Sum ↓ | |
| TOTAL SCORE for system factor | 156 | | 174 | | 156 | | 96 | |
| AVERAGE SCORE (divide by 30) | 5.2 | | 5.8 | | 5.2 | | 3.2 | |

factor for all activities undertaken in a given workplace. The formulae are:

Baseline level of process residual risk

$$= \frac{\Sigma(\text{Activity 1}P_{RR}\text{Score} + \text{Activity 2}P_{RR}\text{Score} + \ldots \text{Activity N}P_{RR}\text{Score})}{N}$$

where $P_{RR}$ is the process residual risk score.

Baseline level of technology residual risk

$$= \frac{\Sigma(\text{Activity 1}T_{RR}\text{Score} + \text{Activity 2}T_{RR}\text{Score} + \ldots \text{Activity N}T_{RR}\text{Score})}{N}$$

where $T_{RR}$ is the technology residual risk score.

Baseline level of physical environment residual risk

$$= \frac{\Sigma(\text{Activity 1}PE_{RR}\text{Score} + \text{Activity 2}PE_{RR}\text{Score} + \ldots \text{Activity N}PE_{RR}\text{Score})}{N}$$

where $PE_{RR}$ is the physical environment residual risk score.

Baseline level of human resources residual risk

$$= \frac{\Sigma(\text{Activity 1}HR_{RR}\text{ Score} + \text{Activity 2}HR_{RR}\text{Score} + \ldots \text{Activity N}HR_{RR}\text{Score})}{N}$$

where $HR_{RR}$ is the human resources residual risk score.

The baselines may be used to identify priority areas and develop strategies, as well as providing the grounds for comparison when future system-wide residual risk audits are undertaken. This method may assist managers to determine whether the strategies they have implemented have been effective in compressing the level of residual risk. Figure 7.15 illustrates how this process of comparison may be applied.

In the figure, two workplaces are shown. In this hypothetical example, the results for the light vehicle workshop and the road haulage route for a transportation company are summarized. The systems residual risk profile results for each system factor, for the year 2001/2002, are listed for these workplaces. The results for processes, human resources, technology and the physical environment for the light vehicle workshop are 5.2, 5.8, 5.2 and 3.2, respectively. The results for these system factors for the road haulage route are 5.95, 4.45, 6.4 and 4.7, respectively.

The company has identified its targets for the compressions of residual risk by 2003/2004 and the effective percentage reduction to be achieved is given below in the figure. Some of the firm's strategies to reduce the risks in each work area are shown. These include, for the light vehicle workshop, the reduction in manual handling load limits to improve processes, decreased employee turnover to enhance the quality of human resources, replacement of the hydraulic jack system with a pit system to compress technological risks, and improvements to work areas. In the

| Workplace: Light vehicle Workshop | | | | Workplace: Road haulage route Chicago to New York | | | |
|---|---|---|---|---|---|---|---|
| Systems residual risk profile results 2001/2002 | | | | Systems residual risk profile results 2001/2002 | | | |
| Processes | Human resources | Technology | Physical environment | Processes | Human resources | Technology | Physical environment |
| 5.2 | 5.8 | 5.2 | 3.2 | 5.95 | 4.45 | 6.4 | 4.7 |
| Target for 2003/2004 | | | | Target for 2003/2004 | | | |
| Target result 5.0 Percentage reduction 3.85% | Target result 5.6 Percentage reduction 3.45% | Target result 4.9 Percentage reduction 5.77% | Target result 3.1 Percentage reduction 3.13% | Target result 5.75 Percentage reduction 3.36% | Target result 4.3 Percentage reduction 3.37% | Target result 6.2 Percentage reduction 3.13% | Target result 4.7 Percentage reduction 0% |
| Strategies:<br>Reduction in manual handling risks using load limits<br>Reduction in employee turnover<br>Replacement of hydraulic jack system with pit to access undercarriage of vehicles<br>Resurfacing of worn floor areas with nonslip finish<br>Improved lighting in the tool storage area | | | | Strategies:<br>Implementation of anti-speeding policy<br>Induction and training for contract drivers<br>Installation of cruise control, new seat belts and driver airbag | | | |
| Systems residual risk profile results 2003/2004 | | | | Systems residual risk profile results 2003/2004 | | | |
| 5.1 | 5.5 | 4.9 | 3.0 | 5.8 | 4.4 | 6.2 | 4.7 |
| % Change achieved 2003/2004 compared to 2001/2002 | | | | % Change achieved 2003/2004 compared to 2001/2002 | | | |
| Reduction of 1.92% | Reduction of 5.17% | Reduction of 5.77% | Reduction of 6.25% | Reduction of 2.52% | Reduction of 3.37% | Reduction of 3.13% | Reduction of 0% |
| Target achieved: Yes/No | | | | Target achieved: Yes/No | | | |
| No | Yes exceeded | Yes | Yes exceeded | No | Yes | Yes | Yes |

**Figure 7.15** Measuring compression of residual risk

haulage area, the company intends to make policy changes to deter speeding, invest in training for contract drivers and install safety equipment on its trucks.

The results for 2003/2004 are given and the percentage change compared to the baseline level of risk (the results for 2001/2002) is calculated. By way of example, in the workshop, human resources residual risks have been reduced from the 5.8 baseline to 5.5, which is a reduction of 5.17 per cent. Using these results, the firm is able to determine whether its targets have been achieved. In this case, all targets have been achieved except the process risks in the light vehicle workshop and haulage area. Regardless of the method that the firm chooses to apply, it is important that where the firm fails to reach its targets, it investigate the underlying causes of this failure. With this knowledge it is better equipped to develop more effective RMS. It may be found, for example, that drivers were exceeding the speed limits because of delivery deadlines. A review of process times would therefore be required.

According to the four-fold strategy, companies have to address residual risks on two levels – through short-term management and long-term compression. To develop appropriate strategies, OHS practitioners and relevant workplace managers need to understand the root causes of residual risk. In the hypothetical example given in Fig. 7.15, the scores shown for the light vehicle workshop may highlight that employee turnover is a significant issue in this work area. In the short term, there is therefore a need for greater supervision and staff training among this group of employees to address current KSA shortfalls. In the road haulage area, the use of double-carriage road trains is a high-risk factor. As an immediate intervention, the firm may need to establish criteria for the selection of contractors and their equipment to ensure that all areas of the operations are legally compliant and that the risks are managed effectively. In the longer term, the principal company may consider strategies that increase the alignment of the contractors' systems and their own by introducing mandatory standards in contract tendering procedures and tighter monitoring of contractor safety.

## Entropy prevention strategies (EPS)

The method described above may be used to quantify the firm's residual risks in relative terms. This method provides managers with a profile of the hazards that are specific to the company's operations and a relative measure of the severity of these hazards. From this, management and compression strategies can be developed. The level of residual risk, because it cannot be compressed in the short term, is a constant source of risk. In contrast, entropic risk is the variable level of risk over and above the residual, as illustrated in Fig. 7.16. This variability means that degradation cannot be quantified in the same way as the level of residual risk. The firm, nonetheless, needs a method to determine the susceptibility of the organization's systems to this type of danger.

In Chapter 1, the following formula for entropic risk was presented. The equation applies the same rationale as the total residual risk score formula by allocating a score of 0 to 10 to entropic risk in each of the four system factors. The score of 0 is only possible in an ideal system where

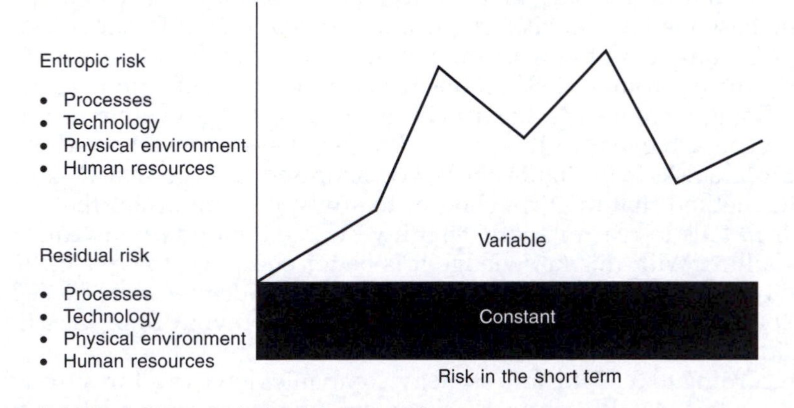


**Figure 7.16** Quantifying residual and entropic risk

there is nil exposure to natural forces of degradation. A score of 10, on the other hand, represented extreme susceptibility. Exposure is an important concept applied to entropic risk. It is not entirely dependent on time as defined traditionally in risk management practices. It is a measure of the system's vulnerability to this type of risk. A firm with a high exposure is very susceptible to degradation and incidents that affect safety and performance. The formula for total entropic risk is:

$$\text{Total entropic risk score} = p \times t \times p_e \times h_r$$

where: p, process entropic risk score; t, technology entropic risk score; $p_e$, physical environment entropic risk score; $h_r$, human resources entropic risk score.

The example given in Chapter 1 compared the vulnerability of two companies to entropic risk. Company A had state-of-the-art technology and a relatively inexperienced workforce. Company B had old technology and very experienced personnel. The former was susceptible to human resources entropic risks and needed to allocate finances to training, whereas the latter had to address the wear and tear of its technology as a priority.

According to the four-fold strategy there are two fronts on which entropic risk has to be addressed. The first is its elimination using corrective action and the second is the development of maintenance strategies to prevent future deterioration. The firm needs to firstly understand its susceptibility to this type of risk in order to formulate appropriate responses. As was the case with residual risk analysis, the entropic risks associated with each system factor need to be quantified independently of each other, thus again applying the principle of separation of system factors.

## Processes

In Chapter 3 it was shown that the sources of entropic risk in processes occur at the interface with other system factors. These are mitigated using JSA, inspections, audits and incident investigations, as summarized in Table 3.1. The firm's vulnerability to degradation depends on the extent and thoroughness with which it uses these activities to take corrective action and to implement maintenance strategies. Fundamentally therefore entropy is prevented through sound management and operational habits. The company can assess its level of 'sophistication' in the use of these measures to determine its exposure to entropic risk. Alternatively, an entropic risk profile can be developed that measures susceptibility to degradation in relative terms.

Entropic risk is variable and is therefore difficult to quantify at any given point in time. It is even more difficult to assess over a period of time because there are innumerable influencing issues that cause the degradation of organizational systems. The firm can, however, use the entropic risk scorers described in this chapter to analyze the impact of the primary risk parameter and risk modifiers on each system factor. These parameters and modifiers are not comprehensive as such, but provide firms with an indication of the nature of the entropic risk they are exposed to. The first of these scorers is the process entropic risk scorer shown in Fig. 7.17.

The scorer indicates that process degradation is most likely to occur in complex work practices. When employees have a simple task to perform, deviations from safe practices are less likely to occur. Complex and/or unfamiliar tasks have a higher level of risk for two reasons. The first is that the requirements of the job can stretch the competencies of the worker; in other words, the task can be beyond the employee's current KSAs. The second reason is that the risks involved may not be fully appreciated.

In the process entropic risk scorer, the first column rates the process by level of complexity which is the primary risk parameter. The more mentally and physically demanding the task, the greater the likelihood of deviation from safe practice. In the second stage, risk modifiers are applied. Adjustments are made upwards if the process has not been standardized. If the process is complex, therefore, it is additionally dangerous if a JSA

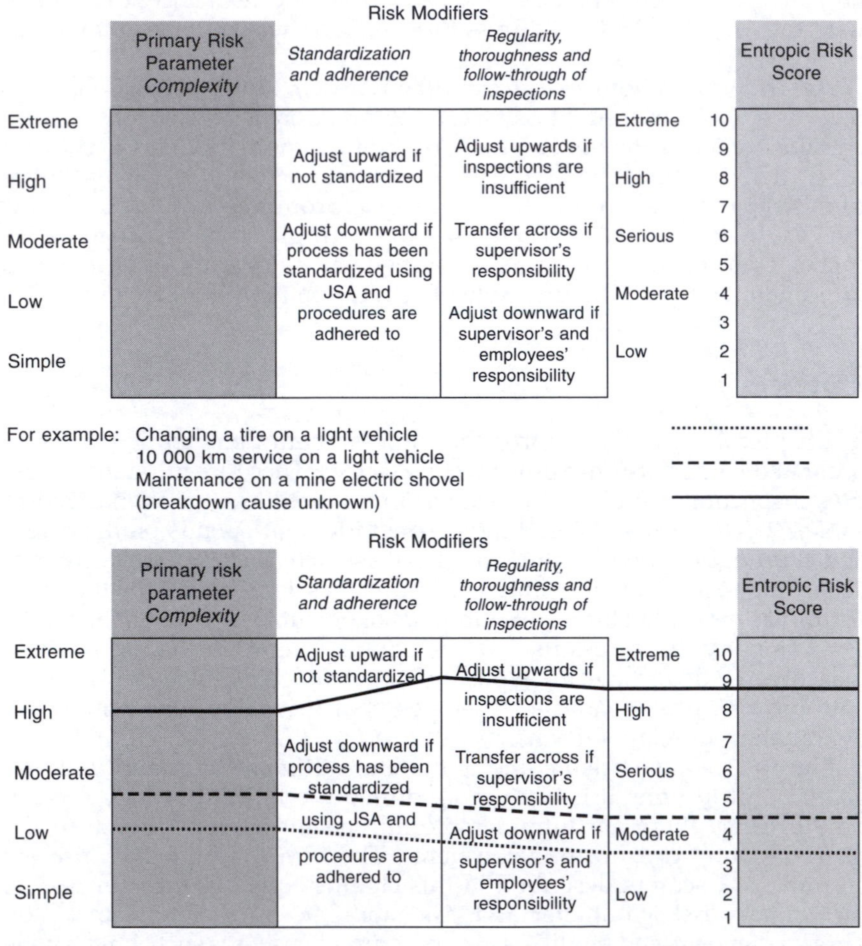

Risk Modifiers

Primary Risk Parameter *Complexity*
*Standardization and adherence*
*Regularity, thoroughness and follow-through of inspections*
Entropic Risk Score

Extreme
High
Moderate
Low
Simple

Adjust upward if not standardized
Adjust downward if process has been standardized using JSA and procedures are adhered to

Adjust upwards if inspections are insufficient
Transfer across if supervisor's responsibility
Adjust downward if supervisor's and employees' responsibility

Extreme 10
9
High 8
7
Serious 6
5
Moderate 4
3
Low 2
1

For example: Changing a tire on a light vehicle
10 000 km service on a light vehicle
Maintenance on a mine electric shovel (breakdown cause unknown)

Risk Modifiers

Primary risk parameter *Complexity*
*Standardization and adherence*
*Regularity, thoroughness and follow-through of inspections*
Entropic Risk Score

Extreme
High
Moderate
Low
Simple

Adjust upward if not standardized
Adjust downward if process has been standardized using JSA and procedures are adhered to

Adjust upwards if inspections are insufficient
Transfer across if supervisor's responsibility
Adjust downward if supervisor's and employees' responsibility

Extreme 10
9
High 8
7
Serious 6
5
Moderate 4
3
Low 2
1

**Figure 7.17** Process entropic risk scorer

has not been undertaken and safe steps have not been established. A downward adjustment is made if the process has been standardized using JSA and the procedures are consistently adhered to. When the risks of a process have been identified and the process modified to manage these risks so that deviations are minimized, complexity is reduced. The method of work is no longer left to the discretion of the worker and thus, provided the JSA was done thoroughly and accurately, it is safer. In addition, when the worker consistently adheres to the safe work procedures it suggests that he appreciates the risks involved. Examples of adherence include knowing when and how to use personal protective equipment and following the correct 'danger tag' procedure when carrying out plant maintenance.

In the third column of the scorer in Fig. 7.17, adjustments are made for the firm's inspection practices. Exposure to entropy increases when inspections are not carried out because corrective action is not as timely. An upward adjustment is required if the firm is 'laissez-faire' in this regard. If inspections are the sole responsibility of the supervisor, the level is transferred across to the entropic risk score. In firms where these practices are decentralized – carried out by supervisors and employees within their scope of work – a downward adjustment is made. Regular inspections are indicative of good risk management practices and the more they are delegated down the hierarchy in addition to being overseen by management, the greater the level of safety consciousness and lower the tendency towards process degradation. The final level in this third column is transferred across to determine the entropic risk score. This scorer indicates that process standardization and regular inspections involving employee participation are positive risk modifiers. They reduce the firm's susceptibility to degradation.

Figure 7.17 contains hypothetical examples of processes undertaken by a mining operation that illustrate the comparative tendency towards entropy in three activities. The scorer gives the firm an indication of its vulnerability to entropic risk in relative rather than absolute terms – the comparative level of risk from process to process. In the first example, changing a tire on a light vehicle, the process has a relatively low level of complexity. The firm has standardized the practice and ensures that procedures are known and adhered to. In addition, inspection practices are decentralized to the workshop level, so the resultant score is 3.5. In the second example, maintenance personnel carry out a routine 10 000-km service on a light vehicle. The process is more complex than changing a tire, but has been standardized as far as practicable. The entropic risk score is 4.5. In the final example involving breakdown maintenance on a mine electric shovel, the process is highly complex and has not been standardized. An upward adjustment is made to account for unforeseen hazards. The example shows that, in this case, the supervisor and workers routinely carry out a JSA prior to undertaking uncertain task to anticipate potential risk sources and to develop a safe method of work. The final entropic risk score is 9.0.

This scorer highlights the danger of complex, unfamiliar tasks, the need for standardization where practicable, and the importance of regular, thorough inspections followed by corrective action and preventative strategies. It also indicates that to avert entropic risk, employees need to be involved in risk management and one of the primary means of operationalizing participation is by delegating the responsibility for inspection and hazard identification to both workers and supervisors.

## Technology

In Chapter 4 it was shown that the sources of entropic risk for the technology system factor are wear and tear and technology/operator interaction. When equipment is used for the wrong process, when the physical environment puts machinery under mechanical strain, and when employees operate it incorrectly, wear and tear is accelerated. The primary variable that affects a technology's exposure to these sources of degradation is its age and usage rate. Technological entropic risk can be quantified in relative terms using the scorer given in Fig. 7.18.

The first step in using the scorer is to estimate the age/usage rate of the technology. New equipment has a lead-time before it begins to show signs of wear and tear. Old technologies, on the other hand, start to reach the point where the costs of maintenance exceed the benefits of continued operation. In the technology entropic risk scorer, the second issue affecting susceptibility to degradation is the level of preventative maintenance. This modifies the risk level either positively or negatively. An upward adjustment is made if the equipment is not maintained regularly and a downward adjustment made if it is. (In Chapter 4 it was also explained that chemicals fall into this category. In this case, the auditor evaluates the age and quality of storage methods and whether these are maintained well. For example, if fertilizer bags are made of heavy paper and become wet then they are in a state of degradation that increases the likelihood of spillage.)

The final determinant of this entropic risk is the regularity, thoroughness and follow-through of inspections. Regular checks allow the firm to identify areas of entropy and to take corrective action. The same rationale is applied as with the process entropic risk scorer. Insufficient inspections result in an upward adjustment. When this responsibility is undertaken by the supervisor alone, a direct transfer of the score is made, whereas a downward adjustment is made when employees are also given responsibility for assessing the condition of technology. The final level is allocated to the entropic risk score. The scorer indicates that preventative maintenance and regular inspections are the main means of modifying entropic risk exposure in a positive direction for the technology system factor.

Figure 7.18 contains examples of three technologies owned by the firm. These are a new light vehicle, a 5-year-old vehicle with low kilometers, and a 7-year-old vehicle with very high kilometers. The scorer shows that the new vehicle is regularly maintained and both supervisors and workers are accountable for routine inspections. Downward adjustments are made in these two columns accordingly. The resultant score is approximately 2.0. The 5-year-old vehicle is above its half-life, is well maintained and routinely inspected. It has a score of 5.0. The oldest vehicle has exceeded its half-life and is also regularly serviced and checked.

ELSEVIER

Risk Modifiers

| Primary Risk Parameter *Age/usage rate* | Preventative maintenance | *Regularity, thoroughness and follow-through of inspections* | | Entropic Risk Score |
|---|---|---|---|---|
| Archaic | | Adjust upwards if inspections are insufficient | Extreme | 10 |
| | | | | 9 |
| Aging | Adjust upward if not maintained regularly | | High | 8 |
| | | Transfer across if supervisor's responsibility | | 7 |
| Half-life | | | Serious | 6 |
| | Adjust downward if maintained regularly | | | 5 |
| Reasonably new | | Adjust downward if supervisor's and employees' responsibility | Moderate | 4 |
| | | | | 3 |
| | | | Low | 2 |
| New | | | | 1 |

For example: New light vehicle ........
5-year-old light vehicle low kilometers - - - - -
7-year-old light vehicle very high kilometers ———

Risk Modifiers

| Primary Risk Parameter *Age/usage rate* | Preventative maintenance | *Regularity, thoroughness and follow-through of inspections* | | Entropic Risk Score |
|---|---|---|---|---|
| Archaic | | Adjust upwards if inspections are insufficient | Extreme | 10 |
| | | | | 9 |
| Aging | Adjust upward if not maintained regularly | | High | 8 |
| | | Transfer across if supervisor's responsibility | | 7 |
| Half-life | | | Serious | 6 |
| | Adjust downward if maintained regularly | | | 5 |
| Reasonably new | | Adjust downward if supervisor's and employees' responsibility | Moderate | 4 |
| | | | | 3 |
| | | | Low | 2 |
| New | | | | 1 |

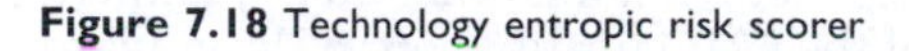

**Figure 7.18** Technology entropic risk scorer

It has a score of 7.0. The scores indicate that the older technology has a greater tendency towards deterioration and, therefore, has a greater likelihood of failure leading to an accident or incident. The scorer illustrates the importance of maintenance and inspections of technologies. If, for instance, the new vehicle was neither regularly serviced nor inspected, its score could become as high as 5. Its usage rate is a key factor in its tendency towards entropy. A possible scenario resulting from lack of maintenance could be the vehicle running out of oil causing the motor to seize whilst traveling along a busy carriageway. The driver's competencies and other factors, such as traffic and road conditions, would influence the consequences.

## Physical environment

The physical environment system factor includes infrastructure and locational conditions. From the physical environment audit, the firm obtains a list of such conditions, which can then be combined with other system factors to generate a profile of business activities. In Chapter 5 it was explained that the physical environment suffers natural degradation and this is accelerated by a poor fit between this system factor and other system factors. For instance, physical conditions deteriorate more rapidly when workers inadvertently cause damage by driving on the eroded edge of the road rather than on the center. Technology also causes deterioration, for example, when emissions lead to the corrosion of infrastructure. Finally, processes can change the environment in undesirable ways, such as when they cause destabilization of natural systems or site contamination.

The rate of deterioration of locational conditions is difficult to quantify. Infrastructure, on the other hand, like technology, tends to have a product life, and therefore measures can be allocated into a range from 'new' to 'half-life' to 'old'. How can locational factors be analyzed? A significant variable in entropic risk exposure in these locational elements is the suitability of the site for the purpose and the quality of site design and planning. In addition, during the life of the operation any design shortcomings that lead to increased degradation tend to become apparent. In open-cut mines, for example, as the excavation becomes deeper, pit wall stability and dewatering become increasingly significant risk management and operational issues. A further example is the 'concrete cancer' and corrosion of infrastructure exposed to sea breezes. The life stage of the operation therefore provides some indication of exposure to entropic risk. The physical environment risk scorer, shown in Fig. 7.19, uses the age of infrastructure or condition of the physical environment as the primary risk parameter.

The scorer shows that there are two risk modifiers which again are preventative maintenance and regular inspections. Maintenance mitigates the degradation of the physical environment. An incremental adjustment is made according to the regularity of the system's upkeep. A downward adjustment is made if the company implements a proactive preventative system for locational and infrastructure conditions. In a mine, for example, this may include road and building maintenance, geotechnical interventions such as using rock bolts to stabilize slopes, repair of barriers and boundaries, and dust suppression methods. An upward adjustment is made if the firm does not carry out regular maintenance of the physical environment and relies on corrective action and reactive maintenance to manage entropic risk.

In the third column of the scorer, a further adjustment is made according to the regularity, thoroughness and follow-through of inspections of the physical environment. As was the case with process and technology entropic risks, inspections allow the firm to identify areas of degradation before they escalate into a high probability of an incident. The firm's exposure to entropic risk is greater if it fails to monitor the physical environment for changes that represent an increase in the severity of

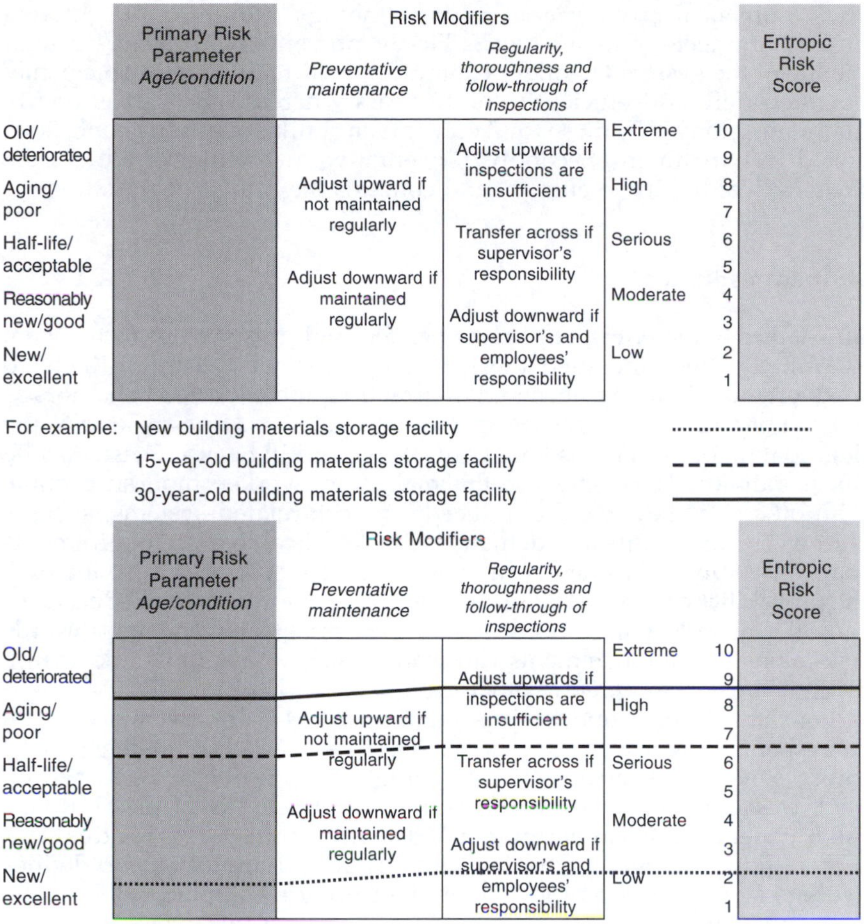

**Figure 7.19** Physical environment entropic risk scorer

hazards. The more decentralized the responsibility for carrying out these inspections, the greater the downward adjustment. The final level is transferred across to obtain the entropic risk score.

Figure 7.19 contains three examples for buildings of different ages at various company operations. The new, state-of-the-art, building materials storage facility has a low level of entropic risk because new infrastructure has a lead-time before it begins to deteriorate. The firm, in this example, does not have a preventative maintenance plan and therefore, an upward adjustment is made. Responsibility for inspections rests entirely with the supervisor so the level is transferred directly across to the entropic risk column, with the final score being 2.0. In contrast a 15-year-old facility owned by the same firm has reached its half-life and scores 6.5. Its 30-year-old facility is aging and if not maintained regularly scores 8.5. In this case, the scorer indicates that the business is failing to take preventative measures to address its exposure to degradation and this is particularly problematic at its older site.

The physical environment and technology scorers highlight two important effects of maintenance. Firstly, preventative practices extend the life of the system factor, for example, a well-maintained vehicle may operate safely and efficiently for 10 years whereas, one that is poorly maintained may only do so for 7 years given similar operating conditions and driver treatment. Secondly, preventative maintenance reduces the firm's susceptibility to entropy and contains the rate of degradation.

## Human resources

In Chapter 6, the entropic risks associated with this system factor were identified. These included the new employee who is more likely to inadvertently deviate from standard practices and expected behaviors as a result of a lack of workplace-specific KSAs. Other issues related to the degradation of human resources are fatigue and ill health. These may be due to individual characteristics that make some workers more susceptible than others or they may be induced by work-related systems, such as shift work, long hours and demanding tasks. Excessive job requirements that cause stress or difficulty in maintaining concentration also increase the probability of an incident. In addition, as supported by earlier accident causation models, some groups of workers have a greater tendency towards risk-taking behavior and this can lead to safety violations and drug/alcohol abuse, compared to other groups of workers.

For the sake of simplicity it is again assumed that older workers are safer than younger workers who have less developed competencies and lower levels of risk awareness. Accordingly, the age of employees within a job classification not only affects their level of residual risk, but also their tendency towards entropic risk. The human resources residual risk scorer, shown in Fig. 7.20, begins with this basic assumption and defines average age by job classification as the primary risk parameter. In the scorer the level of risk is therefore marked against this average age.

In the second column, risk modifiers are taken into consideration. An adjustment is made upwardly for jobs and work practices that place high demands on workers. This includes physically and mentally demanding tasks, long hours, shift work and lengthy commuting times. Some employees who work in remote areas on a fly-in fly-out basis, for example, have to get up very early to catch a plane before starting the first shift of the cycle. Before they start work therefore, there is preparation time, travel time to the airport, flight time, and travel to the work site, before duties are commenced. Similarly, employees who drive long distances between home and work are at greater risk of fatigue than workers who are close to home and, therefore, the former have an increased probability of an accident whilst at work and whilst traveling. Job and travel demands are therefore included as negative risk modifiers.

A second such modifier is the experience level of the employee group. Workplace-specific KSAs reduce the tendency towards degradation. Specifically, these KSAs refer to hands-on experience that allows the worker to understand the hazards that are specific to the workplace, how these are affected by the employee's behavior, and how these should be managed.

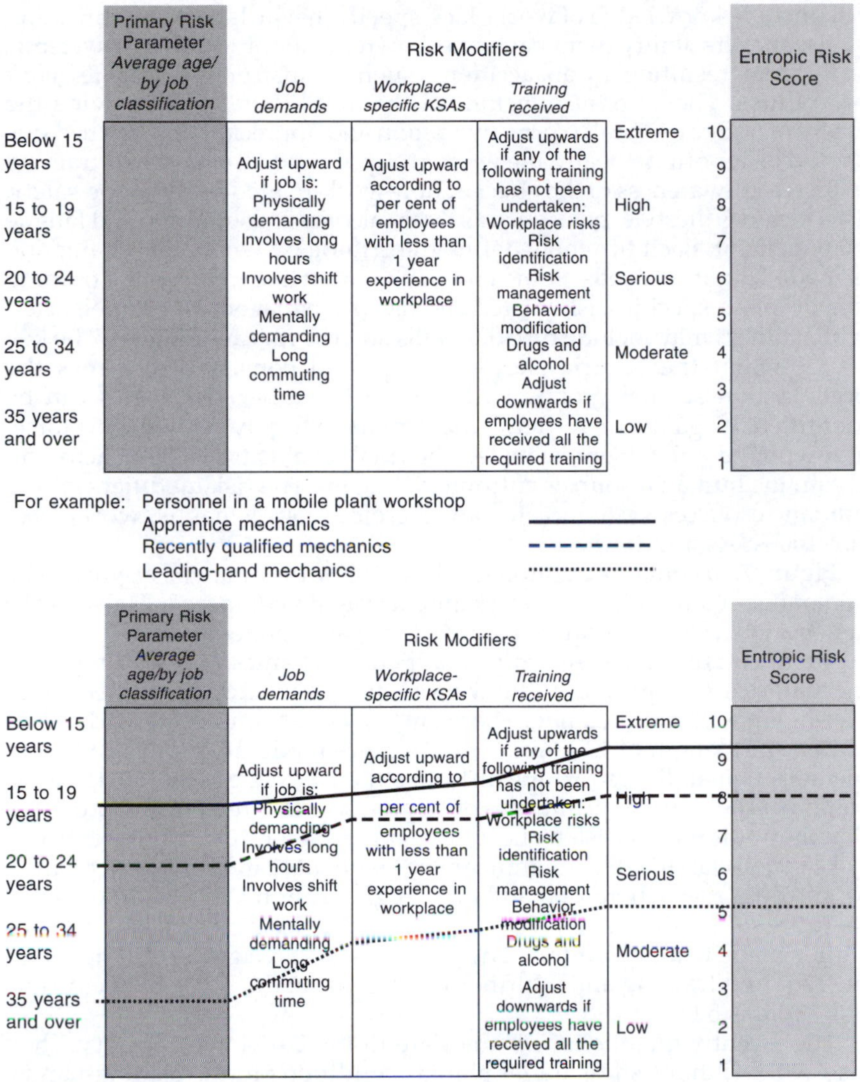

Primary Risk Parameter *Average age/ by job classification*

Risk Modifiers

*Job demands* | *Workplace-specific KSAs* | *Training received*

Entropic Risk Score

Below 15 years
15 to 19 years
20 to 24 years
25 to 34 years
35 years and over

Adjust upward if job is: Physically demanding Involves long hours Involves shift work Mentally demanding Long commuting time

Adjust upward according to per cent of employees with less than 1 year experience in workplace

Adjust upwards if any of the following training has not been undertaken: Workplace risks Risk identification Risk management Behavior modification Drugs and alcohol Adjust downwards if employees have received all the required training

Extreme 10
9
High 8
7
Serious 6
5
Moderate 4
3
Low 2
1

For example: Personnel in a mobile plant workshop
Apprentice mechanics ————
Recently qualified mechanics - - - - - - - -
Leading-hand mechanics ..........................

Primary Risk Parameter *Average age/by job classification*

Risk Modifiers

*Job demands* | *Workplace-specific KSAs* | *Training received*

Entropic Risk Score

Below 15 years
15 to 19 years
20 to 24 years
25 to 34 years
35 years and over

Adjust upward if job is: Physically demanding Involves long hours Involves shift work Mentally demanding Long commuting time

Adjust upward according to per cent of employees with less than 1 year experience in workplace

Adjust upwards if any of the following training has not been undertaken: Workplace risks Risk identification Risk management Behavior modification Drugs and alcohol Adjust downwards if employees have received all the required training

Extreme 10
9
High 8
7
Serious 6
5
Moderate 4
3
Low 2
1

**Figure 7.20** Human resources entropic risk scorer

The scorer makes the assumption that it takes the average worker one year to gain sufficient experience to appreciate and manage workplace-specific risks. Although the assumption is arbitrary, it still allows managers to determine degradation exposure in relative terms. An upward adjustment is made for the percentage of employees within a job classification who have less than one year's experience. The scorer therefore makes the assumption that when the worker is aware of these workplace dangers and is integrated into the organization's safety culture, she will make a conscious effort to be more vigilant to mitigate these risks.

Finally, the scorer makes adjustments for the level of training that the employee group has received. The types of training which prevent entropic risk include knowledge of workplace-specific hazards, risk identification skills, and the ability to manage one's own behavior to prevent systemic risks from resulting in an accident, such as instruction in safe work procedures. The latter also includes behavioral issues like knowing the extent of authority to take corrective action and appreciating when hazards should be reported. Ideally, the work group should also have had training to increase awareness of the effect of non-work-related behavior on safety. This includes lifestyle issues such as drug/alcohol consumption and fatigue management. Each firm may define the criteria by which this evaluation is made to suit its needs. Some companies may like to limit this training to workplace-specific, task-related and hazard management competencies, whilst others may include broader skills such as risk assessment. The key issue is that the criteria should be applied consistently across the organization so that groups with relatively higher risk levels can be identified. If against this criteria, the firm has a highly trained workforce, a downward adjustment is made. The final level is transferred across to obtain the human resources entropic risk score. The risk modifiers for the human resources system factor are therefore job demands, workplace-specific KSAs and training.

Figure 7.20 contains examples of how this scorer may be applied. The personnel in a mobile plant workshop are again compared. The average age by job classification is recorded, as was done for residual risk. Apprentice mechanics, recently qualified mechanics and leading-hand mechanics are employed in the workshop. The leading-hand mechanics are shown with a dotted line. These employees are on a shift work roster with a shift length of 12 hours. The tasks are moderately physically and mentally demanding, so an upward adjustment is made in the job demands column. (The actual adjustment is not a precise measurement and provided that the auditor is consistent, he will be able to determine relative entropic risks by job classification. Significant adjustments can be made in relation to entropic risk where demands are high.) Among the leading hands there are some with less than one year's experience hence a small upward shift is made to allow for this. The group has received some but not all of the required training and a further increase is made. Their final entropic risk score is 5.0.

The recently qualified mechanics are in the 20–24 age category. They also work 12-hour shifts under similar conditions to the leading hands. Their entropic risk level is therefore increased by the same rate. All employees in this group have more than one year's experience and the level is transferred to the 'Training received' column. They are also lacking training in some risk management areas, thus their final score is 8.0.

The apprentices have an average age of 17 years. They only do day shift for 8 hours and are restricted to less demanding tasks in line with their level of competence. As a result, there is only a minor upward adjustment for job demands. Approximately 25 per cent of these employees have less than one year's experience and so an upward adjustment is made. The focus of their training is on task-related abilities and therefore, they have a significant shortfall in risk management skills, as shown by the steeper adjustment in the 'Training received' column. The final score for this group is approximately 9.5. The scorer shows that the experience of the leading hands tends to mitigate the likelihood of entropic risk caused by the demands of the job. When the firm combines the results with other system factors in various activities it may be shown that the company is heavily reliant on the experience and risk awareness of these workers to prevent degradation. The breakdown of the nature of entropic risk in this way may provide clues to the root causes of accidents particularly during incident investigations.

## Total entropic risk scores

Using the entropic risk scorers, a better understanding of the susceptibility of the firm to degradation may be achieved. The first review undertaken may be used to estimate the business' susceptibility to entropic risk. The activities that were developed to analyze residual risk are again used to determine the total entropic risk score. The formula for this, as shown earlier, is:

$$\text{Total entropic risk score} = p \times t \times p_e \times h_r$$

where: p, process entropic risk score; t, technology entropic risk score; $p_e$, physical environment entropic risk score; $h_r$, human resources entropic risk score.

Figure 7.21 proposes a format for the development of activities so that the total entropic risk score for each activity can be calculated.

**(a) Blank table**

| Systems entropic risk profile for workplace: ............................... | | | | | | | | |
|---|---|---|---|---|---|---|---|---|
| Processes | Score | Human resources | Score | Technology | Score | Physical environment | Score | Total entropic risk score |
| | | | | | | | | |
| | | | | | | | | |

**(b) Example**

| Systems entropic risk profile for workplace: Light vehicle workshop | | | | | | | | |
|---|---|---|---|---|---|---|---|---|
| Processes | Score | Human resources | Score | Technology | Score | Physical environment | Score | Total entropic risk score |
| Changing a tire on a light vehicle | 2.5 | Recently qualified mechanic | 7.0 | New tire<br>Jack<br>Wheel brace<br>Tire pressure pump | 2.0<br>2.0<br>1.5<br>3.5 | Light vehicle workshop | 4.0 | 245 |
| | | Apprentice mechanic | 9.0 | New tire<br>Jack<br>Wheel brace<br>Tire pressure pump | 2.0<br>2.0<br>1.5<br>3.5 | Light vehicle workshop | 4.0 | 315 |

**Figure 7.21** The calculation of the total entropic risk score for different activities

A similar process that was used for the residual risk score calculations is used. Each of the system factors in an activity is identified and given a score that is transferred from the respective entropic risk scorers. These scores are multiplied using the above formula to obtain the total entropic risk score for the activity. The example given in the figure calculates this measure for the process of changing a tire on a light vehicle. As was the case with the residual risk score, the highest risk technology is used in the calculation. When this process is undertaken, the exposure of the firm to entropic risk is greater when the apprentice mechanics carry out the work. This does not mean that the apprentices should not undertake the task. The firm needs to look at all activities carried out in the workplace and determine whether the level of risk is acceptable and that in practical terms it can be managed. The results may prompt greater emphasis on supervision and on additional training in manual handling techniques, sound housekeeping and adherence to safe work practices. The results should therefore be used to develop strategies that reduce susceptibility to deviations from safe systems.

The proposed method for developing strategies to address entropic risks follows the same steps as were taken for RMS which were shown in Fig. 7.13. The difference is that step 6, 'Formulate', involves taking corrective action and developing maintenance systems to prevent future degradation. The procedure for EPS development is shown in Fig. 7.22.

In step 1, system factor audits are used to identify potential sources of entropic risk in the workplace. The severity of exposure to this risk for each system factor is determined using the relevant scorer. The activities developed when analyzing residual risks are again used to calculate the total entropic risk score for each activity. These are ranked or graphed to pinpoint areas of greatest vulnerability in the organization's systems. Resources are allocated to higher scoring activities or most vulnerable system factors to address entropic risk in two ways. The first is to take corrective action in the short term to eliminate the risk as far as practicable and the second is to implement maintenance strategies to prevent degradation in the future. The strategies, initiatives, action plans and measures developed should then be included in the productive safety management plan. The results of such interventions may be evaluated at the end of the cycle to determine their effectiveness and to identify opportunities for continuous improvement.

Entropic risk results may be manipulated to provide management with additional information to assist the development of EPS. The firm should determine methods to meet its information needs. Examples of how the results may be used include summing the entropic risk scores for each system factor to identify the factor that has the greatest potential for degradation. The average system factor scores for each workplace can also be compared. This allows managers to more confidently identify those workplaces that have the highest tendency towards entropic risk. Resources for preventative strategies including training, equipment maintenance, housekeeping, and process management may be directed to these areas.

An example of this data manipulation is provided in Table 7.3. In example A, the method of summing system factor scores and averaging

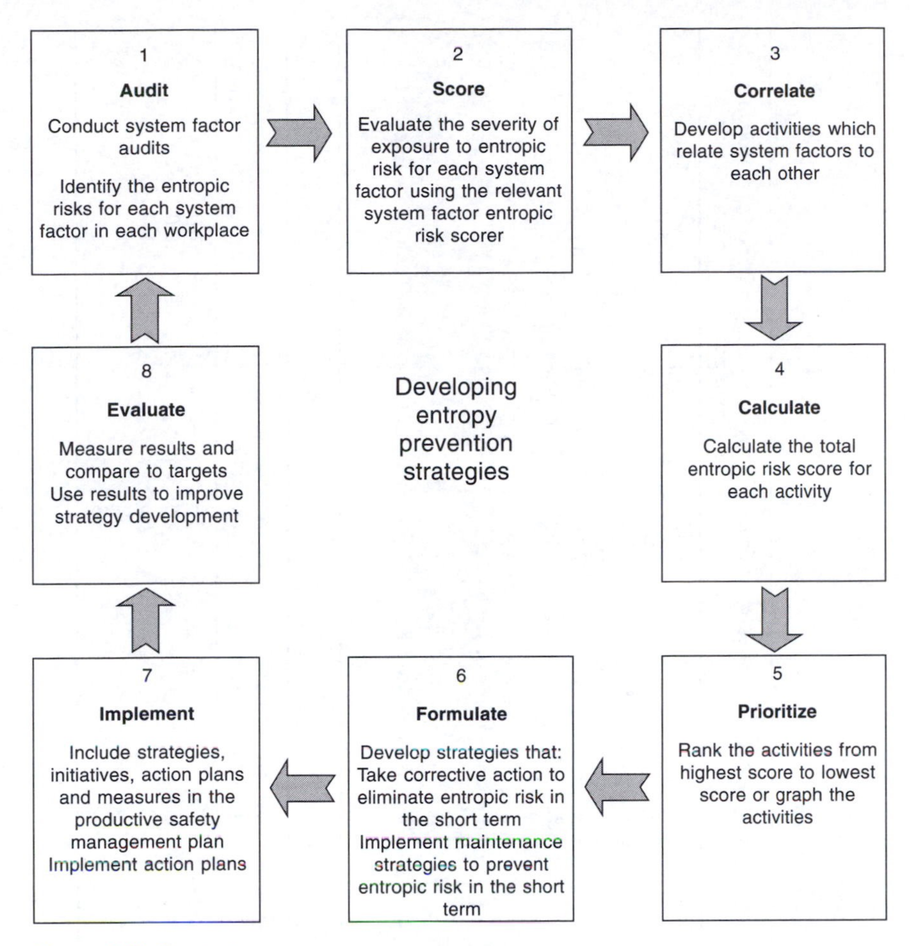


**Figure 7.22** Developing entropy prevention strategies

them is shown. The averages indicate which system factor is the most likely source of degradation within a workplace. Knowing this allows managers to be more vigilant to control these systemic vulnerabilities. These average scores can be used as the baseline level of entropic risk against which future results are compared. The baseline levels of entropic risk are the average of the entropic risk scores for each system factor for all activities undertaken in a given workplace. The formulae are:

Baseline level of process entropic risk

$$= \frac{\Sigma(\text{Activity } 1P_{ER}\text{Score} + \text{Activity } 2P_{ER}\text{Score} + \ldots \text{Activity } NP_{ER}\text{Score})}{N}$$

where $P_{ER}$ is the process entropic risk score.

Baseline level of technology entropic risk

$$= \frac{\Sigma(\text{Activity } 1T_{ER}\text{Score} + \text{Activity } 2T_{ER}\text{Score} + \ldots \text{Activity } NT_{ER}\text{Score})}{N}$$

**Table 7.3** Using the systems entropic risk profile to compare system factor risk levels

*Example A: Systems entropic risk profile for workplace: Light vehicle workshop*

| Processes | Score | Human resources | Score | Technology | Score | Physical environment | Score | Total risk score |
|---|---|---|---|---|---|---|---|---|
| Changing a tire on a light vehicle | 2.5 | Recently Qualified Mechanic | 7.0 | New tire<br>Jack<br>Wheel brace<br>Tire pressure pump | 2.0<br>2.0<br>1.5<br>3.5 | Light vehicle workshop | 4.0 | 245 |
| | 2.5 | Apprentice Mechanic | 9.0 | New tire<br>Jack<br>Wheel brace<br>Tire pressure pump | 2.0<br>2.0<br>1.5<br>3.5 | Light vehicle workshop | 4.0 | 315 |
| Activity 3 to Activity 30 | Sum ↓ | | Sum ↓ | Sum highest score of the technologies used | Sum ↓ | | Sum ↓ | |
| TOTAL SCORE for system factor | 141 | | 210 | | 105 | | 180 | |
| AVERAGE SCORE (divide by 30) | 4.7 | | 7.0 | | 3.5 | | 6.0 | |

*Example B: Average entropic risk scores for four sites at an open-cut mine*

| Workplace | Average score, processes | Average score, human resources | Average score, technology | Average score, physical environment |
|---|---|---|---|---|
| Light vehicle workshop | 4.7 | 7.0 | 3.5 | 6.0 |
| Open-pit mining area | 6.0 | 5.0 | 7.0 | 7.0 |
| Stores office and delivery area | 3.0 | 2.5 | 6.0 | 5.0 |
| Administration office | 5.0 | 3.0 | 3.5 | 1.5 |

where $T_{ER}$ is the technology entropic risk score.

Baseline level of physical environment entropic risk

$$= \frac{\Sigma(\text{Activity 1}PE_{ER}\text{Score} + \text{Activity 2}PE_{ER}\text{Score} + \ldots \text{Activity N}PE_{ER}\text{Score})}{N}$$

where $PE_{ER}$ is the physical environment entropic risk score.

Baseline level of human resources entropic risk

$$= \frac{\Sigma(\text{Activity 1}HR_{ER}\text{Score} + \text{Activity 2}HR_{ER}\text{Score} + \ldots \text{Activity }HR_{ER}\text{Score})}{N}$$

where $HR_{ER}$ is the human resources entropic risk score.

The baseline levels of entropic risk can only be used as a guide in OHS decision-making because of the variability of entropic risk. When used in conjunction with the risk scorers they may provide an additional insight into the underlying causes of incidents. If an accident occurs, assessors may refer to the scorers to gain a better understanding of the factors that potentially contributed to the event. The baseline scores are intended to provide a trigger to investigate how these scores were derived which in turn, may help to identify symptomatic problems and underlying systemic weaknesses. The method aims to prompt the investigator to uncover the organizational conditions and management practices that led to degradation in the first place.

In Table 7.3, example B provides the average entropic risk scores by system factor for four workplaces at an open-cut mining operation. The relative scores indicate areas of greatest vulnerability, however, to derive maximum benefit from the data, assessors need to look at how the scores were reached. In the light vehicle workshop, the two system factors with greatest exposure to entropic risk are human resources and the physical environment. The former may be high because of the low average age of employees and the groups' susceptibility to fatigue resulting from shift work. This would suggest that training and fatigue management practices such as monitoring employee alertness, work schedules and overtime management, are key issues for this area.

This hypothetical case shows that the open-pit area has the highest level of process complexity. In this workplace, drilling and blasting, excavation and ore transport activities are undertaken. The development of safe work practices is therefore critical to risk management, not only in relation to how a particular task is undertaken but how it affects the other processes that are being carried out concurrently. For example, how does the level of risk change when survey work is undertaken adjacent to an area being excavated? How is entropic risk affected when the number of vehicles operating in the pit increases? For instance, on windy days water trucks need to be used to suppress dust levels, whilst other heavy machinery such as trucks and shovels continue to work in the area. On a larger scale, when multiple, complex processes are undertaken in a given area then the vulnerability to entropic risk must also rise.

In addition, in Table 7.3 the scores for other system factors in the pit are also high relative to other workplaces, which indicate that resources should be directed at prevention of entropic risk in this environment. Again it is important to go back to the risk scorers to determine the underlying variables that contributed to these scores. For instance, it may be found that the technology used is aging, so preventative maintenance needs to be a high priority. In addition, if the pit is getting deeper, then vehicle travel distances, and therefore, usage rates may be on the increase, accelerating the rate of degradation of the technology system factor. With this greater depth, dewatering, slope stability and dust suppression become increasingly difficult so the physical environment of the mine has a greater level of susceptibility to entropic risk. Preventative maintenance to address this risk could include increasing the number of site inspections, increasing the use of water trucks to control the dust problem, erection of barriers to prevent access to eroding areas, and implementing a road maintenance plan. The scores allow firms to develop a better insight into the nature of risk in each business unit and therefore devise workplace-specific strategies that make more effective use of limited resources.

A further example illustrates how these scores can be used to develop EPS. In the table, hypothetical results are given for a stores office and delivery area. They suggest that the processes undertaken there are relatively simple and carried out by experienced, mature workers. The technology system factor has the highest entropic risk score of 6.0. Investigation may reveal that forklifts and the other equipment have reached their half-life and have been poorly maintained, so there is a need to improve equipment maintenance in this area. The score for the physical environment is also quite high at 5.0, which may be attributable to insufficient workplace inspections. Specific strategies to prevent degradation would therefore include additional resources for equipment maintenance and the introduction of tighter inspection procedures.

The final work area shown is the administration building which has the lowest exposure to entropic risk, although its processes appear to be more complex and less standardized than the light vehicle workshop and stores office. The nature of this complexity is important. The work is mentally but not physically demanding, so does not pose a significant immediate health and safety risk. This score, however, indicates that the complexity of the tasks may place demands on workers which affect their performance. Strategies to address this issue may include balancing the proportions of complex versus simple tasks and ensuring that stress-related factors are addressed by developing a positive workplace culture.

## Total risk profile

The data from these risk analyses can be used further to develop a total risk profile for the operation based on system factor scores for each workplace. This involves graphing the combined effect of both categories of risk by system factor for each area. The summary profile gives the firm a picture of its total level of risk and also illustrates the impact of residual risk on entropic risk. As discussed in the previous chapter, workplaces that have a high level of inherent danger place additional demands on other system factors. On a construction site, for example, the high residual risk of the physical environment requires the worker to be more vigilant and consistently proficient than in a low residual risk workplace. There is less margin for error. The potential consequences of failing to maintain alertness and competence are hence more severe and concurrently, these constant demands increase the likelihood of employee degradation.

Whilst the firm should develop its own methods for manipulating the data, the following examples may act as a guide. Integrating both residual and entropic risks into one graph allows the auditor to consider the total risk that occurs when system factors interact. In particular, it assists managers to appreciate the demands that the organizational system places on workers. It also provides an indication of the employees' capacity to cope with these demands. A workforce that is mature, experienced and well trained has a lower level of residual risk than does a young workforce. They are thus better equipped to cope with workplace hazards in the sense that they are generally more likely to manage risks by following safe work practices, are able to appreciate changes in the severity of hazards, and contribute to risk management. Other matters that affect susceptibility to entropic risk include work hours and rosters. For instance, there will be less tendency towards degradation if employees work an 8-hour day than if they work 12-hour shifts. The implication is that it becomes increasingly important to minimize human resource risks in operations that have high levels of risk in other system factors. Specifically, this includes businesses with dangerous physical environments, high residual risk technologies and/or complex processes.

Figure 7.23 contains a graph that combines the residual and entropic risk scores for various workplaces in an open-cut mining operation. As shown earlier in Fig. 7.16, the level of residual risk is constant in the short term, whereas, entropic risk is the variable component of the total risk. Management needs to be aware that entropic risk scores are only an estimate of the relative exposure to entropic risk. The potential for deviations in entropic risk cannot be measured precisely as such because there are innumerable contributing variables that may cause changes in this type of risk. There can also be considerable fluctuations within each system factor, for example, two units of equipment are not exactly the same and two workers will never be the same.

In Fig. 7.23, the total risk profiles for a light vehicle workshop and an open-pit area at a mining company site are provided. These indicate that the residual risks in the processes, technology and physical environment are higher in the mining area than in the workshop. In addition, the business has a higher exposure to entropic risk in these system factors in the pit. In the example shown, this may be attributed to the variability and lack of standardization of processes, aging technologies and increasing depth/complexity of the pit, however, to be certain the assessor needs to reconsider how the scores were derived. The results suggest that the firm has experienced, well-trained workers operating in this excavation area. The firm may be relying on the competence and vigilance of its workers to manage the risks associated with the physical environment and

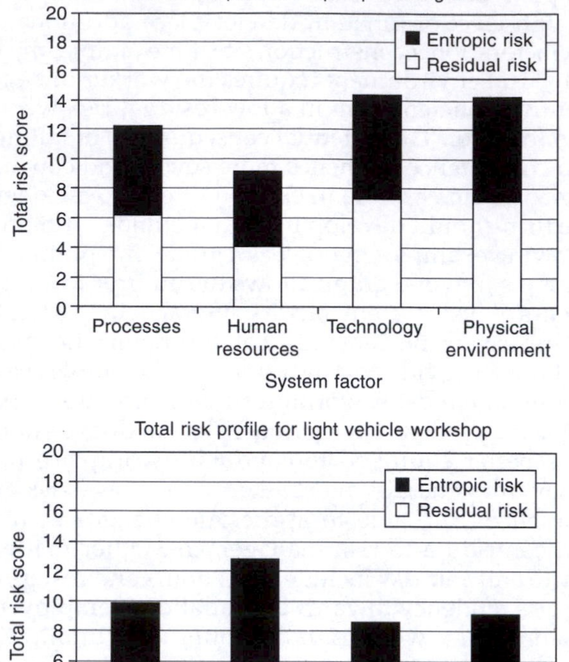


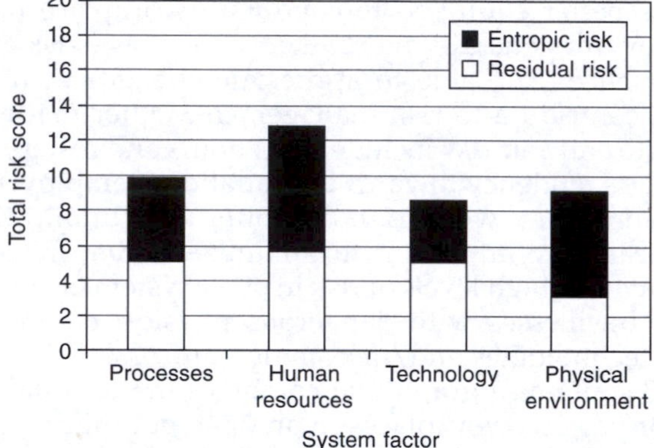


**Figure 7.23** The total risk profile for a business operation

technology, which means that the company cannot afford the entropic risk in the human resources system factor to rise. If this is the case, then strategies need to be developed to ensure that the work and management system do not place additional demands on workers and that such systems allow them sufficient recovery time. This includes effectively managing work rosters, overtime and job demands which potentially increase the rate of degradation.

In the workshop, the residual and entropic risks associated with the workforce are a significant concern. It may be concluded that this is because of the lack of experience of this employee group. A primary focus of risk management in this area would therefore be training and the quality of supervision. There may also be a need to implement more stringent housekeeping procedures and infrastructure preventative maintenance to eliminate the entropic risk associated with the physical environment, which also appears to be high.

Results manipulation like the total risk profile allows the firm to identify priority risk management areas so that it is better able to make effective use of limited resources. The strategies developed as a result of the analysis of the company's residual risk profile and susceptibility to entropic risk should be included in the productive safety management plan. These strategies, for example, 'To eliminate the entropic risk in the physical environment of the light vehicle workshop' may be broken down into specific initiatives such as, 'To develop a building maintenance schedule' and 'To implement a housekeeping policy and procedures'. To make them implementable, initiatives need to be further detailed into action plans that outline specific tasks, those personnel responsible for carrying out these tasks and the relevant dates for completion. Against the initiatives, measures need to be formulated that assess whether the action plans have been completed on time and within budget, and whether the strategy has been effective in reducing the level of risk. (These processes are discussed in detail in Part 4.)

If a firm does not undertake quantitative risk analysis can it develop appropriate strategies? Interventions can be formulated without this level of analysis by using the risk scorecard described in Chapter 3, however, there are two key benefits of evaluating these risks more thoroughly beforehand. The first is to initially improve management's understanding of the nature of the risks the business faces. The second is that the analysis provides baseline levels of risk against which future results can be compared. In this way the firm is able to evaluate the effectiveness of its OHS decision-making and its strategies in terms of risk reduction and control. It can also better appreciate how its risk levels have changed over the life of the operation.

If, however, the company wants to begin controlling risks without undertaking the full auditing process, it has two avenues for strategy development. In the first place it can implement the four-fold strategy whenever and wherever hazards are identified. The second option, which may be applied concurrently, is to develop interventions according to primary risk parameters and risk modifiers. The residual risk and entropic risk scorers identified these for each system factor. The scorers suggest that hazard control begins by reducing the primary risk parameter as far as practicable. For process residual risks, for instance, this parameter is the energy level, therefore, the firm should consider methods which contain or minimize this energy or which reduce the consequences of failures. An example is implementing speed limits in high-risk and heavy traffic areas within the company site. The potential consequences of failures are lessened because speed reduction minimizes the severity of impact and also increases the time that the operator has to react to changes in conditions. To reduce the likelihood of degradation, the firm may limit the numbers of vehicles or personnel allowed in high-risk areas and thereby reduce complexity.

The residual risk scorers also contain risk modifiers. These indicate how risk can be managed in the short term and compressed in the longer term. The technology residual risk scorer, for example, identifies 'in-built safety features' as positive risk modifiers. RMS thus include identifying methods/modifications that make equipment safer, such as attaching tools on elevated work platforms to the infrastructure to prevent injury or damage from falling objects. Other safety features include equipment monitors and warning lights that prevent inadvertent access to dangerous components or hazardous work areas.

Some risk modifiers have a positive impact on hazards, whilst others have a negative impact. The firm needs to consider how these can be influenced to reduce the level of inherent danger and systemic vulnerability. For instance, in Fig. 7.10 which illustrates the physical environment residual risk scorer, exacerbating hazards were identified as a negative risk modifier. RMS may include establishing criteria that define when it is too dangerous to work. In the construction site example given in the figure, this could include allocating ground-level tasks to builders when wet weather and high winds make working at heights unacceptably risky.

The human resources residual risk scorer, given in Fig. 7.11, indicates that 'average age by job classification' is the primary risk parameter. The firm should not and legally cannot discriminate against job applicants based on age. The purpose of the scorer is to allow the firm to identify groups with higher training and supervision needs and therefore make more effective use of training dollars. The physical demands risk modifier indicates that jobs should be designed to fit the limitations of the human body. Appropriate RMS include job rotation, job enrichment, increased use of mechanized aides for manual lifting, and establishment of procedures to reduce the strain of these physical demands on the worker.

This scorer also indicates that employee turnover – expressed for this purpose as the percentage of employees with less than one year's experience in the workplace – is a significant negative risk modifier. This highlights that firms that suffer high turnover should establish strategies to minimize the leakage of workplace-specific KSAs. Turnover is traditionally considered to be an HRM issue, but its direct effect on safety indicates that it should concurrently be considered an area for OHS strategy development.

The residual risk scorers indicate that RMS include two levels of intervention. The first is to address the primary risk parameter as far as practicable. The second is the development of strategies that have a positive impact on the relevant risk modifiers. The same rationale is applied to control entropic risk. The process entropic risk scorer, given in Fig. 7.17, for instance, identifies complexity as the primary risk parameter. This indicates that, where possible, processes should be simplified. Where this cannot be done, it is important to apply JSA to identify potential hazards and to manage these through work practice controls and increased vigilance. The risk modifiers identified in this scorer are 'standardization and adherence' and 'regularity, thoroughness and follow-through of inspections'. The scorer indicates that firms should ensure that JSA is used to develop safe work procedures and that supervisors and employees take responsibility for workplace inspections.

The importance of inspection procedures is evident in each entropic risk scorer as a strategy to mitigate this type of risk. The other modifiers include preventative maintenance of technology and the physical environment. The human resources entropic risk scorer indicates that when the firm begins to address its residual risks in this area by adapting its management strategies to the competence and experience level of its work groups, it also reduces its exposure to entropic risk. For example, the appropriate management style for young inexperienced workers would involve close supervision, on-the-job instruction to enhance KSAs, extensive coaching and support. In contrast, for highly experienced, competent workers, a suitable management style may include the development of autonomous work teams with increased accountability and participation in workplace decision-making. According to this scorer, the modification of work practices to reduce physical demands, the management of turnover, and the implementation of comprehensive training systems contribute to the reduction of degradation in the human resources system factor.

Table 7.4 summarizes the types of strategies that the firm can begin to implement before a full auditing process is undertaken. The table identifies the primary risk parameters and risk modifiers associated with residual and entropic risks for each system factor. Examples of appropriate strategies are also provided. For instance, for the residual risk associated with technology, the primary risk parameter is the energy/toxicity level of the product. The risk modifier for this hazard source is 'in-built safety features'. If the technology is a chemical stored on site, then ideally the toxicity level should be kept as low as possible. It may be stored in unconcentrated form and the storage equipment designed to prevent contamination. Monitoring systems may be installed to reduce the risk of consequential damage from failures and the area isolated to prevent access by unauthorized personnel. The summary in Table 7.4 proposes a strategic direction, additional to the four-fold strategy, for risk management. It is intended to allow the firm, in the first instance, to concentrate on addressing the primary risk parameter and risk modifiers, so that residual and entropic risks are better controlled in organizational systems.

## Summary

The method presented in this chapter proposes an alternative to the use of the risk scorecard and risk assessment code as methods of prioritizing hazards. In particular, the technique attempts to capture a complex set of variables that are the underlying sources of residual risk and degradation in business systems. It is primarily an educational tool to assist managers in their OHS decision-making so that they can more confidently allocate resources to optimize risk reduction and control strategies. In addition, the scorers in particular are intended to act as a prompt for investigators to uncover the underlying systemic weaknesses that lead to workplace incidents.

This is the final chapter of Part 2: Systems change. The following part will present the interventions for behavioral change. These consider issues affecting the organizational culture, such as management commitment and leadership, sharing ownership of productive safety as a key result area, training and behavioral audits. These behavioral change strategies are designed to support the systems change and risk management strategies identified thus far. In Part 4 the initiatives resulting from the risk assessment and cultural development processes will be included in the productive safety management plan.

**Table 7.4** Primary risk parameters and risk modifiers by system factor

| Risk type | System factor | Primary risk parameter | Risk modifiers (potential effect on risk level) | Strategy | Examples |
|---|---|---|---|---|---|
| Residual | Processes | Energy level | Nil | Residual Risk management | • Speed controls<br>• Accessibility controls<br>• Restriction of duties to qualified personnel |
| | Technology | Energy/toxicity level | In-built safety features (+) | | • Purchasing controls<br>• Equipment modifications<br>• Monitoring systems<br>• Isolation of toxic substances |
| | Physical environment | Energy/threat level | Exacerbating hazards (–) | | • Definition of safe work conditions<br>• Monitoring systems |
| | Human resources | Average age by job classification | Physical demands (+ or –)<br>Lack of workplace KSAs (–)<br>Training (+ or –) | | • Alternative duties<br>• Reduced manual handling<br>• Increased supervision<br>• Mentoring<br>• Job rotation<br>• Training plan |
| Entropic | Processes | Complexity | Standardization and adherence (+ or –)<br>Regularity, thoroughness and follow-through of inspections (+ or –) | Entropy Prevention | • Job safety analysis<br>• Procedural training<br>• Increased supervision<br>• OHS representatives and consultative strategies |
| | Technology | Age/usage rate | Preventative maintenance (+ or –)<br>Regularity, thoroughness and follow-through of inspections (+ or –) | | • Preventative maintenance plan<br>• Shared delegation of OHS inspections to engineers (process and mechanical) |
| | Physical environment | Age/condition | Preventative maintenance (+ or –)<br>Regularity, thoroughness and follow-through of inspections (+ or –) | | • Housekeeping plan<br>• Shared delegation of OHS inspection to engineers (environmental and geotechnical) |

**Table 7.4** *(Contd)*

| Risk type | System factor | Primary risk parameter | Risk modifiers (potential effect on risk level) | Strategy | Examples |
|---|---|---|---|---|---|
| | Human resources | Average age by job classification | Job demands (–)<br>Lack of workplace KSAs (–)<br>Training received (+ or –) | | • Competency and value-based recruitment strategy<br>• Mentoring<br>• Increased supervision<br>• Training plan<br>• Training evaluation |

## References

1. Standards Association of Australia (1999) *Risk Management AS/NZS 4360:1999*. Standards Association of Australia, New South Wales, Australia.
2. Stellman, J.M. (ed.) (1998) *Encyclopedia of Occupational Health and Safety*, 4th edn. International Labor Office, Geneva.
3. Mathews, J. (1993) *Health and Safety at Work – Australian Trade Union Safety Representatives Handbook*. Pluto Press Australia, New South Wales, Australia.
4. Four crushed to death in mine (1999) Article dated 25 November. Available on website: http://archive/news.com.au/news/_content/state_content/4368279.htm
5. Department of Mineral Resources (New South Wales) (2000) The aftermath of the Northparkes tragedy. *Mine Safety News*, February. Also available on website: http://www.minerals.nsw.gov.au/safety/msn4/msn4_1.pdf
6. Worksafe Western Australia (1994/5) *State of the Work Environment No. 26: Injuries and Diseases in Young Workers, Western Australia, 1994/95*. Available on website: http://www.safetyline.wa.gov.au/PageBin/injrstat0052.htm
7. National Occupational Health and Safety Commission (2001) *Occupational Health and Safety Management Systems – A Review of their Effectiveness in Securing Healthy and Safe Workplaces*. April 2001. Available on website: http://www.nohsc.gov.au/Pdf/OHSSolutions/ohsms_review.pdf

# PART 3
# Behavioral change

In Part 1, the tools of the productive safety approach – the entropy model and the strategic alignment channel – were presented. In Part 2, the model was applied to the organizational system and the sources of residual and entropic risk within each system factor identified. The quantitative risk management method proposed in Chapter 7, provided an insight into the underlying factors that contribute to incidents. As explained earlier, the intent of this method is to facilitate OHS decision-making that leads to better use of limited resources. In addition in Part 2, the traditional tools of OHS management were restructured to fit the model. These restructuring and risk management strategies were the bases of systems change.

Safety is not, however, purely a structural or administrative process. It also has a social dimension that embraces macro-issues, such as the organizational culture, and micro-issues including supervisor–worker relations and individual behavioral choices. For an OHS management system to be effective, therefore, it has to include strategies to bring about behavioral change. To achieve optimal performance and safety, the people dimension or culture of the company has to be compatible with system modifications. This involves building a management environment that reinforces production, safety and system factor quality as compatible goals, driven by leadership that seeks mutually beneficial outcomes for key stakeholders. Primarily this requires the firm to focus on alignment of its systems and benefactor interests in accordance with the strategic alignment channel.

Whilst the earlier parts of this book explained how the firm can attain external strategic alignment through legal compliance and social responsibility, this part focuses on internal strategic and internal goal alignments. These issues are addressed using HRM systems. In the following part, those systems which develop a positive safety culture are described. One of the key issues discussed is the development of competencies and attitudes, which comprise the behavioral dimension of organizational performance and safety.

Productive safety management is therefore a total management system, which provides both the management systems and the behavioral modeling tools required to pursue production and safety concurrently. The rationale behind integrating the behavioral dimension of safety into the management system can be best explained using an analogy. A purely structural approach is like planning a journey from point A to point B. In the planning stage, management has to benchmark its position in terms of current levels of safety, performance and quality, and identify the target – what it expects to achieve. In reality, however, the attainment of end goals is dependent on the input of each member of the organization and therefore, the journey requires collective effort and commitment. When the goal is not shared and employee buy-in not obtained, the process is more difficult and the objective less attainable. It is hindered by resistance. In addition, there is a danger of developing an adversarial relationship between management and workers if the system becomes quantitative without considering the human factor or qualitative issues. The expedition is therefore not only concerned with reaching the destination, it is equally concerned with how the firm gets there. Whilst systems change focuses on the 'ends', behavioral change focuses on the 'means' – how the business intends to reach its objectives. For this reason, productive safety management, in addition to addressing structural and administrative issues, involves the development of strategies to build a conducive organizational culture that acts as a vehicle for progress.

The primary tool to develop this culture is the strategic alignment channel. The channel indicated that HRM functions such as recruitment and selection, training and development, and reward programs have to be aligned with the OHS system. They have a critical role in improving the quality of human resources and the return on human capital. The entropy model shows that this quality has a direct impact on the firm's level of risk and its capacity to manage hazards associated with processes, technology and the physical environment. In addition, workforce competencies affect whether productivity goals can be achieved and the firm's potential to attain on-going improvement.

Accordingly, OHS and HRM disciplines have to be aligned and this occurs in two ways. The first is that HRM practices have to be compatible with the OHS management system. This involves, for example, implementing training systems driven concurrently by productivity, safety and quality. In addition, position descriptions have to identify accountabilities in organizational roles so that employees gain a sound understanding of their contribution to these outcomes. HRM tools therefore help to operationalize the OHS system. The second means of alignment is for HRM practices to reinforce desirable behaviors and positive results. This includes implementing incentive programs based on production, safety and system factor quality, using a measurement methodology that weights these outcomes to achieve balanced objectives. The balanced approach ensures that short-term gains are not pursued to the detriment of long-term viability. In addition, in the present time frame, production does not override safety and quality.

What implications does productive safety management have for HRM as a discipline? This field of management has emerged over the years from its personnel administrative roots that were closely tied to the payroll function in organizations. With the emergence of downsizing, total quality management and business re-engineering, HRM has had an increased role in decision-making which affects the return on human capital. This has included determining the 'right' numbers and mix of workers to be employed, identifying the competencies the firm requires for future improvement, restructuring, the repackaging of remuneration, and cost control. These organizational development programs brought HRM into the strategic echelons of management. Concurrently, with this 'hard' business approach, HRM sought to develop 'soft' employee-centered initiatives to enhance the productivity and the social aspects of work, such as job satisfaction and intrinsic (nonmonetary) reward factors. The current dilemma over the role of HRM is a consequence of the internal conflict that has emerged from the incompatibility of these 'hard' and 'soft' approaches.

Productive safety management clarifies the role of HRM using the strategic alignment channel, which shows that the strategic level of this discipline is concerned with appropriate resourcing so that internal strategic alignment is achieved. The operational level focuses on developing practices built on common goals and values leading to internal goal alignment. The channel, therefore, explains the duality of HRM and its place within the total management system. Firstly, it assists management to identify human capital needs. HRM therefore actively involves the acquisition of this capital from the marketplace, the investment in training and development to enhance the quality of this capital, and reward systems to recognize the contribution of this resource to organizational success.

The second perspective of HRM is to build an organizational culture that addresses the core needs of both management and workers. This is based on a relationship of mutual benefit; specifically, shareholders' and management's dependence on the workforce for the supply of their competencies and time, and the workforce's dependence on the firm for economic returns and social purpose. It is the author's opinion that HRM practice has been less than fully effective because of weaknesses at the interface between strategy development and operational practice. This is because of lack of custodianship and advocacy of human capital by some professionals in this discipline. Fundamentally, the value of human capital has been undersold. Accounting practices that record labor as a cost and physical capital as an asset have hindered this further. In the future, therefore, there needs to be greater efforts to quantify the financial benefits of HRM programs and the value of human capital.

Hence, the role of HRM must be 'custodian and advocate of human capital'. This involves reinforcing that the quality of human resources has a direct and significant impact on both safety and organizational performance. Primarily the focus of HRM should be to acquire, develop and retain valuable human capital by implementing change processes that bring about positive behavioral modification and alignment of organizational and employee goals and values.

Any successful change strategy needs commitment from management. Training to give supervisors the 'know-how' to implement the system is of primary importance. More significant however, is 'know-why' – appreciating the rationale and philosophy driving the change process and an internal affirmation that the underlying values of the approach are compatible with individual values. A fundamental ethos of productive safety management is that it is important for managers to believe in the system. This leads to appropriate decision-making and management actions that build and sustain a positive organizational culture. It is the basis of sound leadership, and in Chapter 8, the process of gaining management commitment and stewardship will be discussed. In addition, the chapter will describe how to share ownership and accountabilities with employees to make the system inclusive.

Management training is the first stage of productive safety management. It is also the first step in improving the quality of the firm's human capital – expressed as the collective competencies of employees. Under this approach, the HRM training and development function focuses initially on building effective organizational leadership at management level before spreading these skills throughout the hierarchy. In Chapter 9, the training function is discussed further and likened to building a 'reservoir' of competencies. After sufficient investment, the reservoir provides additional returns to the firm through employee contributions. As the firm invests in competency building, it increases its organizational learning potential. Employees become more receptive to learning when they understand the rationale behind the changes that are being made. The spillover of investment in human capital, referred to as 'resourcefulness', leads to better problem solving at both the strategic and operational level.

The productive safety formula is also presented in Chapter 9. It shows a further link between the OHS and HRM disciplines through the relationship between entropy and resourcefulness. It explains that companies that suffer susceptibility to degradation have to rely increasingly on the skills, vigilance and contributions of their employees to take corrective action and to develop preventative strategies. The final chapter of this part – Chapter 10 – explains how to audit the management system from the behavioral change perspective to determine the effectiveness of cultural change strategies. An overview of the elements of behavioral change that will be discussed in this part is provided in Fig. (iii).

| *Behavioral Change* |
| --- |
| Productive safety culture |
| Human resource management systems |
| Shared ownership |
| Safety issues resolution |
| Capacity reservoir – building competencies and vigilance |
| Accountabilities and responsibilities |
| Management commitment and leadership |
| Appropriate supervision/decision-making |
| Productive safety formula |
| Resourcefulness – maximizing learning potential |

**Figure (iii)** Overview of behavioral change elements

In Part 4, the strategies for systems and behavioral change will be consolidated into the productive safety management plan. Central to the plan is the performance management system that rewards achievement of organizational objectives and evaluates the effectiveness of the firm's strategic interventions.

Chapter 8

# Management commitment and leadership

Organizational change strategies require commitment from management to make them work so the first stage in developing a productive safety culture has to involve building management support and leadership. It is important to achieve this before implementing any other element of the approach. If the OHS management system is introduced prior to gaining this commitment, inappropriate managerial behavior can derail the system and also damage the relationship between management and workers. In addition, middle managers may feel that the change is being thrust upon them and this can lead to distrust.

There is compelling research demonstrating that management leadership is the key to high safety performance. Companies with the lowest lost-time injury rates are those with the highest level of management commitment and employee involvement.[1] High-performing firms have a number of traits. Firstly, safety and health efforts are equal to other organizational concerns and are fully integrated.[2] Secondly, their cultural characteristics include specific management behaviors which improve safety and performance such as treating employees with respect, providing positive feedback and encouraging suggestions. These firms also effectively resource the expert knowledge of OHS professionals in the areas of purchasing, processing and other issues affecting employee safety. They therefore have a multidisciplinary approach to risk management facilitated by a climate that values open and explicit communication. In addition, resources are provided for safety training and management development to reflect their priority status as organizational strategies.

## Management training

The process of building management commitment begins with an executive training program that occurs in three stages. Phase 1 is knowledge-based training, referred to as 'know-what', which provides an understanding of the rationale and components of the system. This focuses on explaining the tools and implementation structure. This phase also presents the components of systems change, discussed in earlier chapters, including an overview of the restructuring of traditional OHS practices to fit the entropy model and the development of risk management strategies. Phase 1 provides managers with the knowledge required to implement effectively the structural components of the OHS management system. 'Know-what' lays the foundations of 'system thinking'. The aim of this phase of training is to deliver the first two of six competencies which fit the emerging paradigm of safety leadership identified as appropriate in the current business climate.[3] These core skills have a significant impact on the effectiveness of OHS management systems and the safety culture. The six broad competencies are:

Competency 1: The ability to think in terms of systems and knowing how to lead systems.
Competency 2: The ability to understand the variability of work in planning and problem solving.
Competency 3: Understanding how we learn, develop and improve, leading to true learning and improvement.
Competency 4: Understanding people and why they behave as they do.
Competency 5: Understanding the interaction and interdependence between systems, variability, learning and human behavior; knowing how each affect each other.
Competency 6: Giving vision, meaning, direction and focus to the organization.[1]

These proficiencies have been identified as critical to creating positive behavioral change. Applied to productive safety management, the first competency includes having a sound understanding of the entropy model, the four-fold strategy and the strategic alignment channel. Competency 2 requires that the manager understands that problems are a direct result of ineffective or inefficient parts of the system.[3] Thus, they come to understand that incidents are primarily caused by systems failures not simply human error.

The second phase of the executive training program is value-based training. It provides managers with competencies 3 and 4 which focus on leadership and people skills.

There are a number of steps needed to build values-driven leadership. Firstly there needs to be commitment to values.[4] This is followed through with communication and education. These values also need to be linked to agreed standards of behavior. To create alignment, the systems in the organization, including those for recognizing performance, need to be linked to these values and standards. The rationale is to have a strong underlying framework for behavior and decision-making supported by procedural systems.

Value-based training should be designed to take the executive team through a process of uncovering their underlying beliefs. This can be facilitated through simple learning activities that involve, for instance, identifying those variables that lead to job satisfaction. This is important as it helps managers to identify the dimensions of work that motivate

them. These types of activities tend to generate results that fall into two categories. The first is task-related, for example, achieving a certain goal such as completing a project, making an effective presentation or correcting a technical problem. The second category tends to include relationship-based factors, such as receiving positive feedback from a superior, chairing a meeting in which employees felt enthusiastic about a project, or helping a subordinate to overcome a problem. The identification of both task and relationship components is a significant step in establishing a fit between managers' values and the philosophy of productive safety management.

The process is intended to be a discovery exercise to uncover and analyze the management team's values. In particular, it should help the team determine whether their values are unique to the group or more broadly embraced in the social environment of the firm. They should therefore be challenged to consider whether other company employees hold similar beliefs at a fundamental level – bearing in mind that both managers and shop floor workers are employees of the firm. The aim of this training is to reveal that management and workers share many goals and values. When managers appreciate that there is a common ground, this provides the starting point for a culture based on mutually beneficial outcomes driven by clearly defined behavioral standards. Successful managers and companies place a great deal of emphasis on values.[5] They stand for something, align their beliefs with the external environment and share them internally. Harmony is achieved because the values of the firm are consistent, are aligned with the behavior of its members, and therefore, boundaries that define appropriate actions also tend to be consistent. This harmony is important because it will:

(1) Provide a sense of common direction for all staff and guidelines for their daily behavior.
(2) Provide the social energy and esprit de corps that moves the organization into action.
(3) Permit upper management to influence employee behavior without being present physically.
(4) Provide a framework for managerial decision making.
(5) Provide a sense of stability and continuity in a rapidly changing environment.[5]

Examples of common values include safety, autonomy, involvement, respect for people and self-improvement. The linking of these to productive safety management, as both an ideology and a management system, provides the executive team with 'know-why'. By identifying their own values, appreciating that these values are broadly held within the workforce, and realizing that this management system is driven by these common beliefs, acceptance of the approach as an appropriate means of pursuing organizational change and improvement is more likely to be developed and sustained.

Once phase 2 has been completed, the management team should be ready to commit to change. The approach should be accepted as both an appropriate methodology for pursuing production, safety and quality (as an 'end') and as a vehicle for developing a positive safety culture (as a 'means'). At this point, they are ready to operationalize productive safety as a key result area (KRA) of the business.

Key result areas define the firm's business activities into core focal points that are critical to the success of the firm. A KRA is a subdivision of an organizational area which, regardless of its nature, has a common purpose that directly relates to the operational goals and objectives of the firm.[6] Productive safety is a linking KRA in that it affects all branches of the business including manufacturing, marketing, financial management and product development. After phases 1 and 2, the management team should have foundational knowledge of this approach and be able to identify the link between the system and their own values. The training discussed thus far is therefore knowledge-based and value-based, as summarized in Fig. 8.1. The figure illustrates the management training pyramid which shows conceptually that management skills development needs to occur in a serious of phases with one built on to the next. These early steps center on understanding systems, appreciating values and motivation, before the necessary behavioral competencies are acquired.

The completion of phase 2 requires managers to translate values-based learning into something tangible – something that they can realize in the workplace. In the next step, the management team should be given the opportunity to instil meaning into this KRA by expressing it in a value statement. This statement defines the values that are tied to productive safety and expresses the intent with which it will be operationalized. In other words, while the KRA defines what the firm intends to achieve, which is an 'end', the value statement defines the 'means'. An example is as follows:

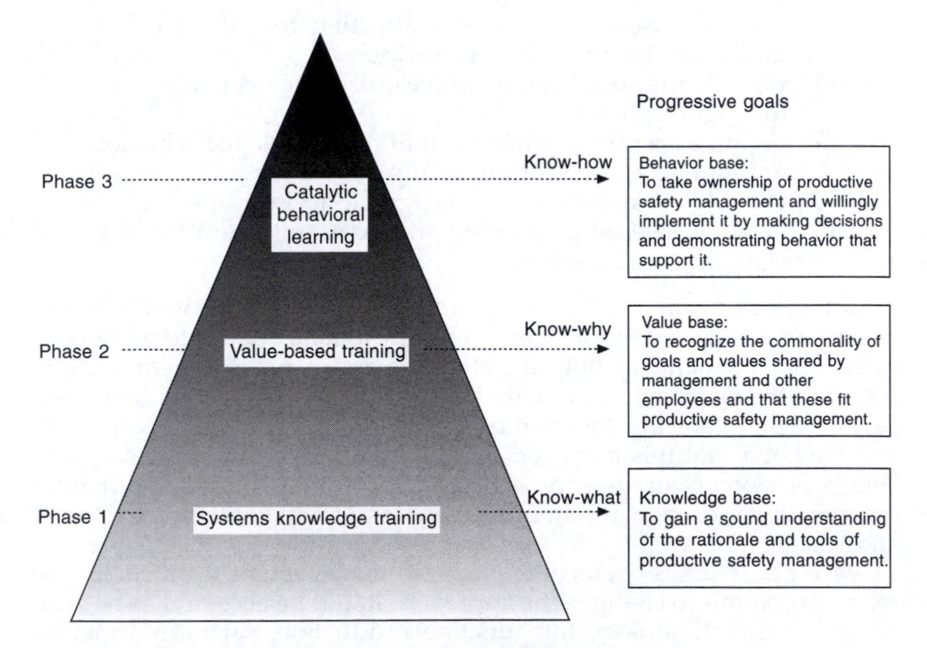


**Figure 8.1** The management training pyramid and progressive goals

> Key Result Area: Productive Safety
> Our company achieves legal compliance and pursues socially responsible outcomes through the effective, integrated management of occupational health and safety. We believe that safety and company performance are inextricably linked and ensure that our actions reflect this belief. Our employees are provided with a safe workplace and their contributions to improved safety, productivity, systems quality and our operation's financial accomplishments are valued and rewarded.

This key result area can be aligned with both policy and value statements. For example, the policy can express a pledge to protect the environment and health and safety of employees. It can also extend to the company's responsibilities to the users of their products and the communities in which they operate.[8] The scope of the policy should reflect the firm's depth of commitment to legal compliance and social responsibility. A value statement can accompany this policy and express the company's approach to corporate ethics and leadership, embracing its responsibilities to its key stakeholders. The most important criterion in writing such statements is that they annunciate core values.

In effect, the value statement iterates the underlying principles of the total management system into which OHS is integrated. It provides a reference point for decision-making and strategy development. The more specific it is, the more it provides a compass for organizational behavior at strategic and operational levels, and the more effective it will be in shaping management actions. It is, therefore, not sufficient for the firm to define what it does, it also needs to state how it intends to do it, to overcome the tendency for each member of the management team to have different perceptions of how safety and production issues should be handled. Hence, the value statement is a standard against which to check those perceptions. It is also important in the sense that it assists management to apply consistency and certainty to complex situations. In addition, it enhances organizational synergy – where the whole is greater than the sum of its parts[6] – by facilitating a common purpose.

As shown in Fig. 8.1, the final phase of executive training is referred to as 'catalytic behavioral learning'. It involves training managers to convert their knowledge and values into behaviors that achieve positive safety and performance outcomes thus providing them with 'know-how'. After this stage they should have greater confidence in their decision-making ability and the competence to demonstrate appropriate behavioral skills. The channel and derived tools, which will be presented in the following discussions, are used in this process.

This final stage of training provides managers with competencies 5 and 6. Competency 5 was an understanding of the interaction and interdependence between systems, variability, learning and human behavior and how these affect each other. In particular, it should help them to appreciate the impact their own behavior has on the organizational system and culture. The final competency is having the ability to give vision, meaning and focus to the business. This requires them to acquire the 'know-how' to implement change and to demonstrate the leadership skills required to disseminate this change to all levels of the firm.

The three phases of the management training program – systems knowledge, value-based and catalytic behavioral learning – make up the 'management training pyramid'. It illustrates how these generic competencies are developed and the progressive goals of each training phase. The training should be designed as an incremental process of consecutive steps until the individual takes ownership of the system and willingly adjusts his behavior to model the desired culture. This willingness stems from the congruence between his values and the philosophy of the system. The process is not 'training' in the traditional sense but 'learning', which is the basis of positive, sustainable behavioral modification. This approach fits the current rationale of the Australian national training reform agenda which focuses on the development of specific, industry-based competencies. It defines competency as:

> what is expected of an employee in the workplace rather than on the learning process and embodies the ability to transfer and apply skills and knowledge to new situations and environments.[8]

For supervisors to have the ability to transfer newly acquired competencies into the workplace requires them to break incongruent habits and reframe their behaviors according to values-based standards. This may be achieved through catalytic behavioral learning, which must be applied in the two key areas of management, illustrated by the channel. The first is the strategic perspective that involves allocating resources to achieve internal strategic alignment. As explained in Chapter 2, resourcing decisions result from the analysis of cost-benefit and other quantitative and qualitative information such as feedback from the external environment, from the management system itself and from internal sources. These resourcing decisions focus on balancing physical, financial and human capital. The resultant mix and level of resource allocation is often used by employees and outsiders as a barometer of management commitment. For instance, when management does not provide sufficient funds for OHS management, it is generally perceived as indicative of lack of commitment to safety. On the other hand, appropriate resourcing tied to measurable targets, shows that management has taken an approach that balances short-term gains and long-term sustainability with a genuine commitment to employee health and safety.

The second area of management in which behavioral competencies need to be demonstrated by the company's leadership, is the development of systems, practices and behaviors, to achieve internal goal alignment. This involves translating strategies into workplace activities and is affected by the organizational culture and the subcultures of different work areas. Examples of cultural factors directly within management's control include the structures of formal communication systems, work pace, employee involvement,[10] the extent of co-operation versus conflict over resources and the degree of flexibility in work systems. Operational decision-making is therefore the second barometer of management commitment. Consequently, well-developed leadership skills are demonstrated by sound strategic and operational decision-making, supported by behaviors framed according to shared values.

## The reasonableness test

How can managers feel confident about their resourcing and operational decisions when systems are complex and there is seldom one 'right' answer? Managers have to deal with uncertainty that results from incomplete information, variability and their own limited KSAs. In addition, a given problem may have a number of viable alternatives and without the benefit of hindsight it is usually impossible to appreciate the full implications of a particular choice. For this reason, managers have to make decisions based on available information, their evaluation of the situation based on current knowledge and past experience, and select an option which best fits the case. Management decision-making is therefore inherently imperfect by nature, and by implication, management systems have a residual risk constrained by information and cognitive limitations.

In terms of overall organizational performance, however, each decision has a positive, neutral or negative impact. Each decision has strategic implications. It is important therefore, for decisions to be related in some way to the overall goals of the firm. This is done using the value statement as a test to assess the appropriateness of the decision to the overall 'ends' and 'means' of the firm. The rationale of this approach is that when decisions are verified against a fixed criterion, certainty increases. This assessment process is referred to as the 'reasonableness test'. Figure 8.2 illustrates how it can be applied to ensure that resourcing decisions lead to internal strategic alignment.

The figure shows that the manager uses information from the external and internal environments to assess the situation and to formulate alternative strategies. These are evaluated using quantitative and qualitative analysis. Resourcing decisions are then made which potentially affect the balance of physical, financial and human capital. The 'reasonableness test' is applied to determine whether or not the decision fits the value statement. Those that do not are rejected while those that do are aligned with the productive safety culture and are suitable for implementation. As an example, the decision to introduce a communication strategy involving JSA meetings prior to major plant overhauls and the provision of walkie-talkies to maintenance personnel would contribute to both safety and team performance, and therefore, fits the value statement given earlier. The decision to increase production by reducing the number of fire and evacuation drills carried out per year would not pass the test and should not be implemented. Managers can therefore increase their level of confidence in their strategic decisions when such choices fit the value statement and, as a consequence, the firm can be sure that its leadership team is focused on sustainable outcomes.

The reasonableness test can also be applied to ensure that operational decisions are aligned with the productive safety culture, to achieve internal goal alignment. As explained previously, the value statement reflects the

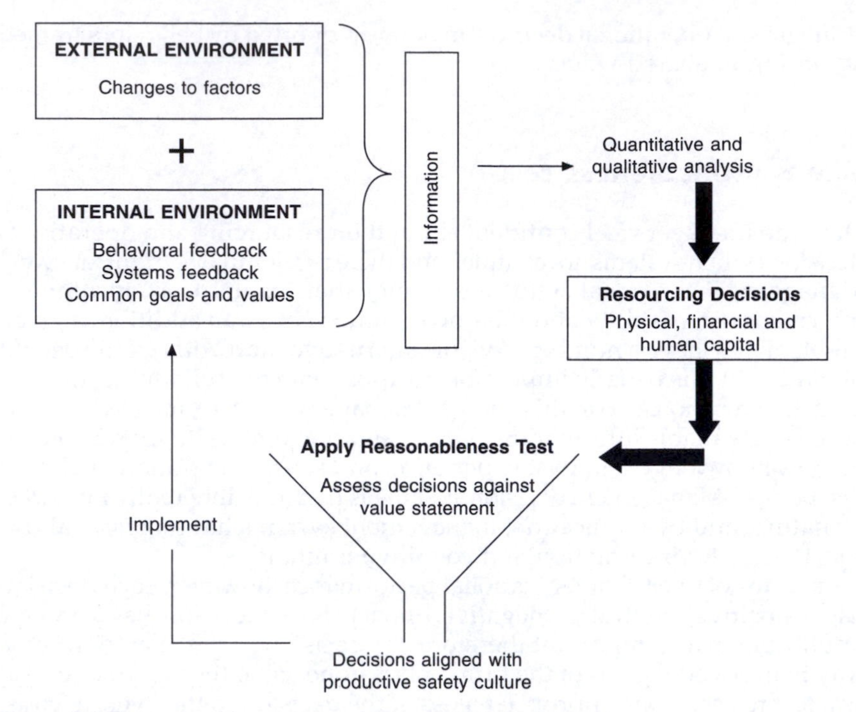


**Figure 8.2** The reasonableness test and internal strategic alignment

core values jointly held by management and employees. The development of the statement was part of the management training process so therefore it is primarily 'owned' by the supervisory team. It is also beneficial for employees to be involved in developing their own team ethic statement that fits the company value statement. These can be developed through small working parties of employee representatives under the facilitation of the supervisor or HRM professional. The team ethic statement is designed to give meaning to and reinforce the value statement at an operational level. Examples include:

Plant maintenance workshop: team ethic statements

(1) We maintain our equipment and workplace to make it safe and efficient.
(2) If it's broken, we fix it.
(3) We do our work safely, efficiently, professionally and in a spirit of co-operation.
(4) We are vigilant.
(5) We receive recognition and rewards for our achievements.

Figure 8.3 illustrates how the value statement can be applied to operational decision-making. It shows that the supervisor uses information from the internal environment, such as company objectives, current conditions and the team's ethic statement, to assess the situation. Options are formulated as a result of this situational analysis. When an operational decision is chosen, it is tested against the value statement to ensure that

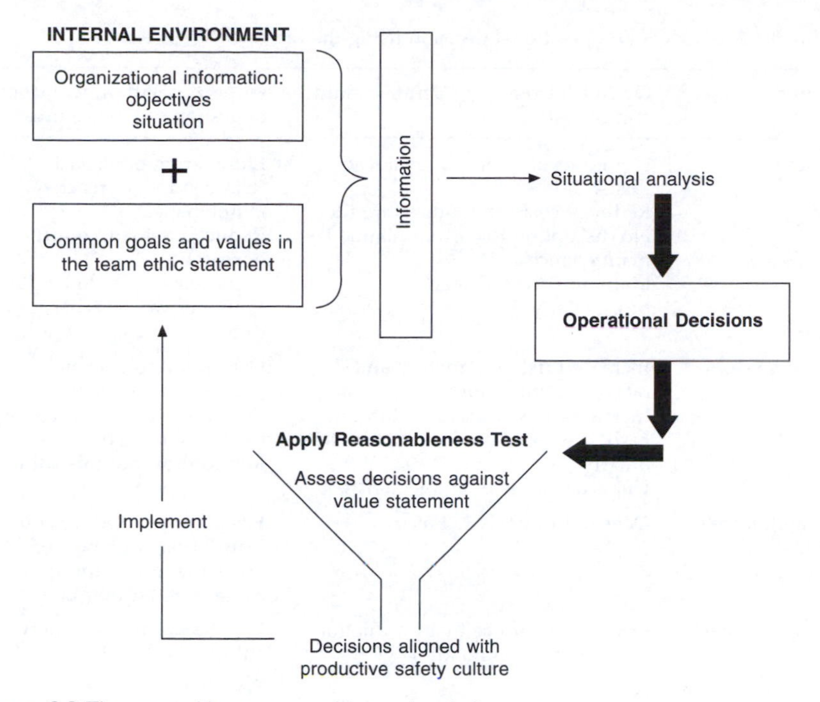


**Figure 8.3** The reasonableness test and internal goal alignment

this final decision fits the culture so that only appropriate courses of action are implemented.

The suggested process does not necessarily ensure that the 'best' option is taken. It is impossible to determine the 'optimum' decision without the benefit of hindsight. It does, however, ensure that decisions fit the productive safety culture and are aligned strategically. For example, a truck used in an open-cut mining operation may have badly worn tires. The shift supervisor has to make a decision whether to continue operating to pursue additional movement of material or to withdraw the truck from operation and have the tires replaced. He has to weigh up the consequences and benefits of the two alternatives. In addition to organizational output objectives, the team's ethic statement has to be taken into consideration and the decision assessed against the reasonableness test. The analysis of the two options is given in Table 8.1.

Under normal work conditions the supervisor does not have the time to undertake this type of analysis, however, the example illustrates the mental process of taking appropriate operational actions. The two options – to continue operation until the end of the shift or to withdraw from operation to replace the tires – are shown. The benefits and consequences of each are given. The gains from continued usage include higher output for the shift, longer tire life, minimization of disruption to current maintenance and continuation of the driver's productivity. The consequences are increased likelihood of a puncture requiring reactive maintenance and the risk of an incident. The benefits of replacing the tires are the elimination of a hazard, reduced probability of reactive

**Table 8.1** Analysis of an operational decision using the reasonableness test

| Options | Option 1: Continue operation until end of shift | Option 2: Withdrawal from operation to replace tires |
|---|---|---|
| Benefits | Higher level of output achieved this shift<br>Reduced cost from longer tire life<br>No disruption to current planned maintenance<br>Maintains truck driver's productivity | Elimination of hazard<br>Reduced risk of reactive maintenance<br>Reduced demand on truck operator<br>Replacement has to be done soon anyway – earlier return to productive operation |
| Consequences | Increased risk of puncture and reactive maintenance<br>Increased risk of incident through higher demand on driver and poor braking<br>Unknown costs of an incident | Disruption to current planned maintenance<br>Reduces driver productivity<br>Cost of shorter tire life<br>Reduced output this shift |
| Fit with team's ethic statement | Does not fit goals 1, 2 and 3 | Fits all goals. Reduces our team's output, however, it helps the next team to achieve better output |
| Reasonableness test | Does not reinforce that production and safety are compatible goals | Fits the productive safety culture |
| Implement yes/no | No | Yes |

maintenance, timely intervention and control of the demands on the operator. When option 1 is checked against the team's ethic statement, it does not fit statements 1, 2 and 3 listed earlier. For instance, it does not fit, 'We maintain our equipment and workplace to make it safe and efficient.' When the reasonableness test is applied, option 1 does not reinforce that production and safety are compatible goals and is therefore not a valid decision. Option 2, on the other hand, passes the test and is suitable for implementation.

The application of the reasonableness test helps the supervisor to achieve greater certainty and confidence in decision-making. This assuredness is attributable to knowing that his final choice is aligned with the system. The workforce can observe that the initiatives taken by the supervisor are compatible with this culture thus demonstrating that management is committed to achieving optimal performance and safety concurrently. What management does, rather than what management says, determines their actual commitment to safety,[7] and this will influence the workforce's perceptions in such a way that shapes the organizational culture.

In the example given in Table 8.1 the better option is to change the tires during the current shift. This eliminates the hazard and allows production to be maintained during forthcoming shifts; in other words, the problem is not left to the next team. This approach creates better teamwork and encourages interdependencies that result in positive outcomes for the firm and for employees. Making a decision such as this in the interests of safety and overall organizational performance would be much more difficult in a firm that is solely production-driven and where safety is rhetorically referred to as an organizational goal.

The reasonableness test shows that all decisions, both strategic and operational, should be assessed for alignment with the organization's goals and desired culture. The test can also be applied to evaluate the appropriateness of current systems. For example, if teams are rewarded on shift-based output and encouraged to compete for these rewards, the decision discussed above – to change the tires – would be hindered because there is an incentive for the supervisor to maximize production on his shift and keep the truck operating. It is important therefore to ensure that all areas of the management system pass the reasonableness test. Alignment is thus concerned with the compatibility and congruence of organizational systems with the overall management philosophy and direction.

The reasonableness test is a fundamental tool of management-driven behavioral change. It allows supervisors to modify their decision-making processes and interactions with subordinates to pursue production and safety simultaneously. The test increases team interdependencies by filtering out decisions that stimulate internal competition. At the operational level the likelihood of conflict between supervisors and workers is reduced because the test integrates the team ethic statement as well as the firm's overall value statement. This helps to reduce the role conflict traditionally experienced by supervisors when they are held accountable for production and safety yet organizational systems are skewed towards output-driven goals.

The true values of a company are brought to the fore when supervisors have the authority and support to act in the interests of all parties in a manner that balances short-term gains and long-term benefits. In some cases this may include shutting down a production line for safety reasons. If the team leader has to explain personally to the manager why he did so then he is likely to think three or four times before taking action[5] unless the justification has a solid values-driven basis supported by senior management. The hierarchy must therefore endorse sound decision-making based on such principles. Further, when employees are able to observe, from the decisions made by supervisors, that management is committed to these principles, this encourages employee buy-in. Ownership is thereby disseminated from the executive through to the operational level.

## Appropriate leadership at management and supervisory levels

What leadership style is expected to emerge in a productive safety environment? Management should consistently demonstrate leadership that is 'appropriate' to the circumstances. This concept of 'appropriateness' is akin to situational leadership discussed in management literature. It is compatible with the management grid developed by Blake and Mouton. This grid has two axes – concern for people and concern for production.[10] The ideal style is 9,9 team management that demonstrates the highest concern for both people and production.[10] These dimensions are the same as the task component and the relationships component of the job that the management team identified as sources of job satisfaction during the value-based phase of the management training pyramid. 9,9 Team management entails work accomplished by committed people and interdependence through a 'common stake' in organizational purpose, underlain by relationships based on trust and respect. The style is based on a number of principles including open communication. In addition, conflict must be solved by addressing underlying causes and carried out in a spirit of co-operation. This requires employees to demonstrate a high level of maturity, which is only possible though shared ownership of power and authority. Other principles include managing using defined goals and rewarding employees on the basis of merit. Both triumphs and tribulations are considered part of the organization's learning process, and critique is undertaken in a context which defines the required standards of behavior and performance.

Productive safety management embodies the participative principles of 9,9 team management, but goes further to explain why in some circumstances, although the team-based model is highly desirable overall, the collaborative approach to decision-making can be ill-advised under certain conditions. How then is appropriateness determined? This depends on a number of situational factors. The entropy model helps to identify those cases where it is more appropriate for management to be authoritative rather than consultative.

The entropy model indicates that the level of degradation of organizational systems and the inherent danger in system factors affects the overall level of risk. Where degradation is rising at a significant rate or where there are catalysts that can trigger residual risk into imminent danger, management needs to act promptly with authority. The appropriate style is 'assertive leadership' that involves taking corrective action with maximum efficiency. The moderator that prevents workers from perceiving the manager's actions as dictatorial is the application of the reasonableness test in the decision-making process. Even though the manager may act without consultation under conditions of high risk, the decision is made in the best interests of employees and the organization. As a result, the decision is compatible and aligns with the productive safety culture and is perceived by employees as fair.

There may also be situations where employees initiate corrective action by informing the supervisor of rising entropic risk or that residual risk is an impending hazard. In these circumstances, the appropriate style is 'responsive leadership'. The supervisor acts promptly and authoritatively on the information provided by workers. Where remedial action is required quickly, therefore, an assertive response is required. In low to moderate risk situations, in contrast, employee participation should be maximized and this requires 'collaborative leadership'.

Figure 8.4 illustrates the appropriate leadership style according to these situational variables. It indicates that management commitment and leadership are prerequisites for effective decision-making. At the supervisory level, decisions fall into two categories based on resource usage or operational practice which affect internal strategic alignment or internal goal alignment, respectively. It is shown that sound decision-

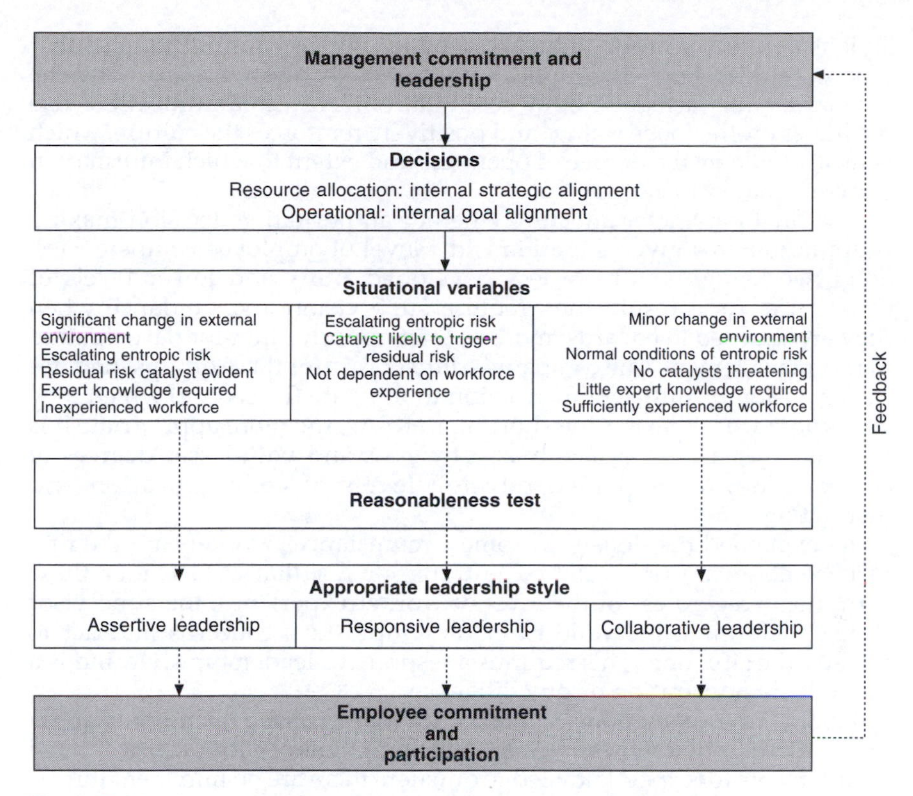


**Figure 8.4** Appropriate leadership at management and supervisory levels

making involves assessing situational variables and applying the reasonableness test. This process leads to the selection of the appropriate leadership style, positively influencing employee commitment and facilitating the most effective level of employee participation given the situation.

There are a number of situational variables that determine what leadership style should be applied. For example, when there are significant changes in external environmental factors such as amendments to the legal framework, assertive leadership responses may be required. This is because the full implications of legislative changes take time to filter through to the workforce level and will sometimes necessitate re-education of employees. Landmark legal judgements, that prove the enforceability of particular statute provisions, can act as a trigger for authoritative change in organizations. In these cases, management tends initially to drive the change to ensure compliance with legal requirements and then follows through with employee training and consultative practices.

A further parameter which indicates that an assertive leadership style may be appropriate is where expert knowledge is required to remedy a hazard. Where decisions need to be made using highly technical KSAs, nonparticipative processes may be more effective. It is important, however, that any modifications which have an effect on the level of risk to which workers are exposed, are explained to them. When an assertive leadership style is applied, it needs to be followed up with communication so that workers understand the implications of the changes and also why the supervisor has acted without consultation. This communication also facilitates on-the-job learning[9] and positively reinforces the culture, which is measurable by the degree of openness and extent to which information is disseminated to employees.

The final parameter affecting whether authoritative decision-making is applicable in a given situation is the level of employee competencies. Some worker groups have less developed KSAs and lower levels of experience. As a result, they require supervision and guidance when they are exposed to hazards, more so than opportunities for participation in risk management. The company's duty of care for these groups includes the provision of sufficient instruction to mitigate the risks to which they are exposed. The greater the shortfall in KSAs, the more appropriate it is for the supervisor to act with assertiveness and with lesser degrees of consultation, accompanied by high levels of communication and instruction.

As explained previously, in some circumstances workers are the first to become aware of escalating entropic or residual risk. Under these conditions, regardless of the level of worker experience, the supervisor should determine the validity of the employee's concerns and act to remedy the situation. This requires a responsive leadership style and is a tangible demonstration of due diligence.

Finally, where situational variables are characterized by minor changes in the external environment, manageable conditions of entropic risk, limited requirements for expert knowledge or where there are no imminent threats from residual risk sources, collaborate leadership should be applied. These consultative processes are more effective when employees' KSAs are well developed. However, provided that workers have a basic understanding of the nature of risk and satisfactory task competencies, participation will be beneficial. It will provide workers with an additional opportunity to develop their understanding of risk and to enhance their appreciation of the need for safety consciousness.

In summary, Fig. 8.4 illustrates that the appropriate leadership style is dependent on situational factors. Once such variables have been taken into account, all decisions that pass the reasonableness test are compatible with the productive safety culture. It is, therefore, not the type of leadership style that is applied but the appropriateness of the style for the situation that affects the level of employee commitment. (Management can gauge the effectiveness of its leadership style from workforce feedback and this will be discussed in Chapter 10.)

What is the correlation between employee commitment and safety performance, and what are the factors that determine this commitment? The variables that employees have reported are:

(1) The degree to which individual employees believe themselves to be an important part of the organization.
(2) Organizational support and value for personal and professional development, education and training.
(3) A safe workplace.

(4) Systems and methodology for problem reporting.
(5) Management/system responsiveness to reported problems.
(6) Performance-based consequences: recognition and reward for positive performance.
(7) Performance-based consequences: confrontation and correction of poor performance.
(8) Employee involvement, participation and input.
(9) Job security (mergers, takeovers, layoffs, wage reduction).[11]

At the heart of these matters is the extent to which the management system concurrently supports organizational and employee objectives or is centered on common goals and values. The other core issue affecting employees' commitment to organizational output, safety and system factor quality goals, is the power balance between management and workers. This balance becomes particularly evident in relation to management's willingness to respond to reported problems, to recognize positive performance and to share decision-making power by increasing employee participation. This is an important matter in achieving industrial harmony and maximizing the return on human capital. Maintaining the balance of power does not, therefore, necessitate that management relinquishes its control as such. Instead, sufficient internal feedback mechanisms and trust are required to secure strategic control while concurrently increasing the degree of self-supervision and ownership at the shop floor level, as illustrated in Fig. 8.5.

The figure shows that there are a number of prerequisites for effective self-supervision. The first is a context where decision-making leads to

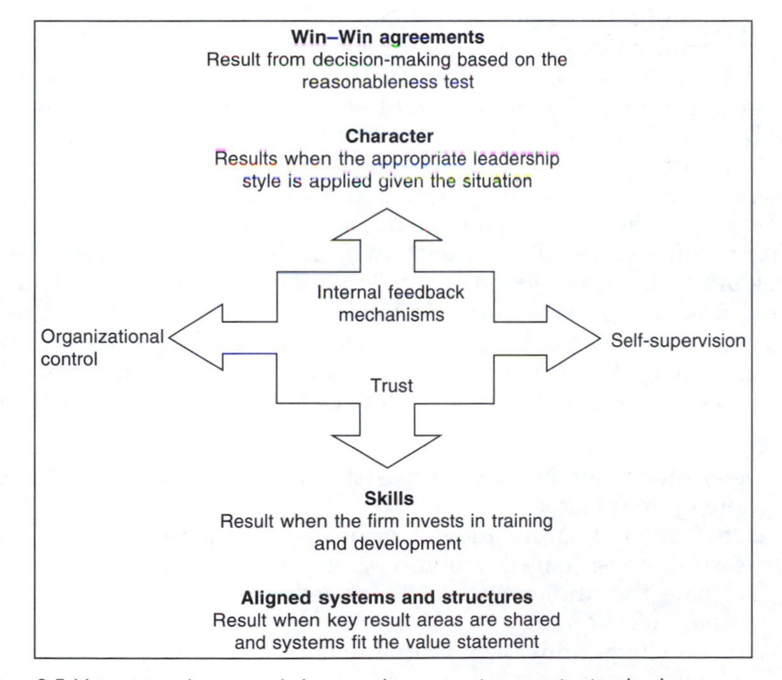


**Figure 8.5** Maintaining the power balance and a supportive organizational culture

win–win agreements based on common goals and values that comply with the reasonableness test. As a consequence, regardless of who takes the action, either the supervisor or workers, the result is in the best interests of both the firm and the workforce. The second is leadership expressed as 'character', which is determined by an appropriate style for a given situation, as illustrated in Fig. 8.4. The use of situational leadership accompanied by effective information-sharing reinforces trust and thereby keeps power and control in balance. In this way, management can allow higher degrees of autonomy in the knowledge that employees have the KSAs and information available to make sound decisions within the scope of their jobs. As a prerequisite, the firm has to invest in the development of employee skills so that they are equipped with the competencies to manage their jobs safely and efficiently while incrementally gaining a better understanding of the 'big picture' challenges faced by the company. Balance is also maintained when management systems and structures are aligned with the value statement and responsibility for the achievement of key result areas is shared. Figure 8.5 shows that central to this power/ control balance is a supportive organizational culture founded on trust and on-going internal feedback. In practice, this means that not only must supervisors keep employees informed of changes but they must also be receptive to their input.

## Resolving safety issues

From time to time, the power balance in organizations becomes destabilized leading to conflict between management and the workforce. Management systems, communication, trust and relationships are hence also subject to degradation. The same rationale of corrective action and maintenance strategies, proposed by the four-fold strategy, needs to be adopted to remedy such situations. The failure to deal with conflict effectively can have a negative impact on the company culture and lead to long-term damage to management–employee relations. To address these issues effectively, conflict resolution procedures need to be established so that confrontation is solved directly with an underlying intent by both parties to seek understanding, agreement and co-operative effort.[11] In relation to the resolution of specific issues affecting safety, output and system factor quality, there are ten steps to be taken. These steps require that management adopt a collaborative leadership style and provide a forum for problem solving in which employees' representatives actively participate. The steps are:

(1) Determine whether the issue raised indicates degradation of one or more system factors.
(2) Determine the nature and extent of any degradation.
(3) Assess the residual risks involved or present.
(4) Evaluate the implications of these risks on safety, efficiency and system factor quality.
(5) Assess whether this issue is an indicator or warning of underlying systemic weaknesses.

(6) Use the four-fold strategy to develop alternative courses of action.
(7) Ensure that chosen courses of action are compatible with the team's ethic statement if the problem is limited to the workplace and manageable at shop floor level. If the issue has strategic implications, ensure that representatives of other affected business units are informed and involved in the resolution process.
(8) Apply the reasonableness test to ensure that the chosen courses of action are appropriate.
(9) Develop action plans that identify time frames, resourcing and accountabilities and ensure that stakeholders are in agreement.
(10) Establish a review process.

The intent of this approach is to focus on the quality of the system and its impact on safety and performance. The manager's responsibility is to steer the conflict resolution process to ensure that all parties remain centered on pursuing a solution that improves the organizational system and is compatible with the culture. His role is therefore to prevent the group from being distracted by any nonconstructive dynamics that may arise between individuals during the process.

The manager should maintain a positive and objective position that facilitates a constructive problem-solving and negotiation process. In the first place, personal interests of the parties should be separated from the problem so that the focus is on areas of benefit, not on positions and power plays. To do this the problem firstly needs to be clarified and agreed upon. After this, with an emphasis on mutual gain, a variety of options should be considered and then tested against objective criteria.[13]

These basic principles indicate that primarily, disputing parties should focus on system factor quality issues that are the source of the problem. Fundamentally, it is in the interests of both groups to improve organizational systems and to maintain effective working relationships. With this approach there should be fewer tendencies for parties to resort to allocating blame. In addition, the value statement and established common goals and values may be used as a point of reference during the conflict resolution procedure to reinforce the mutual benefit inherent in the employer–employee relationship. Once the problem has been clearly defined, the four-fold strategy provides the objective criteria for the development of remedial plans. To ensure that the choices made are for mutual gain, the selected strategies can be assessed against the reasonableness test. The productive safety tools therefore provide an objective framework for resolving conflict, particularly those that result during the process of change implementation as a result of residual habits that historically allowed production to over ride safety and quality goals.

It is important to handle dispute negotiations effectively because of the potential impact on the company's culture, which can be influenced in a number of ways. Firstly, it will be affected by how the resolution process is undertaken. Is it, for example, done in a spirit of co-operation or are the relevant parties dogmatically pursuing their own agendas? Secondly, the results and how the workforce perceives them will temper the organizational climate. Does the outcome indicate that management is committed to employee safety or are they paying 'lip-service' to it?

Ultimately, the dispute resolution process should return the power balance to an equilibrium in which both management and the workforce are comfortable with their levels of control and influence. In contrast, where either or both parties are dissatisfied and trust has been eroded, the power balance will be in a state of flux that leads to on-going conflict because adversarial relations have been established.

In some cases, conflicts that arise between supervisors and workers cannot be resolved through internal negotiations using the steps described above. This can occur for a number of reasons. Disputing parties may have different perceptions of the problem or the level of risk and therefore may not be able to reach agreement over the root cause of the matter. In addition, the issue may not be limited to the workplace in which it was raised and may have wider implications for the organization, thus requiring the involvement of technical experts or higher levels of management. To ensure that the firm is well prepared to address these complex issues, a written procedure outlining the avenues for negotiations is needed. As a safety net for such issues OHS legislation provides for external adjudication by regulators. For instance, the Occupational Safety and Health Act 1984 (Western Australia) states:

> 24. (1) Where an issue relating to occupational health, safety or welfare arises at a workplace the employer shall, in accordance with the relevant procedure, attempt to resolve the issue with –
> (a) the health and safety representative;
> (b) the health and safety committee; or
> (c) the employees,
> whichever is specified in the relevant procedure.
> (2) For the purposes of subsection (1), 'the relevant procedure' means the procedure agreed between the employer and the employees as applying in respect of the workplace concerned or, where no procedure is so agreed, the procedure prescribed for that purpose in the regulations.[13]

The Act goes further to refer any unresolved issues where there is a risk of imminent and serious injury to an inspector who is authorized to take any action considered appropriate. An example of a safety issue resolution flowchart is provided in Fig. 8.6.

The legislation above requires management and employees to be involved in the development of a resolution procedure. Although not specifically stated, the implication is that it is also the firm's responsibility to ensure that its workforce is aware of the process and their rights to raise safety issues. In Fig. 8.6, after employees raise their concerns with the supervisor they undertake collaborative analysis using the ten steps above. Issues that are unresolved are referred to the safety officer and departmental manager, who are then involved in the second round of analysis. Where the matter remains unresolved, the inspector is notified in all high-risk cases. This prompts an investigation by the regulator and the implementation of appropriate action, which depending on his assessment of the risk, may include serving a return to work order, an

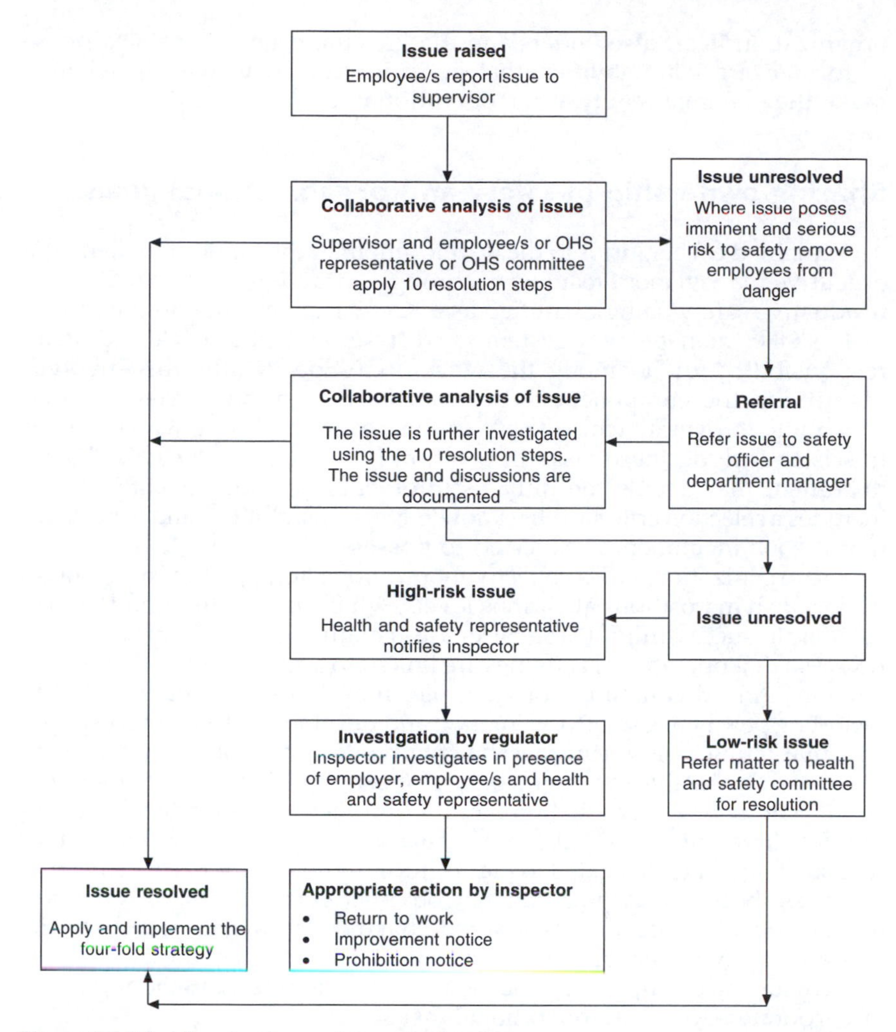


**Figure 8.6** Workflow for the resolution of safety issues

improvement notice or a prohibition notice. Minor-risk issues may be temporarily delayed and referred to the health and safety committee for resolution.

The value statement, reasonableness test and safety resolution procedure, in effect, establish guidelines that define acceptable management behavior and provide the foundations upon which a productive safety culture can be developed. Once supervisors have the 'know-how' to apply these tools consistently then they should be more confident that their decisions are aligned with the strategic objectives and philosophy of the firm. The primary objective of such training is to enlist management commitment to the productive safety approach. Once this commitment has been attained, systems changes can be implemented. Ownership of the system can also be spread to the lower levels in the firm using HRM tools. At this point, management is ready to create behavioral change throughout the organization. They also understand that it is important in the next phase to formalize accountabilities, that is, to identify who will do what to make the system a reality at the operational level.

## Sharing ownership of safety and organizational goals

As explained earlier, prior to the development of the value statement, the executive management team went through a training process that led to productive safety being identified as a KRA. The next step in the rollout of the OHS management system is to 'fragment' this KRA and share responsibility for it among the workforce. Specifically, this involves identifying the component duties that each position carries out that contribute to overall achievement of this result area. One way of doing this is to include these tasks in position descriptions (PDs). A PD is a document that details the duty requirements of a job. Usually it also includes a selection criteria of the knowledge, skills, abilities and experience that the job incumbent is expected to possess.

The organization's PDs collectively provide a total picture of the tasks undertaken in positions at various levels, which are required for the firm to operate successfully. These tasks are usually derived from the firm's KRAs so that operational activities are linked to and aligned with business strategy. An oil company, for example, may have a number of KRAs including productive safety, exploration and development, customer supply and marketing. These can then be fragmented into 'outcomes' or core activities. Productive safety outcomes, for example, include 'occupational health and safety', 'production', and 'equipment maintenance'. During the development of the PDs, these outcomes can be distributed so that responsibilities are delegated to the appropriate level. Under each outcome the tasks specific to the position may be described. By using this method to develop PDs, the firm is assured that the tasks undertaken at the operational level contribute to the strategic achievement of the firm. Figure 8.7 provides an example of a proposed structure for the strategic alignment of corporate KRAs with operational tasks.

At the top of the figure is the strategic level of the firm. Most companies develop a mission statement that is an overarching description of their purpose and philosophy. The firm's KRAs are the next level in the strategic hierarchy. In the example given, it is shown that a value statement should accompany the productive safety KRA. As explained earlier, the KRA describes the 'ends' while the value statement describes the 'means'. There are specific areas of measurement that apply to this KRA which are captured by the alignment indicators. These indicators are the perspectives in which measures are developed within the performance management system, to ensure that a balanced approach is taken. The alignment indicators are:

(1) Productivity;
(2) Safety;
(3) System factor quality;
(4) Financial and customer;

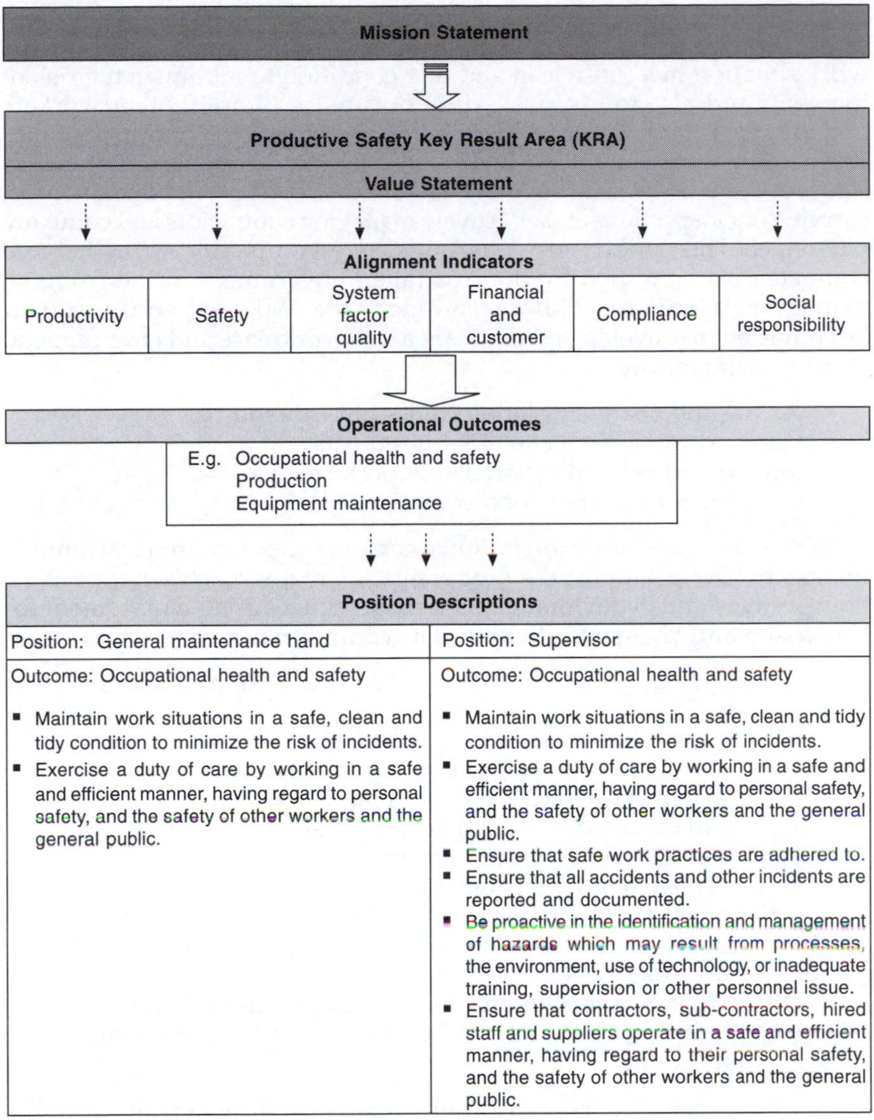

**Mission Statement**

**Productive Safety Key Result Area (KRA)**

**Value Statement**

| **Alignment Indicators** | | | | | |
|---|---|---|---|---|---|
| Productivity | Safety | System factor quality | Financial and customer | Compliance | Social responsibility |

**Operational Outcomes**

E.g. Occupational health and safety
Production
Equipment maintenance

| **Position Descriptions** | |
|---|---|
| Position: General maintenance hand | Position: Supervisor |
| Outcome: Occupational health and safety<br>▪ Maintain work situations in a safe, clean and tidy condition to minimize the risk of incidents.<br>▪ Exercise a duty of care by working in a safe and efficient manner, having regard to personal safety, and the safety of other workers and the general public. | Outcome: Occupational health and safety<br>▪ Maintain work situations in a safe, clean and tidy condition to minimize the risk of incidents.<br>▪ Exercise a duty of care by working in a safe and efficient manner, having regard to personal safety, and the safety of other workers and the general public.<br>▪ Ensure that safe work practices are adhered to.<br>▪ Ensure that all accidents and other incidents are reported and documented.<br>▪ Be proactive in the identification and management of hazards which may result from processes, the environment, use of technology, or inadequate training, supervision or other personnel issue.<br>▪ Ensure that contractors, sub-contractors, hired staff and suppliers operate in a safe and efficient manner, having regard to their personal safety, and the safety of other workers and the general public. |

**Figure 8.7** Sharing ownership of productive safety as a key result area

(5) Compliance; and
(6) Social responsibility.

(Later in Part 4 it will be explained how to develop such measures to ensure that the firm does not compromise safety for production and sustainable business outcomes are pursued.)

In the figure, the productive safety KRA is fragmented into 'operational outcomes'. These are distributed to the relevant PDs and the tasks specific to this outcome are described. The general maintenance hand's PD shown in the figure, like all other PDs, contains the 'occupational health and safety' outcome. The incumbent's responsibilities are to maintain the work situation in a safe, clean and tidy condition to minimize the risk of incidents and also to exercise a duty of care for himself, other workers and the general public. The supervisor has a higher level of responsibility and accountability for OHS. His PD also includes enforcing adherence to safe work practices, ensuring that incidents are reported and documented, carrying out inspections and effectively managing contractors and company personnel. The supervisor's OHS tasks are compatible with the core competencies identified by the Australian government in the frontline management initiative (FMI). According to the FMI, a supervisor should be competent to develop and maintain a safe workplace and environment. Specific elements are:

Access and share legislation, codes, and standards.
Plan and implement safety requirements.
Monitor, adjust and report safety performance.
Investigate and report non-compliance.[14]

In total there are eleven operational competencies required by frontline managers. These support the productive safety approach's emphasis on both systems and behavioral management with a strong requirement for leadership and change facilitation. The competencies are:

(1) Manage personal work priorities and professional development.
(2) Provide leadership in the workplace.
(3) Establish and manage effective workplace relationships.
(4) Participate in, lead, and facilitate work teams.
(5) Manage operations and achieve planned outcomes.
(6) Manage workplace information.
(7) Manage quality customer service.
(8) Develop and maintain a safe workplace and environment.
(9) Implement and monitor continuous improvement systems and processes.
(10) Facilitate and capitalize on change and innovation.
(11) Contribute to the development of a workplace learning environment.[14]

The PD is an effective tool to formalize shared ownership of KRAs because it defines the tasks and competencies required to perform the job safely and efficiently. In addition, it is readily accessible to the position's incumbent and often referred to during the course of employment, for example, during recruitment, goal-setting and performance appraisal. (The use of PDs to set goals at the supervisory level is explained further in Chapter 12.)

Under productive safety management, the PD is structured into three key areas. The first of these is a list of competencies required to undertake the position. These are used as criteria against which the knowledge, skills and experience of job candidates are tested during the recruitment process. These selection criteria also provide the basis for gaps analysis – comparing the competencies held by the employee against those required to perform the role. Any gap identified should be rectified through training. The second section of the PD contains 'achievement indicators' based on the four system factors. Every employee's PD states the importance of undertaking processes, using technology, operating in the physical environment and being part of a team of human resources in a manner that is efficient, safe and quality assured. An extract of a sample job statement showing these achievement indicators is given in Fig. 8.8.

The figure also shows the 'operational outcomes for this position'. The outcomes in the figure apply to a supervisory role and include the

**Achievement Indicators** (achievement indicators apply to all positions within the organization.):

A high standard of achievement for each position outcome is demonstrated by:

*Processes:*
- ❑ undertaking tasks efficiently
- ❑ carrying out tasks in a safe manner

*Technology:*
- ❑ using equipment safely and efficiently
- ❑ maintaining equipment in a safe and efficient condition
- ❑ using resources effectively and minimizing wastage

*Physical environment:*
- ❑ maintaining the work area in a safe, clean and tidy condition

*Human resources:*
- ❑ working effectively in a team environment
- ❑ taking ownership of production, safety and system factor quality
- ❑ responding to customer needs
- ❑ contributing to organizational improvement

| **Operational outcomes for this position**<br>(operational outcomes identify the key responsibilities of the position and contribution to the productive safety key result area) | |
|---|---|
| *Outcome: Technology and environmental management*<br>▪ Ensure that the work team maintains tools and equipment in good working order and act on any faults reported by them.<br>▪ Ensure that necessary routine maintenance checks are carried out on vehicles, machinery and tools and that servicing is undertaken to maintain efficiency and safe operation.<br>▪ Ensure that the work area is well maintained. | *Outcome: Occupational Health and Safety*<br>▪ Maintain work situations in a safe, clean and tidy condition to minimize the risk of incidents.<br>▪ Exercise a duty of care by working in a safe and efficient manner, having regard to personal safety, and the safety of other workers and the general public.<br>▪ Ensure that safe work practices are adhered to.<br>▪ Ensure that all accidents and other incidents are reported and documented.<br>▪ Be proactive in the identification and management of hazards which may result from processes, the environment, use of technology, or inadequate training, supervision or other personnel issue.<br>▪ Ensure that contractors, sub-contractors, hired staff and suppliers operate in a safe and efficient manner, having regard to their personal safety, and the safety of other workers and the general public. |
| Outcome: Production. . . | Outcome: Internal customers. . . |
| Outcome: Work team and contractor supervision | Outcome: Administration. . . |

**Figure 8.8** Extract of a PD showing achievement indicators and outcomes

compulsory OHS outcome as well as others that support and are derived from the productive safety KRA such as 'technology and environmental management', 'production' and 'work team and contractor supervision'. Achievement indicators are included in the PDs for all organizational roles for a number of reasons. Firstly, they outline the expected criteria of performance for operational activities. Secondly, the indicators describe behavioral factors under 'human resources' including working well as part of a team and taking ownership of safety and production outcomes. In addition, it captures other key business performance indicators such as customer service. Finally, the achievement indicators reinforce the importance of maintaining high quality systems as a means of managing risk and improving organizational performance.

## Aligning human resource management systems

Including these indicators in the PDs also links them strategically to the performance evaluation cycle thus aligning operational tasks with the total management system. The four system factors therefore run as a 'theme' from the productive safety KRA in the strategic plan to business unit plans and into the goal-setting carried out by supervisors in their annual performance appraisal. This alignment within the HRM systems is important to ensure that there are no internal contradictions. It also communicates a consistent set of behavioral standards to employees. Figure 8.9 illustrates how these strategic themes connect the various levels of the hierarchy and management system.

In the figure the arrow on the left runs from the strategic level of the firm to the PDs at the operational level. The arrow is two-headed indicating that the linkage flows both ways – top-down and bottom-up. The business' strategic plan contains the mission statement – an enduring statement of purpose that addresses the basic question 'what is our business?'.[15] It also contains KRAs and value statements. Strategies, initiatives and action plans are derived from these KRAs (as will be explained in greater detail in Chapter 11). At the next level down in the hierarchy and management system are the business units that undertake core activities such as production and marketing. Business unit KRAs are directly linked to corporate KRAs. Productive safety is common to both of these strategic levels and ownership is shared among all business units. Other KRAs, for example, exploration and development may be 'owned' by a specific business unit.

The achievement cycle is the performance management system used in the productive safety approach and is explained fully in Chapter 12. As an overview, it contains the business alignment scorecard and cultural health scorecard that measure/monitor performance at the organizational level. (Scorecard provide balance as explained by Kaplan and Norton.[16] The system presented in this book is designed specifically to resolve the safety versus production dilemma in hazardous industries.) The business alignment scorecard contains the alignment indicators that quantify the firm's achievement in the areas of production, safety, system factor quality, financial and customer, compliance and social responsibility. The balanced

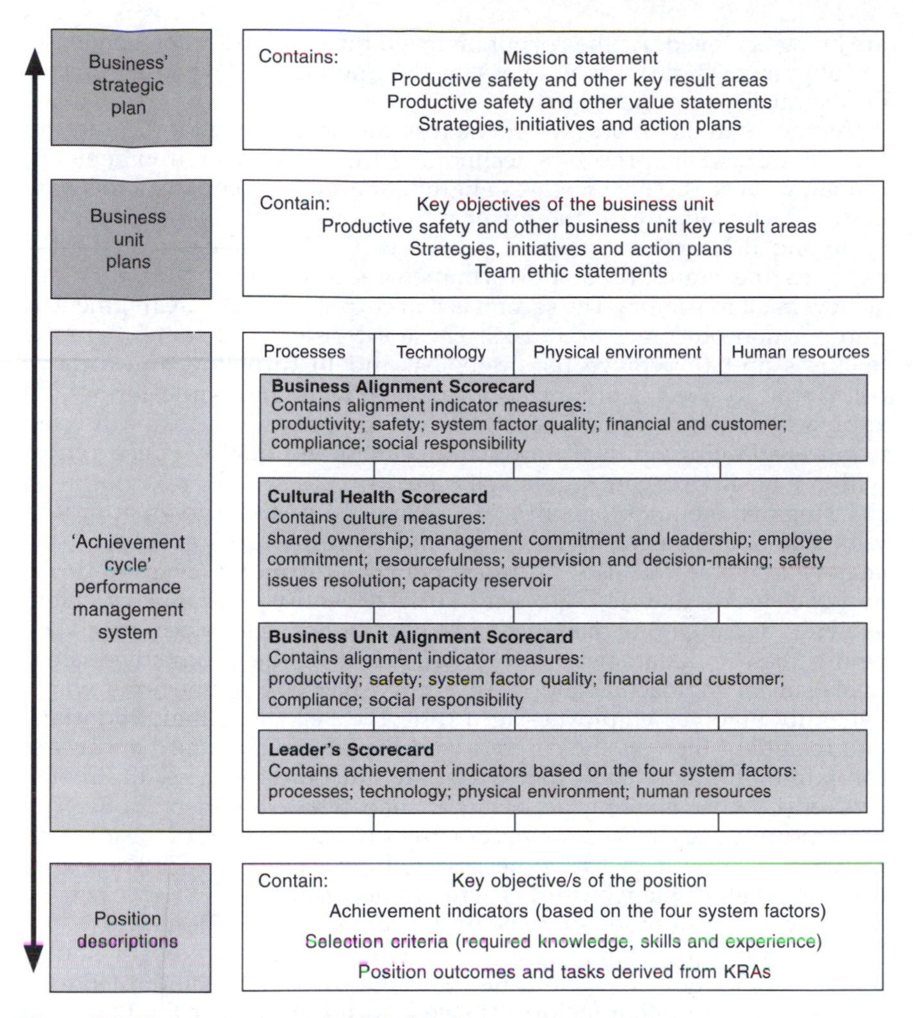


**Figure 8.9** Aligning management levels and tools

use of these measures ensures that production does not compromise safety and system factor quality, and also that short-term goals do not over-ride long-term sustainability. The cultural health scorecard complements the business alignment scorecard by measuring organizational leadership and other parameters that affect the implementation of change and maintenance of improvements.

The achievement cycle also contains business unit alignment scorecards so that each unit can contribute to overall organizational success. Again, at the business unit level, the alignment indicators are used as measures of performance. At the operational level, the leader's scorecard is used for goal-setting by managers and supervisors. Each team leader establishes goals, action plans and measures that contribute directly to improving the quality of system factors. This multi-tiered performance management system is used to determine whether the objectives in the strategic plan are being achieved. At the corporate level, the emphasis is on achieving the alignment indicators, whereas, at the operational level, the focus is on the quality of system factors.

As shown in the figure, system factors and achievement indicators are the same. These are processes, technology, the physical environment and human resources. This increases alignment and the company's focus on system factor quality. In the lower part of the figure, the achievement cycle and PDs are also aligned in two ways. The first is that the PD explains the standard of performance required by including these achievement indicators. The second is that corporate KRAs are fragmented into position outcomes and tasks. These lower-level outcomes help the business unit to achieve its strategies, and in turn flow upwards to contribute to overall organization achievement. In simple terms, the approach ensures that employee efforts and company resources are directed towards attaining optimal production, safety and quality concurrently, within a positive organizational climate.

Using this methodology, PDs are developed in a top-down approach with line management involvement and with the focus on goals to be achieved and not on lists of duties being performed.[17] The top-down approach means that strategic issues such as identifying productive safety as a KRA and attaining management commitment need to be undertaken before these job statements are modified to fit the productive safety management approach. A bottom-up approach, used by some firms, which generally involves employees recording the tasks that they undertake and including these as the core areas of the job, is ill advised because it can fail to link these tasks to appropriate outcomes. As a result, the PD can reflect what is being done rather than what should be done. The bottom-up approach does not necessarily prompt analysis of the job such as whether specific tasks are necessary, how well the work should be done or whether the job itself is actually necessary.

There are a number of reasons for developing PDs early in the implementation program. The first is to give employees a sound understanding of their responsibilities, their duty of care and role in the maintenance of system factors. The second reason is that the PD is the main document used in the recruitment and selection process. It is important to ensure that new employees have the competencies and attitudes which are compatible with the desired culture. The PD is drawn on extensively during the recruitment process. The need for the outcome – 'occupational health and safety' – derived from the productive safety KRA becomes particularly evident at this time. From the outcomes and specific tasks in the PD, the selection criteria are determined. They define the competencies and experience that new recruits are expected to possess that allow them to perform the role safely and efficiently. During the selection process, applications are assessed against these criteria to determine the candidate who is the best fit with the job requirements. In the recruitment process, therefore, the emphasis on safety as a competency increases when the PD contains the OHS outcome.

It is particularly necessary to screen candidates for their safety knowledge and attitudes when recruiting supervisors or when employees have to work in high-risk areas. Senior staff recruitment is critical because, as explained earlier, management commitment and leadership are required for the effective implementation and continuity of an OHS management system. The employment of team leaders who, for example, consider 'worker error' to be the primary cause of accidents can damage the culture. Any differences or gap in the competencies and attitudes of new recruits, and those needed to perform the role effectively and to fit the organizational culture, need to be addressed through training. To minimize the costs of training, the closer the match between the job candidate's KSAs and those required by the firm, the better. For this reason, value-based recruitment is strongly advocated. A study of risk-taking behavior in the Western Australian mining industry found that value-based selection procedures led to success in employee retention and improved the organizational culture.[18] For hazardous industries, such as the meat industry, recruitment with a greater focus on safety behavior and attitudes is highly recommended.[19]

How does value-based recruitment work? When screening for a supervisory position, for instance, the interviewers can ask the applicant to describe an experience in which he was involved in an accident investigation. The ideal answer would identify what happened, why and the action taken. If the candidate identifies 'human error' as the cause, a further question asking whether the organizational system played any part in the incident should be asked. This would ascertain whether or not the applicant understands the significance of underlying systemic problems in accident causation. If the interviewee appears not to understand these matters, an interviewer should explain some of these issues to determine whether the candidate is receptive to a systems perspective. The purpose of such inquiry is not to expose the applicant's lack of knowledge and cause embarrassment or discomfort, but assess the level of willingness to acquire the knowledge, attitudes and behaviors required to fit the productive safety culture. Table 8.2 shows some possible responses and their implications.

The applicants for a supervisory position may demonstrate a range of attitudes to accident causation ranging from 'human error' to lack of supervision and training to systemic weaknesses. The table proposes a number of possible implications of these responses. A candidate who 'blames the worker' may be unwilling to take management responsibility for incidents and may be reluctant to consider underlying issues. This would indicate a possible poor fit with the culture and the need for extensive training. An applicant who takes an organizational view of accidents is more likely to fit the culture and require less training to adapt to the productive safety approach. The responses and implications given are hypothetical but serve to illustrate the behavioral issues that can affect the organizational climate and the cost of integrating a new recruit into the firm.

According to the responses given in Table 8.2, the candidate who takes a systems perspective of accident causation appears to be the best fit. This does not mean, however, that the applicant is necessarily the best person for the job. The overall match of KSAs and attitudes to the position requirements determines this. If a person with some competency weaknesses is employed, then the firm should be prepared to address

**Table 8.2** Applicant attitudes to accident causation
Objective of interview question:
Assess the applicant's experience in incident investigation and attitudes to accident causation.

| Applicant response | Possible implications | Fit with culture | Training needs |
|---|---|---|---|
| 'Accidents are caused by people making mistakes.' | Responds by blaming the worker. May be reluctant to take management responsibility for accidents or to consider underlying systemic issues. Alternatively, may not have developed knowledge of the systems view of accident causation. | May be a poor fit. Use additional questions to determine receptiveness to a systems perspective. | May need extensive training to create a fit. |
| 'Accidents are caused by lack of supervision and training.' | May be acknowledging that he/she has a role to play in preventing accidents. May indicate a concern for worker safety and a willingness to take a leadership role. | Possible fit. Use additional questions to determine level of ownership and receptiveness to a systems perspective. | Likely to need some training to shift to a systems perspective and to create a fit between the applicant and productive safety management. |
| 'Accidents are caused when things go wrong with the system.' | May be acknowledging that he/she is part of the system and has a role to play in preventing accidents. May wish to focus on maintaining a safe system or may wish to dilute responsibility away from him/herself and the team. | Possible fit. Use additional questions to determine level of ownership and specific knowledge of systemic weaknesses. | May need training to enhance the fit between the applicant and productive safety management. |

this gap. (It should be noted that the company should never use an applicant's accident history as a selection criterion. This perpetuates the myth of the 'careless worker' and leaves the company open to claims of discrimination.)

The purpose of the value-based approach to recruitment and selection is two-fold. The first is to assess the candidate's KSAs against those required to perform the job satisfactorily. The second is to evaluate the candidate's underlying values. These determine the applicant's willingness to action those behaviors that fits the culture. For employees to demonstrate desired responses, therefore, they need both the required competencies and appropriate motivation.

The recruitment and selection process is important because it is the first step in acquiring the human capital that the firm needs to achieve its objectives. Once a new employee has joined the business, steps are then taken to develop their competencies and attitudes through employee training and development, performance management and rewards. These are discussed in detail in the following chapters.

## Summary

In this chapter it has been explained that management commitment and leadership are prerequisites for the implementation of an effective OHS management system. Leader behaviors that contradict the philosophy and values of the system erode employee buy-in and prevent the desired organizational culture from developing. Without accompanying positive behavioral change, the system is in danger of becoming a pile of paperwork. The initial executive training process is therefore extremely critical. Once managers and supervisors resolutely identify the compatibility between their own values and those underlying productive safety management, they are ready to be advocates of the approach. When they have the 'know-why' and 'know-how', it is the appropriate time to disseminate the system by sharing ownership of it with the workforce using PDs, other HRM tools and using participative processes, with the intention of enlisting commitment at the shop floor level. To sustain the change it has to be driven by the leadership and 'salesmanship' of the management team and reinforced through the employee training program. The following chapter addresses operational training and explains how to develop competencies and attitudes that maximize employees' contributions to safety, productivity and system factor quality.

## References

1. Most, I.G. (1999) The quality of the workplace organization and its relationship to employee health. *Abstracts of Work Stress and Health March 1999, Organization of Work in a Global Economy*. American Psychological Association/National Institute for Occupational Safety and Health Joint Conference, Baltimore, USA, p. 179.
2. Erickson, J.A. (1994) The effect of corporate culture on injury and illness rates within the organization. *Dissertation Abstracts International 55 (6)*. Doctoral Dissertation, University of Southern California, USA.

3. Pierce, F.D. (2000) Safety in the emerging leadership paradigm. *Occupational Hazards*. June 2000, Penton Media, Inc in Association with The Gale Group and LookSmart. Available on website: http:www.findarticles.com/cf_dls/m4333/6_62/63825876/p1/article.jhtml?term=pierce
4. Harmon, F.G. (1996) *Playing for Keeps*. John Wiley & Sons, New York, USA.
5. Hitt, W.D. (1988) *The Leader-Manager Guidelines for Action*. Battelle Press, Columbus, USA.
6. Banki, I.S. (1997) *Dictionary of Professional Management*. Systems Research, Los Angeles, USA.
7. Manuele, F.A. (2000) *On the Practice of Safety*, 2nd edn. John Wiley & Sons, New York, USA.
8. Blewett, V. and Shaw, A. (1997) *Best Practice in OHS Management*. CCH Australia Limited, Sydney, Australia.
9. Glendon, A.I. and Stanton, N.A. (2000) Perspectives on safety culture. *Safety Science*, No. 34, Elsevier Science, New York, USA, pp. 193–214.
10. Blake, R.R. and Mouton, J.S. (1981) *The Versatile Manager: A Grid Profile*. Richard D. Irwin Inc., Illinois, USA.
11. Gardner, R.L. (1999) Benchmarking organizational culture: organizational culture as a primary factor in safety performance. *Professional Safety*, American Society of Safety Engineers, USA, March, pp. 26–32.
12. Covey, S.R. (1990) *Principle-Centered Leadership*. Simon and Schuster, New York.
13. *Occupational Safety and Health Act 1984 (Western Australia)*. Western Australian Government Printers, Perth, Australia. Available on website: http://www.austlii.edu.au/au/legis/wa/consol_act/osaha1984273/
14. Line Management Pty Ltd, *Frontline Management Leadership Development Program*. Outline available on website: http://www.linemanagement.com.au/FLMhome.htm
15. David, F.R. (1995) *Strategic Management*, 5th edn. Prentice Hall, New Jersey, USA.
16. Kaplan, R.S. and Norton, D.P. (1996) *Translating Strategy into Action – The Balanced Scorecard*. Harvard Business School Press, Boston, Massachusetts, USA.
17. Stone, R.J. (1991) *Human Resource Management*. John Wiley & Sons, Queensland, Australia.
18. Mines Occupational Safety and Health Advisory Board (1998) *Risk Taking Behavior in the Western Australian Underground Mining Sector – Risk Taking Behavior Working Party Report and Recommendations*. Available on website: http://www.dme.wa.gov.au/prodserv/pub/pdfs/riskreport/PDF
19. Correll, M. and Andrewartha, G. (2000) *Positive Safety Culture – The Key to a Safer Meat Industry – A Literature Review*. Available on website: http://www.workcover.com/documents/meatCultureLiteratureReviewV81/pdf

# Chapter 9

# Training – building the organization's capacity

Productive safety management involves systems and behavioral change to shift the firm towards optimum performance, safety and quality. Those changes discussed so far – the restructuring of traditional OHS practices to fit the entropy model, building commitment and leadership among supervisory staff, and identifying accountabilities – have primarily been management-driven and management-centered changes.

## Gaining employee commitment

The next step is to disseminate the philosophy and values of the approach to the operational level so that it becomes fully integrated into the total management system and the foundation of the organizational culture. Productive safety management is not intended therefore, to be an 'off-the-shelf' management system but an organizational 'lifestyle' that embodies the company's sense of identity, its standards, its goals and its social dynamics. Both the economic and social purposes of the firm are considered to be important. A business' contribution to the community is not purely transaction; that is, limited to financial factors such as providing goods and services, employing people and generating profit. The firm is also a microcosm of the society and thus has a significant impact on social outcomes. The strategic alignment channel illustrated this interdependency. In particular, the safe behaviors learned by employees at work have an impact on community health and safety and vice versa. In addition, when employees act safely outside the workplace, the likelihood of absenteeism due to nonwork-related injury and sickness also declines. The benefits for the community and for the firm are therefore mutual.

The consideration of the firm as both an economic and social entity is highly pertinent in the current changing business environment. The traditional view of the company as a body motivated almost entirely by economic parameters, the neo-classical model, is being surpassed by a social-economic (SE) perspective.[1] The SE model sets the high-bar for firms entering the new economic age:

> The ideal (or Z-) firm is the SE firm that has developed all these internal organizational capabilities as much as possible, thereby reaching its highest human potential. The Z-firm is thus both highly competitive and highly responsible. It is not only outstanding in responding to competitive challenges related to its basic business activities, but it has also found fully responsible ways to deal with its external and internal stakeholders and society as a whole.[1]

The SE model identifies the characteristics of the ideal firm, but acknowledges that actual firms ordinarily manifest some degree of inefficiency and irresponsibility. This is a view supported by the entropy model which attributed this to the residual risks within the firm's system factors and its tendency towards chaos. From the social perspective, these management weaknesses stem from the risks within the human resources system factor such as incomplete KSAs. Training or investment in human potential can reduce these deficiencies and enhance the quality of management. The significance of competencies and cultural issues in creating effective OHS management systems indicates that both the economic and social perspective of the firm require attention concurrently. These areas are addressed using tools such as the reasonableness test, to influence the internal dynamics of work relationships positively, as well as the achievement cycle to measure production, safety and system factor quality outcomes, which are directly linked to the 'bottom-line'.

There is, therefore, a strong case for investment in human potential to acquire competencies that contribute to the economic performance of the firm and to develop attitudes that create an organizational climate in which social dynamics are conducive to success. Once management has been provided with the skills to implement the approach, the next priority is to provide employees with training that enhances these competencies and attitudes. This chapter explains the core areas of training and the returns the firm can attain from this investment.

The entropy model indicates that to minimize risk, the quality of system factors including human resources has to be maintained at a high standard. The firm's ability to improve workforce quality is not entirely dependent, however, on its training budget. The extent to which training creates real change – training is transferred into the workplace and translates into tangible results – is also dependent on the willingness of employees to learn new competencies and to demonstrate behaviors that fit the desired culture. Hence, companies have to gain employee support for the change program.

How can this support be won? A prerequisite is a sense of ownership. As explained in the previous chapter, sharing KRAs using PDs assists employees to have a clearer understanding of their responsibilities and contribution to organizational achievement. This, however, is mostly procedural rather than motivational. It was also explained that the behavior of management is a very significant parameter in acquiring and sustaining employee commitment and that organizational leadership is an influencing agent of change. This, however, is only half the picture because management effort alone does not necessarily generate initial commitment among workers. There are many other variables in play including the level of trust and the firm's history of change management. From the worker's point of view, management's 'salesmanship' is an external factor and does not guarantee employee buy-in.

'Real' change occurs when workers make an internalized decision to accept the corporate ethos based on the perceived compatibility between this philosophy and their own values. The probability that an action will be taken or a behavior will be followed increases dramatically when people develop a belief in its value.[2] This means that the training provided to enlist employee commitment has to have the same rationale that applied to training of the management team. This is to deliver 'know-what', 'know-why' and 'know-how'. The basis of commitment and behavioral change is belief, which must then be upheld by management behavior that reinforces this belief as a shared ideal.

> Basic to management behavior is its value system because, . . . behavior is the expression of values. When there is congruence between management's values and those of the employees, employees will feel that they 'fit' in the organization and safety performance improves. However, when there is an incongruence between the value systems of management and employees, safety performance is lower.[3]

Employees can be encouraged to develop a belief in the value of the management system through a learning process. The traditional approach has been to provide training 'to bring a person to a desired state or standard by instruction and practice'.[4] Learning, however, is much more pervasive, continuous and variable in its structure than formal training. It occurs through different contexts such as on-the-job discussions, group problem solving and observations. Management's behavioral modeling, whereby supervisors demonstrate appropriate attitudes and values, plays a significant part in employees' learning processes. Workers absorb social-contextual information during their job activities and this can either reinforce or erode any formal training that has been provided. Management should be conscious of and not underestimate the impact of these influencing factors. For example, as explained in the previous chapter, an appropriate leadership style depends on situational variables including the level of risk. When a supervisor acts in a directive manner in low-risk conditions and denies a highly skilled team its autonomy, this can contradict any training aimed at increasing ownership at the operational level and reduce the return on this investment.

To prevent supervisors from making responses that are counter-productive, management leadership has to be developed prior to any behavioral instruction for the general workforce. By so doing, unintended sabotage that results when the training given to workers is contradicted by nonconstructive supervisory behavior, should be avoided. The importance of leader behaviors as a source of employee learning has been explained in the path–goal theory of leadership developed by House and Mitchell.[4] This theory reinforces that the manager's leadership style has to be appropriate to the situation. Figure 9.1 illustrates the effect of leaders' behaviors on subordinate attitudes within a general workplace context.

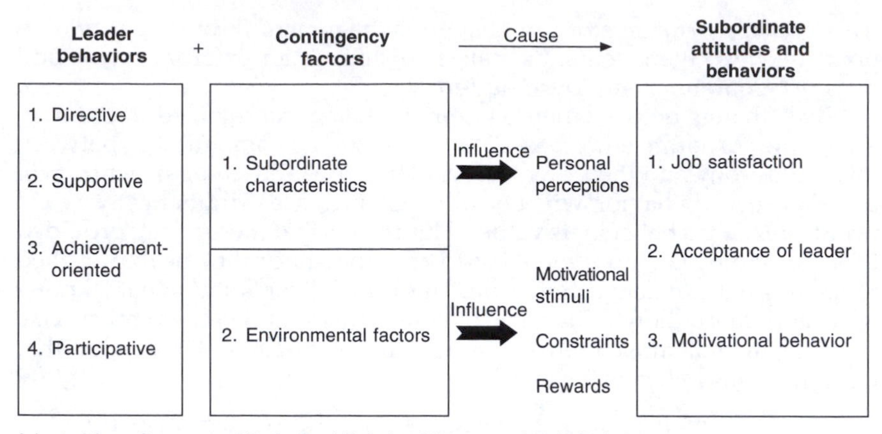


Adapted from Summary of Path–Goal Relationships, page 633 in Ref. 4.

**Figure 9.1** Workplace relationships and the path–goal theory of leadership

Overall, the theory conveys that workers' attitudes and actions are affected by leaders' behaviors and this, in turn, determines the individual worker's motivation to learn and to change. The manager can choose from a number of leadership styles depending on the conditions. The styles shown are directive, supportive, achievement-oriented and participative, which relates to the assertive, responsive and collaborative styles presented in Fig. 8.4 (all of which were both achievement and relationships oriented). The path–goal model indicates that giving directives is appropriate and has a positive impact on satisfaction and expectancies of subordinates when tasks are ambiguous.[4] There is a negative correlation with satisfaction when workers undertake clear tasks. The implication is, that when risks are known, operators are aware of them and sufficiently skilled to mitigate them, the leader should not use a directive approach. Supportive leadership fits conditions where subordinates work on stressful, frustrating or dissatisfying tasks. This would include situations where the risks are high and the implications of these risks are well-appreciated by workers. Achievement-oriented leadership encourages employees to strive for higher standards of performance and can be reinforced by gearing financial incentives to safety and production outcomes. According to the theory, the participative approach leads to greater clarity of the paths to various goals.[4] Collaboration therefore clarifies the 'means' by which objectives are to be achieved. As a general management technique, when objectives are selected according to employee values, and autonomy is increased, this stimulates a desire to perform that originates from the free will of the worker rather than being imposed by the supervisor or the organizational system. In other words, the theory indicates that the participative approach generates a higher level of motivation.

The figure illustrates that the chosen leadership style is filtered through contingency factors, which are subordinate characteristics and environmental factors. These contingencies imply that there are two dimensions of organizational management that are important. The first is the recruitment and selection process, which has a direct impact on the characteristics of employees entering the firm. This reinforces the need to use value-based techniques to evaluate job applicants. The second is the organizational culture. This is a significant environmental parameter that shapes the norms of the work group and which supports the formal authority system.

The path–goal theory reinforces that appropriate leadership is linked to the context. In previous discussions it was explained that the level of risk is the primary determinant of the degree of authority that the leader should use, particularly in hazardous workplaces. Hence, the greater the risk, the more appropriate it is to use directive leadership. In a dangerous situation, where the worker understands the potential consequences of the risk, nonconsultative corrective action by the supervisor is more likely to be perceived as acceptable and constructive. Using assertive leadership in hazardous conditions can therefore have a positive impact on employee job satisfaction and also increase support for the leader. In low- and moderate-risk conditions, it is appropriate for the employee to be part of the solution and involved in the process of risk management. Figure 9.2 illustrates further the continuum of assertive versus collaborative leadership as it relates to the level of risk. (The figure applies the Tannenbaum and Schmidt model[4] to risk management.)

The figure defines the appropriate use of supervisory authority according to the severity of risk in a given situation. It shows when the manager should use legitimate power in the interests of employee health and

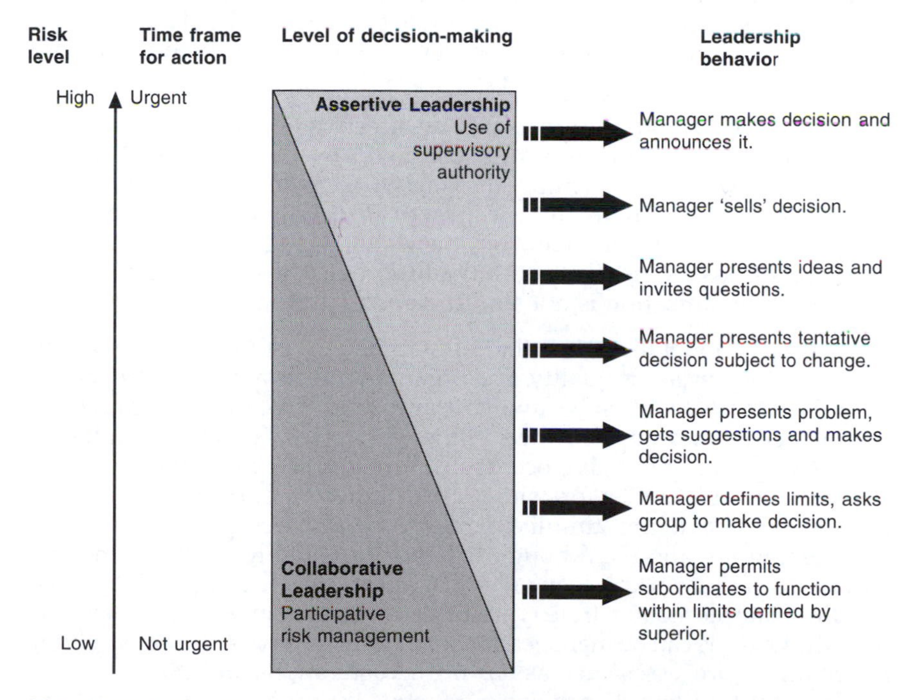


Adapted from Continuum of Leadership Behavior, page 652 of Ref. 4.

**Figure 9.2** Risk and the appropriate use of assertive versus collaborative leadership

safety and the protection of company resources. When the risks are not urgent then the responsibility for health and safety should be disseminated to lower levels of the organization. The result is that employees are given the opportunity to problem solve and learn on-the-job without the manager actually relinquishing control. The supervisor can hold subordinates accountable for the results and therefore in effect defines the limits according to the group's competency level. In contrast, high-risk situations require urgent action, therefore it is appropriate for the manager to act decisively without consultation. The right-hand side of the figure shows how the leader's behavior may be adjusted to facilitate the most effective level of employee input.

## Employee training

In addition to applying the 'best fit' leadership style, the firm has to provide training so that workers understand the nature of risk, how it changes, its repercussions, and how it is influenced by their behavior. Training is used therefore to develop the worker's ability to evaluate and adapt to situational factors. Higher levels of employee KSAs should also increase the worker's tolerance of centralized authoritative responses in high-risk conditions and allow decentralized participative management in lower risk circumstances to be more effective.

Training is a critical component of OHS management because the quality of human resources has a significant impact on both residual and entropic risk levels. Training, in itself, however, is not the core solution to risk management:

> education and training can only be effective where the desired forms of behavior are not discouraged by the nature of the work. For example, simply making workers aware of dangers and tutoring them in safe practices in a work situation characterised by chaotic work peaks and dangerous physical conditions . . . is likely to have little effect if more effective work organization is not simultaneously introduced.[5]

The entropy model illustrates that human resources are only one of four factors in total systems quality. The organization, as a multi-dimensional organism, therefore must be favorable and conducive to allow training to be transferred and implemented in the workplace. In behavioral terms, positive transfer of learning occurs from the classroom to the workplace when the stimuli in the work situation and the responses required are as close as possible to the stimuli experienced and the responses learned in the training situation.[6] Although it is not possible to replicate the risk situations of the workplace in an information session, it is possible to use real life examples to illustrate types of risk. The examples can also explain how the hazards can be managed by behavioral adjustments and applying the required processes such as hazard investigations and JSA.

There are four keys to achieving effective training. The first is to make the learning meaningful by providing material within a context that facilitates understanding, such as using hands-on experiences. Secondly, the theory and practice can be integrated by highlighting the common principles underlying the system. For instance, workers can be shown how the entropy model applies in real life by identifying the different sources of system factor risks for an activity, like climbing a ladder to cut an overhanging branch of a tree. A third option is for trainees to be involved in discovery learning by working through a situation to identify the problem, evaluating the level of risk associated with it, and developing alternative courses of action, which are then assessed against the reasonableness test. The final means of increasing learning transfer is to motivate participants by accentuating the benefits to them from acquiring the relevant KSAs and structuring the program so that trainees work to short-term, realistic goals.[6]

How can the firm determine what training is required? Productive safety management uses a concept called the 'capacity reservoir' to represent the total competency pool of the firm's human resources system factor. Initially, it involves applying traditional approaches to training needs determination using gaps analysis to identify the shortfall between the KSAs that the workforce has and those required to work safely and efficiently. When a new employee starts with the firm, he brings a set of KSAs, values and attitudes into the workplace. There is often a gap between this set and that required to perform the job to the required standard. In addition, a degree of incongruence usually exists between the recruit's values and attitudes and those espoused by the productive safety culture. To address these disparities and to clarify expectations, new recruits should be put through induction. The importance of this orientation process is widely recognized, particularly in relation to fulfilling the employers' duty of care and as a result, many organizations take a legalistic view of induction:

> The employer has a duty to inform employees of occupational health and safety rules in the organization, equal employment opportunity policies and general work rules that govern conduct in the organization. The formal induction is a convenient forum for this activity and provides a record that employees have received that information in case of litigation.[6]

### Induction

Apart from fulfilling these legal obligations, induction has two other purposes that have implications for the organizational culture. The first is to clarify the economic nature of the employment relationship; that is, the exchange of time and labor for remuneration. This includes accountabilities, hours of work, leave entitlements and other issues centered on compensation for effort. The exchange of this transactional information during induction leads to realistic expectation of extrinsic (financial) rewards. The economic nature of this relationship has two components. These are a task component and a performance component, as detailed in the job descriptions described earlier. The primary aim of induction is to assist the new employee to achieve a high standard of safety and performance in as short a time as possible. This assists the firm to manage the risks associated with the new recruit's incomplete KSAs and accelerates the time taken to achieve an acceptable level of job performance. Induction is, therefore, both an important human resources risk management strategy and productivity strategy. Figure 9.3 illustrates the process of clarifying the economic nature of the employment relationship through induction.

Prior to the orientation program, the remuneration and labor/time exchange is not clearly defined, except for those matters contained in the letter of offer or employment contract. During the induction, task requirements and performance standards, included in the PD, are explained. The new recruit is also put through safety training to develop an awareness of safe work practices required to manage workplace hazards. Induction also provides opportunities for the incumbent to clarify uncertainties concerning entitlements. This transactional clarification process is the first stage of integration into the organization.

Economic interdependence, however, is only one dimension of the employment relationship. The second is the social perspective referred to as organizational socialization.[7] The newcomer has to learn the values, norms and behavioral standards that are required by the firm. This involves assimilation into the company culture and addresses issues including the

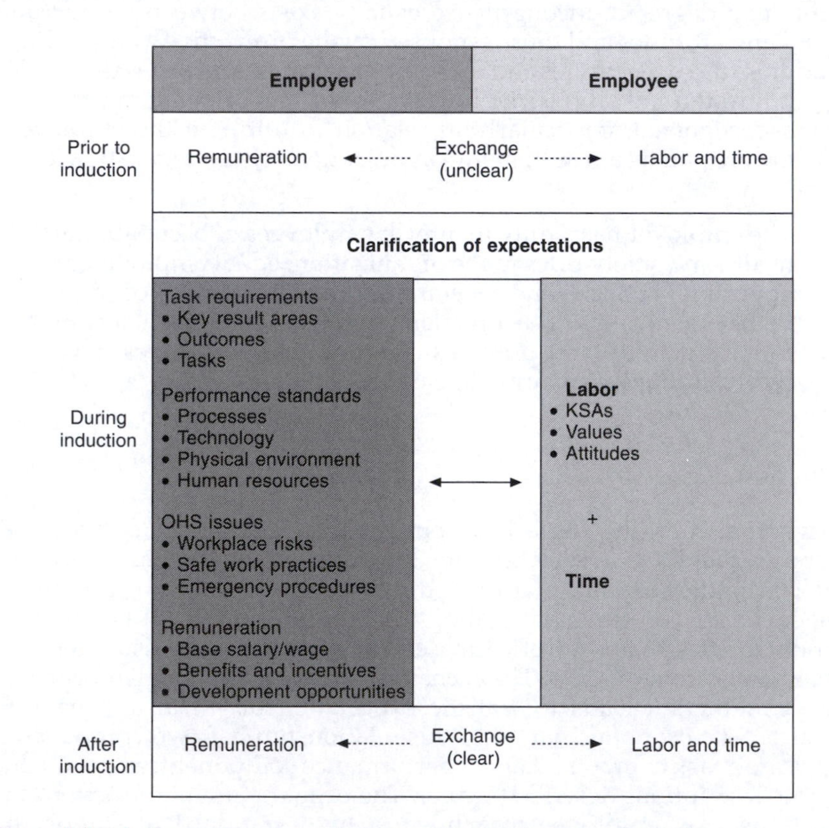


**Figure 9.3** Induction and the economic nature of the employment relationship

driving goals of the organization. Whilst these end goals are important, the means by which they are achieved is also critical to company success. In addition, employees need to understand their roles and responsibilities. This means that the company has to have a clear set of rules or principles which maintain the identity and integrity of the organization[8]. The rules, derived from values, serve to define the behaviors of organizational members and begin the integration of the employee into the company.

Induction is the first step in this integration process because it involves introducing the new employee to the workplace's social dynamics, as shown in Fig. 9.4. Initially, there is an information barrier between the new worker, who feels like an outsider, and the existing organization, composed of insiders. During the formal induction, information is provided which falls into two categories. The first is structural information that explains where various roles and incumbents are in the hierarchy that, in turn, defines patterns of communication and relationships. The structure shows how positions within the department and workplace are interdependent and provides the new employee with a social orientation.

The second is cultural information communicated explicitly that includes documented goals, values, policies and the overarching philosophy. The new recruit also gathers information implicitly from observations that provide an insight into the social climate and peer-group dynamics. The relational dimension of the induction process allows the incumbent to express personal goals and values, and assess the fit between them and those of the firm.

What difference does this make to organizational performance? The induction process erodes the information barrier thus integrating the new employee into the company. Both economic and social assimilation put a new recruit on a learning curve that shifts her towards optimal performance and safety, thereby accelerating the time normally taken for

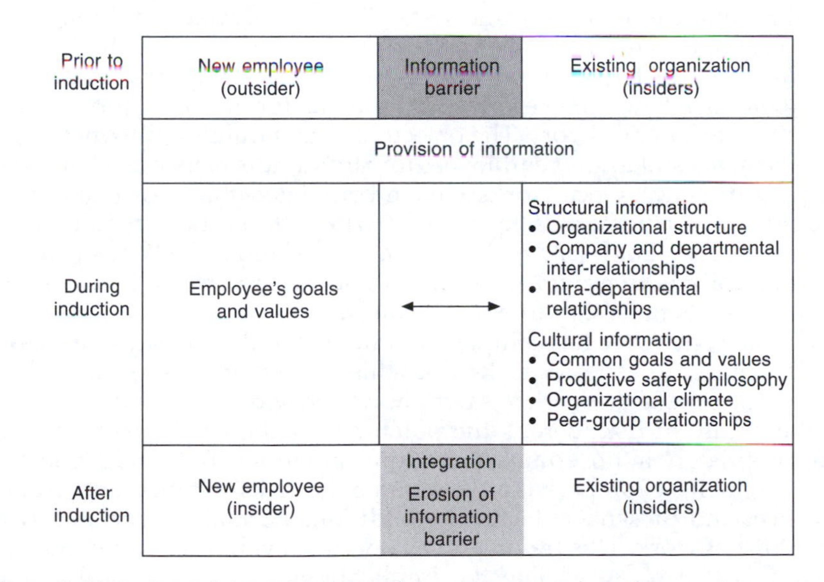


**Figure 9.4** Induction and the social nature of the employment relationship

a new employee to make positive contributions to organizational achievement. In addition, induction plays a significant role in risk compression and control. Through the orientation program, a new employee becomes an insider with goals and values more closely aligned with the culture. Hence, it is critical for the integration process to be successful because of the impact on both the individual and the business. At stake are:

(1) The new employee's safety, performance, satisfaction and commitment to the firm;
(2) The work group's safety, performance and satisfaction;
(3) Start-up costs invested in the new employee;
(4) The likelihood that the employee will remain with the firm and therefore, the overall level of turnover and retention of competencies;
(5) The cost of replacement of employees that resign.

## The capacity reservoir

Induction is the first step in integration and alignment of new employees. It provides an initial in-flow into the capacity reservoir, which stores the collective competencies of the workforce employed by the firm. There are six competency builders that fill it, as shown in Fig. 9.5. The first of these are the KSAs developed by the new recruit through previous work experience and training, which the firm sources from the labor market. The business obtains these proficiencies through the recruitment and selection process. After the employee joins the company, the firm begins to enhance such skills to meet its human resources needs and the first stage of this in-house training is induction.

In addition, to operate safely and efficiently, the employee must have a full contingency of job-specific KSAs. These are developed through on-the-job support and training or, where the firm cannot provide these in-house, by contracting external trainers to deliver the required instruction. Competencies that need certification, such as forklift operators' tickets, fall into this latter category. The objective of this training is to meet legal requirements and to shift employees towards a satisfactory level of safety and performance. Once workers have completed this instruction, they should have a solid base of 'know-how' within the scope or requirements of the role they hold. Under duty of care legislation, the firm's primary responsibility is to provide the training needed to ensure that workers have the skills necessary to operate safely.

Once an employee has completed induction and achieved satisfactory competence in job-specific tasks, does this mean that he is a safe worker? These KSAs make the worker 'safer', however, without an understanding of the nature of risk there is the potential for the employee to behave inappropriately in uncertain, nonroutine situations. Both induction and task 'know-how' prepare the worker for the risks within standardized practices and present under those conditions normally associated with the scope of work. This training does not, however, provide the skills to identify hazards, appreciate the implications of those hazards, or to

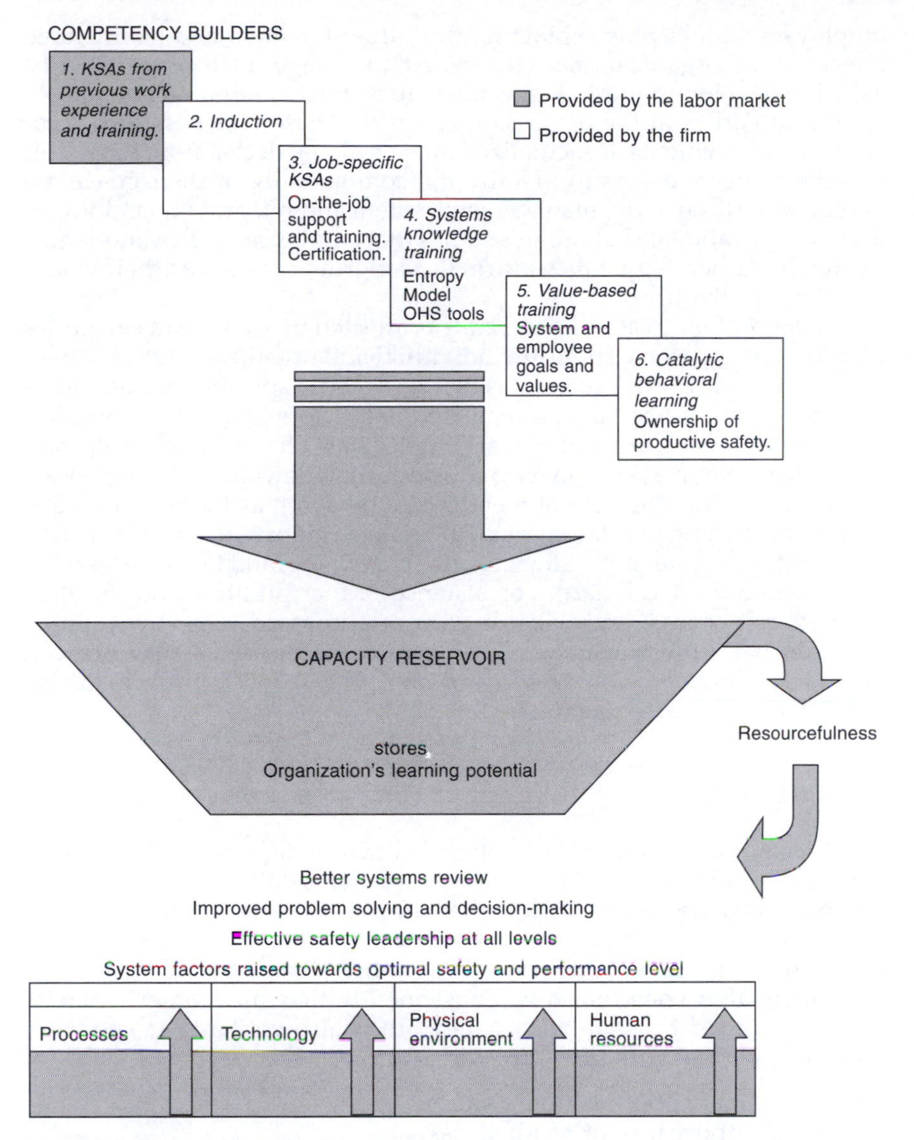


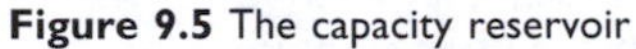

**Figure 9.5** The capacity reservoir

understand the impact that worker behavior has on the level of risk. This competency gap is addressed by providing the 'know-what' of productive safety management, which as explained in the previous chapter, covers the nature of risk based on the entropy model and the tools used in the OHS management system. The instruction delivered should therefore be similar to the knowledge-based training received by the management team, but more operational rather than strategic in its thrust. This phase of competency building is shown at point 4 in Fig. 9.5.

The next stage of training has to focus on gaining employee commitment. This involves linking the philosophy and values of the approach to employees' values plus explaining the value statement and its intended effect on the organizational culture. At this stage, teams may also be asked to develop workplace-specific values in team ethic statements. As explained earlier in the discussion about the reasonableness test, these workplace statements are actualized in operational decision-making. This process allows workers to identify the commonality of their goals and values with those of the management system and integrate them into the system. The rationale behind this step is that employees will be motivated by the change strategy if they affirm the congruence between their values and those of the firm.

The level of motivation that results is affected by two broad categories of influencing factors. These are individual and organizational variables. The former describe the attributes that are worker-specific such as ability, experience and personality. Organizational factors include the physical work environment, organizational climate, team climate, leadership, and the human resources system used to select, train, reward and compensate individuals.[9] Training cannot therefore be relied on as the sole source of managing behavior-related risk. It is also important to ensure that organizational systems are aligned with the training, and facilitate learning transfer into the workplace. For instance, if the conditions on the shop floor do not change and hazards are not addressed, then the training program has little substance. To generate the desired behaviors and attitudes, therefore, a strategic and comprehensive approach to risk reduction and management is required.

### The safe worker

The final phase of training is catalytic behavioral learning, which shows employees how they can positively influence safety and performance outcomes by adopting behaviors that support the system and that build the desired organizational culture. This learning involves taking ownership of productive safety as a KRA, with the totality of the acquired knowledge leading to safety consciousness. Consequently, the 'safe worker' not only has the required KSAs to work safely but is also vigilant. As shown in Fig. 9.6, the 'safe worker' is therefore someone who:

(1) Is educated about residual risk;
(2) Is vigilant because of residual risk;
(3) Is kept informed of changes in entropic and residual risk;
(4) Works safely and efficiently to keep entropic risk low;
(5) Has the competencies and opportunities to make suggestions that contribute to improved safety, performance and system factor quality.

## Employee empowerment and developing resourcefulness

The reason for depicting these competencies as a reservoir, and employee development as the means of filling it, is to show the importance of

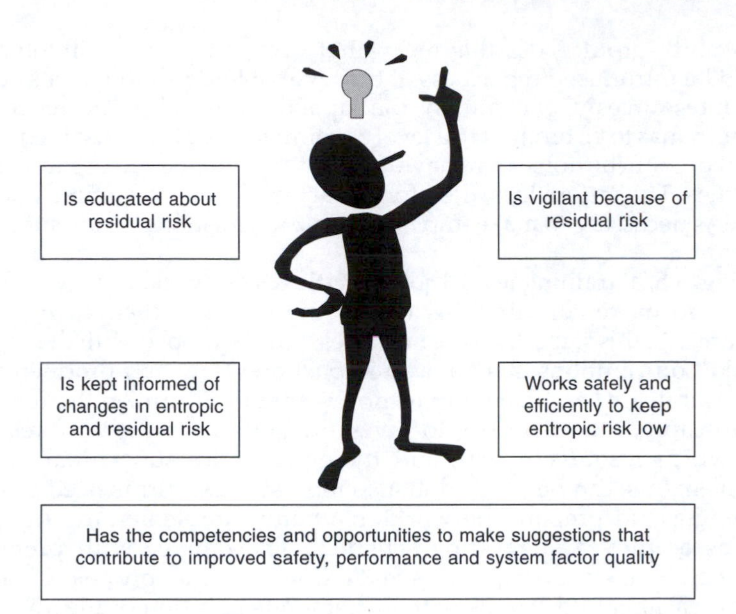

**Figure 9.6** The productive safety worker

investing in the organization's 'learning potential'. As the workforce's proficiencies are enhanced, the level of residual risk associated with the human resources system factor decreases and so does the firm's susceptibility to entropic risk normally associated with incomplete KSAs. In addition, workers have a higher level of safety consciousness that translates into better risk management through appropriate behavioral responses at the interface with other system factors. Workers also acquire the skills to contribute productively to the system's review, problem solving and decision-making. The 'over-flow' of employee competencies that leads to improved system factor quality is referred to as 'resourcefulness'. When management systems encourage this resourcefulness by providing avenues for employee input, the company is better able to pursue continuous improvement.

There are a number of contextual variables that determine the 'flow' of resourcefulness. In particular, the health of the culture is critical to this release. It provides the environment and climate in which employees can confidently make contributions to make their jobs safer and more productive.

There is a strong link between workplace consultation involving management and their workforce and successful organizational change. These arrangements make employees more inclined to become involved in continuous improvement.[10] This contributes to greater efficiency within their own jobs that translates into enhanced productivity for the firm. There is also greater commitment to organizational goals. A more cooperative work environment is created with underlying values based on mutual trust, a desire for job satisfaction, and emphasis on common goals such as safety and production.

Employee participation is thus a key issue affecting system factor quality, safety and output. Does this mean that opportunities for involvement should be introduced regardless of the level of KSAs of the firm's current human resources? Conceptually, the capacity reservoir indicates that the workforce has to be brought to a level of competency at which it is equipped to make contributions that have a positive impact on organizational outcomes. The rationale is, therefore, that employees must firstly acquire the KSAs necessary to make implementable suggestions before such input is sought.

In Fig. 9.5, at training level 3, job-specific KSAs, workers have sufficient abilities to make contributions within the scope of their own job and work area. At this stage, they are not necessarily equipped with the capacity to make contributions that have an impact on safety and productivity at a more strategic level. The implication is that to maximize the return on human capital, the firm has to invest in building a capacity reservoir from which resourcefulness, and therefore, more substantial systems improvements, can be derived. It also indicates that firms need to invest in training and prepare the workforce before implementing programs such as employee suggestion schemes. The danger with premature introduction of such programs is that employees may propose ideas that are not feasible. In these cases, management's rejection of the ideas may be perceived by workers as lack of commitment to safety and lack of genuine interest in employee input. This is particularly so if management fails to provide adequate feedback concerning the resultant courses of action.

There are therefore a number of prerequisites for the implementation of participative or empowerment strategies. Workers need to possess adequate motivation, ability, resources and self-efficacy to successfully perform such expanded roles. This requires them to undertake training and to acquire relevant skills.[11] In addition, managers have to develop a satisfactory level of 'maturity' in terms of appropriate decision-making before they are able to empower employees effectively. Without this there is the danger that supervisors will erroneously view collaborative processes as a relinquishment of managerial authority,[12] and therefore, be reluctant to 'let go'. In contrast, the constructive view of empowerment is to see it as a method of enabling managers to obtain maximum leverage from the talent and motivation of their team members.[13]

The capacity reservoir illustrates that human resources have to reach satisfactory proficiency in task-related, risk management and interaction skills to achieve consistently effective contributions within participative processes. At the point where the majority of employees within a team meets these minimum competency standards, the firm can expect returns on its investment in training through joint problem solving and innovative solutions.

How can the firm measure the gains of its investment in the capacity reservoir? The most direct way is to measure changes in safety and productivity. As employee KSAs and safety consciousness are raised, these results should improve concurrently. In addition, when employees interface with other system factors, the entropic risk that they introduce should also decline because of a heightened level of safety consciousness. This means, for example, that the quality of maintenance practices such as housekeeping, the care of equipment, and adherence to safe work practices should improve. A further means of measuring yield on investment is by assessing the benefits of employee suggestions. Figure 9.7 summarizes the returns attainable from filling the reservoir.

Resourcefulness is optimized when employees contribute a large number of suggestions of high quality that result in a high implementation rate. On the left-hand side of the figure, the overflow of the reservoir in a context that facilitates employee input leads to the release of resourcefulness. On the right-hand side, investment in training raises the level of safety consciousness that allows better risk management. Employees become aware of the hazards in the workplace and are vigilant to reduce the probability of an undesirable event. The combined effect of resourcefulness and enhanced safety consciousness improves system factor quality, giving the firm better positioning to pursue continuous improvement. Collectively, this accelerates the firm's learning curve, which is comprised of the sum of each employee's learning curves.

As explained earlier, conceptually the reservoir stores the organization's learning potential. Through programs such as employee suggestion schemes, OHS representative groups and autonomous team structures, the firm capitalizes on this learning potential. It converts it to some form of tangible financial value. Without a conducive, organizational culture, this potential remains dormant and unprofitable.

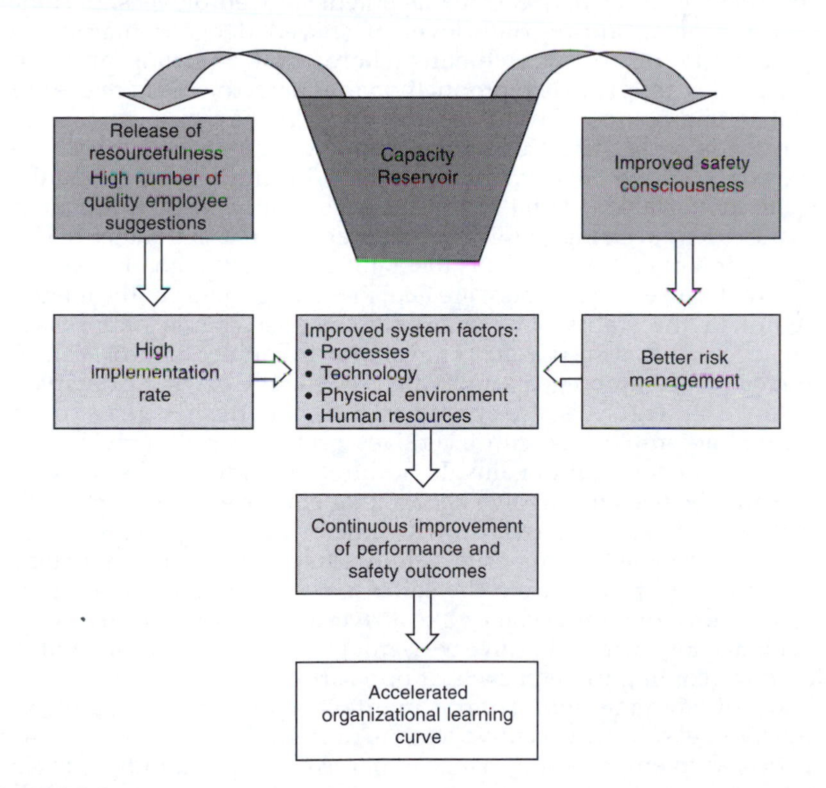


**Figure 9.7** Optimizing returns from the capacity reservoir

The final method that can be used to measure the benefits of training investment is to compare the results of an auditing cycle with the findings of the previous audit. This should confirm that improvements in risk management have been made in areas that are directly affected by the quality of the human resources system factor such as housekeeping standards and a reduction in equipment damage resulting from inappropriate use.

## Overcoming resistance

Does investment in the reservoir guarantee returns from employee contributions? Management needs to be aware that resourcefulness can be blocked. The main causes of obstruction, shown in Fig. 9.8, are distrust, resistance, insufficient management commitment, poor supervision or decision-making and past inaction. The figure also describes alternative solutions to nullify these blockers, for example, distrust can be remedied by emphasizing common goals and values using the value statement as a shared point of reference. Management must therefore be committed to applying the statement and using the reasonableness test to ensure that actions taken are equitable and also balance short- and long-term goals. When there is trust between management and employees, a context facilitative of an appropriate level of shared decision-making and responsibility result. In a climate where trust is strong and input opportunities are provided, productivity and effectiveness are greater.[10]

The second source of blockage identified in the figure is resistance, which tends to be a fact of human behavior. Commonly, good ideas create hope.[14] Some people respond to this with enthusiasm whilst those who are comfortable with the status quo become anxious. It is this anxiety that causes fear of change, which becomes apparent as resistance of either an overt or a covert nature. This tends to sabotage the good idea.

The root causes of resistance are fear, low energy, inertia (the tendency to maintain the status quo), memories of past change failures, and percentage – what is the payoff.[14] The solution to counter these obstructions is to focus on common goals and values as the underlying principles of the change strategy. It is also appropriate to share the benefits of change by providing employees with incentives geared to productivity, safety and system factor quality gains. In addition, the greater the resistance, the greater the need for involvement so that employees can be part of the transition and take ownership of it. Inherent in all group situations, however, there will be a minority of individuals who will be reluctant to embrace the shift away from the status quo. When a strong corporate culture is developed it becomes like a river of change and carries these individuals along with it. Positive peer-group pressure can play a significant role in overcoming these pockets of opposition.

Lack of management commitment is also a major blocker of resourcefulness. If the executive team is reticent then employees cannot be expected to embrace the change either. As explained in the previous chapter, this obstacle can be overcome through management training.

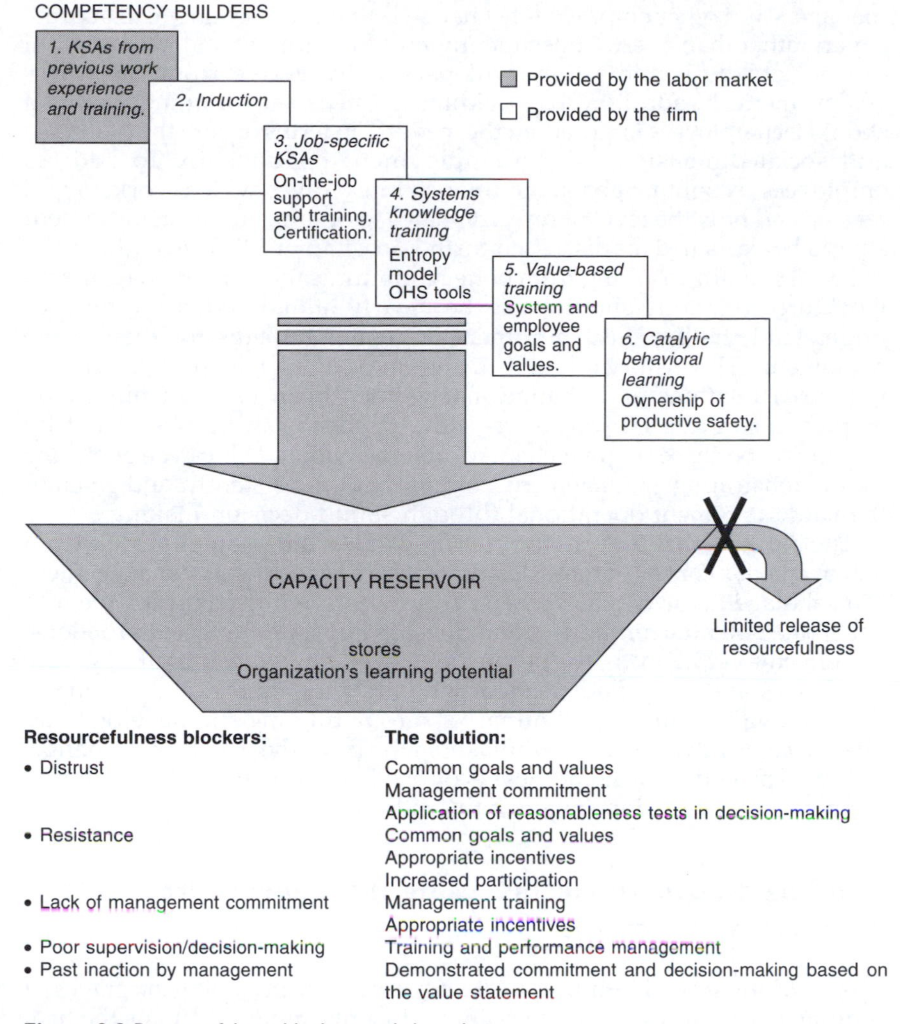


**Figure 9.8** Resourcefulness blockers and the solution

The core objective of the training is to provide 'know-how', underscored by the belief in the fit between the system's philosophy and the management team's values. It is also appropriate to reward these senior employees with incentives when they deliver the required improvements through effective leadership and decision-making. (The achievement cycle performance management system discussed in Chapter 12 links leaders' achievements to rewards.)

Poor supervision and decision-making can hinder resourcefulness and are usually symptoms of lack of management commitment or 'know-how'. If the firm does not correct these weaknesses, employees will be reluctant to make suggestions if they perceive their supervisor to be unreceptive. The quality of leadership has therefore a direct impact on the willingness of workers to 'add-value'. The difficulty with 'forcing' managers and other employees to change is that the resistance can become covert rather than overt. Consequently, employees must understand, from the outset, the benefits they will personally derive from the firm's achievement. In addition to providing the necessary training, the firm should let employees know about the 'payoff', in terms of both the economic and social dimensions of the employment relationship. To address employees' economic goals, the financial incentives which workers will receive and how the level of reward relates to organizational achievement should be explained (as discussed further in Chapter 12). The nonfinancial or social benefits resulting from better communication and management–workforce relations should also be clearly stated. When employees understand the benefits, then there is greater impetus for them to be resourceful. The history of the firm and, in particular, its past patterns of industrial relations and change management, can have an impact on employee contributions. The free flow of ideas can be obstructed by previous experiences of poor change implementation. This can be overcome by demonstrating management commitment consistently and making the value statement operational through sound decision-making.

The blockers that prevent the overflow of learning potential are within management's control. If these obstructions are left unchecked management can sabotage its own achievement. It is therefore important for them to anticipate potential hindrances and develop appropriate remedial actions. An effective strategy is to 'permeate' lower ranks by identifying core employees and shop floor leaders who are progressive, and involving these individuals in the preliminary stages of training and development. They become advocates or pathmakers who pave the way for the change. This also provides for higher levels of participation and collaboration in the pre-implementation stages of the rollout.

## Building a positive culture using human resource management systems

Figure 9.5 illustrated building a capacity reservoir involving six stages of competency building. At some point in this investment strategy, the firm has to determine that the reservoir is sufficiently full to introduce an employee suggestion program. Independently of this formal structure, the firm should also run information sessions such as toolbox meetings, to provide a forum in which safety, performance and quality issues can be raised. These meetings provide 'grass roots' information and knowledge-enhancement opportunities that relate directly to the risk factors dealt with on a daily basis at the operational level.

There are no hard and fast rules concerning when resourcefulness should be released using a suggestion scheme, however, it is essential that employees have the job-specific KSAs to work safely and also a sound understanding of the nature of risk. It is also helpful for common goals and values to have been widely communicated and accepted so that all parties have a shared understanding of the 'means' by which objectives are to be pursued. Effectively, this suggests that the majority of core employees should have completed stage 4 of competency building training. Once the environment is right in terms of this capability, the suggestion program may be used to obtain higher degrees of return on human capital investment. Concurrently, at the shop floor level, the management team should be working towards increased devolution of decision-making to lower levels in the hierarchy, which involves expanding the opportunities for participative decision-making and allowing greater autonomy in work teams.

Much of the value-based training, stage 5, and catalytic behavioral learning, stage 6, in the competency building process can occur through on-the-job experiences rather than formal training. The suggestion program and communication strategies, such as toolbox meetings and team-based problem solving, are intended to contribute towards the latter stages of the capacity reservoir. Formal training is not the only vehicle used to extend the skills pool. In fact, there are a number of HRM functions that contribute to capacity development, as shown in Table 9.1. The first, which was discussed earlier, is the recruitment and selection function. It is used to source the labor market for the KSAs the firm needs to achieve its objectives. A strong emphasis is placed on safety competencies and attitudes as selection criteria, in addition to task-related KSAs.

Induction also adds to organizational capacity and is multipurposed, not simply a means of covering the firm's legal obligations. While assisting the company to meet its duty of care, it also clarifies the economic and

**Table 9.1** Using human resource management functions to build the capacity reservoir

| Human resource management function | Strategies |
|---|---|
| Recruitment and selection | • Apply value-based recruitment practices<br>• Use job-specific KSAs, productivity and safety competencies as criteria for assessing job candidates |
| Induction | • Carry out induction and assess learning when completed<br>• Assess the effectiveness of the induction program using new starter feedback |
| Mentoring | • Provide a mentor for on-the-job support and training, particularly in hazardous areas<br>• Inform co-workers of effect of new employee on residual and entropic risks |
| Training needs analysis | • Assess current KSAs for fit with position tasks/requirements<br>• Assess attitudes for fit with the productive safety culture<br>• Carry out a gaps analysis to identify KSA shortages |
| Training and development | • Provide formal training:<br>Using findings of gaps analysis<br>Systems knowledge training<br>Values-based training<br>Catalytic behavioral learning<br>• Evaluate learning and identify future training needs |
| Performance management | • Implement goal-setting and reviews using the achievement cycle<br>• Identify training required to achieve these goals |
| Reward systems | • Develop and implement incentive systems which reward productivity, safety outcomes and system factor quality |

social nature of the employment relationship and begins the integration of new recruits into the company culture. The HRM practitioner's role is to deliver key components of induction, to evaluate learning and to assess the effectiveness of the program. The HRM department may elect to develop mentoring arrangements whereby an experienced employee acts as a 'buddy' for the new recruit. This is particularly appropriate in hazardous work areas to mitigate the risk associated with the new employee's lack of workplace-specific KSAs. The senior employee's role is to provide on-the-job care including information about hazards, safe work practices and required standards of behavior. This approach follows the recommendations in the WorkSafe plan published by WorkSafe Western Australia in 1999, which identifies training as a primary indicator of an effective OHS management plan. It also states that the required training may be provided using formal courses, mentoring and on-the-job training,[15] and as a result, the firm can take a multipronged approach to learning about hazards and risk management.

The HRM branch plays an important role in identifying competency shortages that affect the level of the capacity reservoir. At the time when the worker begins employment with the firm, a training needs analysis should be carried out. This is to ascertain the fit between current KSAs and position requirements, and the fit between the new recruit's attitudes and the productive safety culture. The gaps analysis assists the firm to provide training targeted specifically at areas of weakness in human resources capacity and thereby maximizes the return on its training dollar. To ensure that this training is effective, learning should be measured to determine whether objectives have been achieved. If for example, the program does not lead to enhanced KSAs as expected, the training strategy will need to be revised to increase effectiveness by either improving the content, changing the mode of delivery, or using multiple contexts in which to apply the learning practically. According to the WorkSafe plan training effectiveness indicators, the key criteria are:

(1) Management has identified the requirements for training under occupational health and safety laws.
(2) Employees understand the requirements for training under occupational health and safety laws.
(3) Management has identified the occupational health and safety training needed by each of their employees.
(4) Training is planned in a systematic way using information from a training needs analysis.
(5) Occupational safety and health training has clear objectives.
(6) Appropriate occupational health and safety training is provided.
(7) Occupational safety and health training is evaluated.
(8) All employees are able to follow emergency procedures.[15]

Table 9.1 also indicates that the performance management system has to be aligned with and support OHS management in order to build the required competencies. It does this by helping to identify KSA shortfalls when managers and team leaders undertake goal-setting. As a manager formulates objectives and action plans to improve the quality of system factors, it may be found, for instance, that she requires better project management skills to implement a structural improvement plan for the workshop. The training that comes out of the performance management system, the 'achievement cycle' (presented in Chapter 12), is thus directly geared to assisting the supervisor to reach her goals. The cycle gives employees the opportunity to increase their effectiveness, contribute to organizational success and enhance their skills. During this process, it is the firm's responsibility to provide the training and support to address any competency shortfalls that need to be bridged to make these goals obtainable. This outcome-driven training also increases the level of proficiency in the capacity reservoir.

Table 9.1 summarizes the key HRM functions that should be aligned with OHS within the total management system. For change implementation to be effective, these organizational systems must be congruent and consistent. An area that can introduce considerable nonalignment or weakness is the mismatch of the overall change strategy with reward systems. This has particularly been the case in companies that have tried to improve their safety but continued to operate incentive schemes in which rewards are primarily geared to output. A report by the New South Wales (Australia) Minerals Council in 1997 found that most of the incentive schemes surveyed were based on production rather than productivity. This was believed to encourage risk-taking behavior and it was recommended that alternative schemes which rely on performance measurement and flat salary as the base of pay be used to manage risk better.[16] The consequence of production bonuses is the reinforcement of behavior that is contradictory to the aspired organizational culture in which production and safety are compatible goals. (The development of appropriate incentives will also be discussed in Chapter 12.)

Productive safety management endeavors to develop synchronicity in organizational systems that directly affect the capacity of human resources. At the individual level, the approach aims to create genuine change by shifting the competencies, safety consciousness and resourcefulness of the worker to a high level, which in turn, raises the human resources system factor as a whole, towards optimal safety and performance. As explained earlier, the stages of training illustrated in the capacity reservoir are intended to enhance KSAs and improve the fit between the employee and the firm so that the relationship of mutual dependence inherent in the employment relationship is strengthened. At the core of this fit are individual perceptions, which as explained earlier, affect an employee's commitment. The process of developing the individual in terms of these perceptions can be mapped out as illustrated in Fig. 9.9.

The diagram summarizes the stages of creating a fit between the employee's perceptions and the organizational culture, so that the worker can ultimately define 'what I do', 'how well I do it', 'how I achieve it' and the desired 'outcomes'. These steps are needed for the worker to achieve optimal performance and safety. The steps also match the firm's expectations of what should be done, how well it should be done, how it is to be achieved and the results. The worker's perceptions are developed within the economic and social framework of the organization as a result of the firm's investment in training and integration.

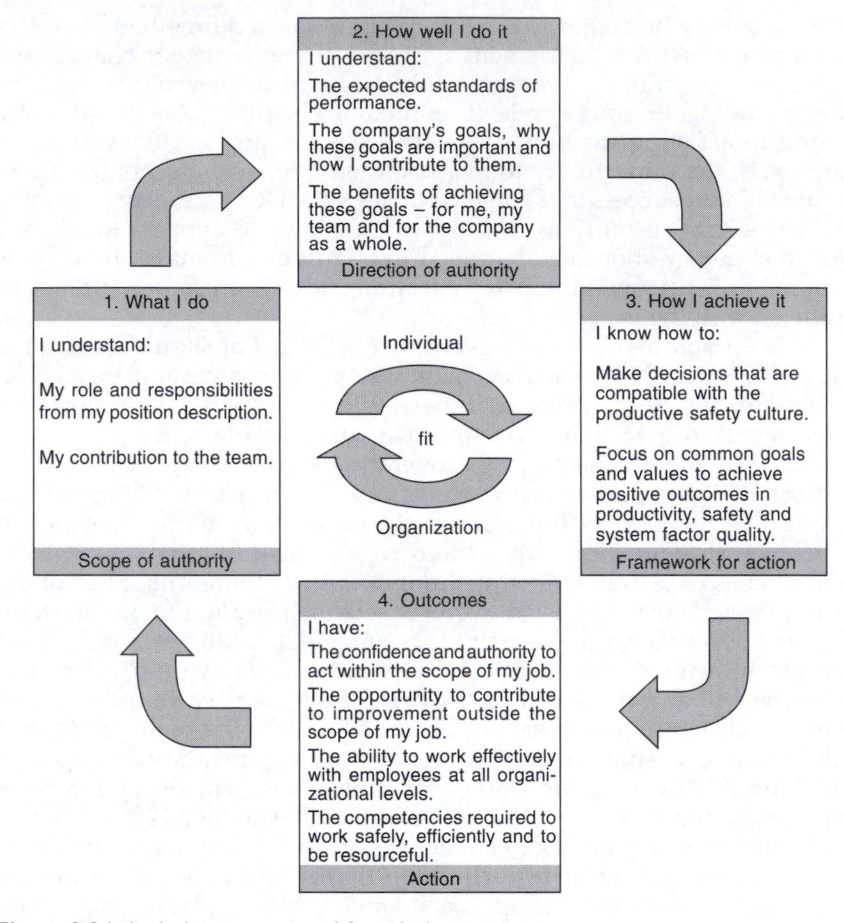


**Figure 9.9** Individual perceptions and fit with the organization

At point 1, the employee begins to fit the organization when core responsibilities and his contribution to the work team are understood. This understanding matches the worker's 'scope of authority' or simply knowing 'what I do' in terms of the 'duties of employees' and the task components of the job. The second stage of integration is to appreciate the importance of performance or how well the work has to be done. This requires accurate or shared perceptions in three areas. The first is the expected standards that define 'satisfactory' performance. In other words, for the employee to fit into the company, he has to understand what level of effort and outcomes have to be achieved to meet the employer's minimum requirements. The second is the company goals, the importance of these goals and how individual achievement contributes to their attainment. These perceptions relate to how the employee fits into the business as an economic entity or his contribution to the 'bottom-line'. Finally, the individual worker needs to be informed about the potential payoff of working well – the exchange of rewards for effort. This effectively links worker effort with profit and other corporate goals, and the sharing of the benefits that come from that success. The understanding of these performance-based issues provides 'direction for authority' in the workplace. It gives managers the legitimate right to set performance standards for the benefit of all parties. For example, when the company becomes more profitable, gains a larger market share or is more efficient, employees should have greater access to rewards.

The third point – 'how I achieve it' – is centered on the 'means' described earlier. It provides the 'framework for action' – standards of behavior or the social perspective of integration into the firm. This framework is important for keeping social order within the business. It defines the boundaries of acceptable behavior and desirable behavior. When the worker makes decisions that are compatible with the culture and focuses on common goals and values to achieve positive results, this reinforces the fit between himself and the company. It demonstrates a willingness to conform to organizational norms.

Finally, as a result of the training and integration process, an individual should have the confidence and authority to act within the scope of the job. On operational issues, therefore, work teams should effectively have a sense of ownership of those system factors within the workplace and be motivated to take responsibility for them. This fit is further enhanced when the operator is allowed to contribute to overall improvement beyond the scope of his job, for example, through the suggestion scheme. The management system thereby permits individuals to develop their learning potential. Overall, the fit between the individual worker and the organization is optimized when the worker has the competencies to work safely and efficiently and to be resourceful. At this point, the firm gains maximum return on its human resources investment.

Over the longer term, building the reservoir should lead to less centralization of authority in the business to the extent that is appropriate for the effective management of risk and for the optimization of performance. In high-risk workplaces, particularly, this is relevant because workers need to feel a strong sense of proprietorship over the job and of the system factors for which they are responsible in order to minimize human factor risks. In addition, the channel indicated that a characteristic of the current labor force is a desire for greater control at the operational level. There are therefore two drivers for greater ownership on the shop floor – those changing employee expectations and the need for better risk management.

The firm can attain numerous benefits from this process of competency building and devolution. In summary, the advantages of employee involvement in the OHS management system are:

Employees are the ones in contact with potential hazards and will have a vested interest.

Group decisions have the advantage of the group's wider field of experience.

Research shows that employees are more likely to support and use programs in which they have had input; employee buy-in for the needed changes is more likely.

Employees who are encouraged to offer their ideas and whose

contributions are taken seriously are more satisfied and productive.
The more that employees are involved in the various facets of the program, the more they will learn about safety, what is causing injuries at their site, and how they can avoid being injured. The more they know and understand the greater their awareness will be and the stronger the safety culture of the organization will become.[17]

Given these benefits, how can employee input be increased? Various contexts and forums can be established for workers to be actively involved in decision-making. These include joint employee–management committees, special OHS committees and advisory groups. They can also participate in site inspections, hazard analysis using JSA, developing safety rules, and training new recruits. Collaborative opportunities can be strategic not just operational, through organizational decision-making, accident investigations, change analyses, and pre-commissioning evaluations. Some members of the workforce may also be trained to be safety observers and safety coaches.[17]

## The productive safety formula

The channel highlights that firms have limited resources, so is it necessary for all firms to invest in the capacity reservoir? How can they determine the level of investment required? Resource allocation for organizational learning depends on the firm's need for safety consciousness and resourcefulness. Issues to be considered include:

(1) Is the organization subject to high residual and/or entropic risk?
(2) Are system factors currently degraded?
(3) Are effective maintenance strategies in place and adhered to?
(4) Is the company currently an industry leader in productivity and safety outcomes?
(5) Is the company consistently operating at high levels of performance and safety?
(6) Does the company face strong competition from rival firms?
(7) Have the strategies used to date effectively pursued production and safety concurrently, addressed systemic weaknesses, and balanced short- and long-term goals?

The higher the level of risk faced by the firm, the greater the need for investment in resourcefulness. Firms with high-risk processes, technologies and/or environments need their employees to be more safety conscious than those operating in lower risk industries. To manage hazards effectively, greater vigilance and better problem solving are required. In addition, the further behind the company is from industry leaders in terms of productivity and safety, the greater the incentive for the firm to improve underlying variables that affect longer-term sustainability. These include the quality of its management systems, the competencies of its supervisors and the intensity of its strategic focus. In effect, the failure to address these underlying issues can lead to a vicious cycle that forces the firm to settle for second best. For instance, if it does not invest in its workforce by providing them with training and opportunities for participation, the turnover rate is likely to be higher and therefore, the level of residual risk in this system factor will also tend to rise. In addition, this instability will be inclined to push entropic risks upward.

The relationship between risk and resourcefulness can be shown using the productive safety formula, given in Fig. 9.10. It states that the change in system factor safety and performance equals the change in resourcefulness less the change in risk.

Risk in the formula refers to the combined effect of residual and entropic risks. Inherently more hazardous operations have a tendency for system factors to operate at lower performance levels because hazards act as a hindrance to efficiency and safety. In these workplaces, the demands on human resources are greater and therefore the susceptibility to entropy is also higher. In businesses where the residual risk is comparatively lower, but where system factors are poorly maintained, the total level of risk can be high due to degradation. The formula indicates that companies that have a high degree of entropic risk exposure also need to have employees who are more safety conscious and resourceful than those operating in state-of-the-art industries.

The formula also suggests that firms which optimize resourcefulness are able to achieve improved safety and performance, particularly if, concurrently they pursue a reduction in the level of risk. To enhance system factor quality, therefore, companies have firstly to compress and control risk, then increase employee participation and ownership. In addition, the need for these two variables is not static over the life of an operation. In firms where, as a result of the aging of physical factors, residual risk rises, for example, as an open-cut mine becomes deeper and more complex, the importance of mitigating risk using resourceful also

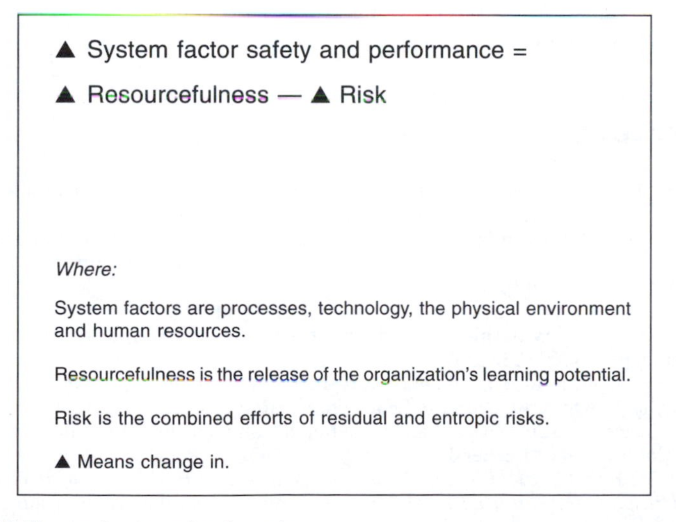


**Figure 9.10** The productive safety formula

rises. Firms therefore need to prepare for these increasingly hazardous conditions by expanding employee involvement in risk management. Finally, the formula indicates that risk, resourcefulness, the quality of system factors, and safety and performance outcomes should be measured.

This equation links the reservoir to the entropy model. Building the reservoir increases the quality of the human resources system factor. It shifts human resources towards optimum safety and performance. Concurrently, this provides opportunities to enhance the quality of other system factors through better standards of workmanship and operation, thorough hazard identification, and more effective problem-solving and decision-making. The firm should therefore expect a return on its training investment in real terms measured by safety, productivity and system factor quality results. These measures are discussed further in Part 4.

## Summary

This chapter has highlighted the need for employee training as both a means of better managing risk and pursuing continuous improvement of organizational systems. The rationale of the reservoir provides a structured, objective approach to the training and development of employees, which shows that the firm, after sufficient investment, can attain positive returns on its training investment, provided that concurrently it provides a social environment that encourages employee participation and contributions.

In the following chapter, which is the final chapter of Part 3, behavioral auditing is discussed. The purpose of this audit is to measure the return on this investment in employee training. Measures are also developed in the core areas of behavioral change that have been discussed thus far. This includes training evaluation, competency and attitudinal changes, and the health of the organizational culture. The behavioral audit results contribute towards the cultural health scorecard, which is one of the tools used to monitor corporate achievement within the performance management system presented in Chapter 12.

## References

1. Tomer, J.F. (1999) *The Human Firm – A Socio-economic Analysis of its Behavior and Potential in a New Economic Age*. Routledge, London and New York.
2. Topf, M.D. (2000) Including leadership in the safety process, in *Occupational Hazards*, March. Available on website: http://www.findarticles.com/cf_dls/m4333/3_62?63269469/print/jhtml
3. Erickson, J.A. (2000) Corporate culture: the key to safety performance, in *Occupational Hazards*, April. Available on website: http://www.occupationalhazards.com/full_story.php?WID=1876.
4. Hampton, D.R., Summer, C.E. and Webber, R.A. (1978) *Organizational Behavior and the Practice of Management*, 3rd edn. Scott, Foresman and Company, Illinois, USA. Figure 9.1 adapted from Summary of Path–Goal Relationships page 633 and Figure 9.2 adapted from Continuum of Leadership Behavior page 652.
5. Quinlan, M. and Bohle, P. (1991) *Managing Occupational Health and Safety in Australia, a Multidisciplinary Approach*. MacMillan Company of Australia, Melbourne.
6. Smith, A. (1992) *Training and Development in Australia*. Butterworths, Sydney, Australia.

7. Harris, D. and De Simone, R. (1994) Employee orientation, chapter 7 in *Human Resource Development*. Dryden Press, Harcourt Brace Publishers, Orlando, USA, pp. 203–234.
8. Schein, E. (1988) Organizational socialization, *Sloan Management Review*. Fall, Massachusetts Institute of Technology, Massachusetts, USA, pp. 53–65.
9. Neale, A. and Griffin, M.A. (1999) Developing a model of individual performance for human resource management. *Asia Pacific Journal of Human Resources*, 37 (2), Australia, 44–59.
10. Gollan, P.J. and Davis, E.M. (1999) High involvement management and organizational change: beyond rhetoric. *Asia Pacific Journal of Human Resources*, 37 (3), 69–91.
11. Heslin, P.A. (1999) Boosting empowerment by developing self-efficacy. *Asia Pacific Journal of Human Resources*, 37 (1), 52–63.
12. Burke, W.W. (1986) Leadership as empowering other, in S. Srivasta and Associates, *Executive power: How executives influence people and organizations*. Jossey-Bass, San Francisco, California, USA.
13. Covey, S. (1989) *The Seven Habits of Highly Effective People*. Simon and Schuster, New York, USA.
14. Robbins, H. and Finley, M. (1996) *Why Change Doesn't Work – Why Initiatives Go Wrong and How to Try Again – and Succeed*. Orion Business Books, London.
15. WorkSafe Western Australia (1999) *WorkSafe Plan – Western Australia's Assessment of Occupational Safety and Health Management Systems*. WorkSafe Western Australia. Available on website: http://www.safetyline.wa.gov.au/PageBin/bestplan0001.htm
16. Mines Occupational Safety and Health Advisory Board (1999) *A Review of Incentive-Based Remuneration Schemes in the Western Australian Mining Industry – Report and Recommendations from MOSHAB Incentive-Based Remuneration Working Party*. Available on website: http://www.dme.wa.gov.au/prodserv/pub/pdfs/remun.PDF
17. Occupational Safety and Health Association (2001) *Strategic Map for Change and Continuous Improvement for Safety and Health – Employee Involvement*. Other section used – 'Management Leadership'. Available on website: http://www.osha slc.gov.SLTC/safetyhealth_edcat/mod4_strategicmap.htm

Chapter 10

# Behavioral audits

In Chapters 8 and 9 the key elements of behavioral change were discussed. Primarily these were concerned with building a safety culture driven by management leadership and high quality human resources. It was explained how HRM practices can be used to support the OHS management system by sharing ownership of company KRAs, selecting new employees on the basis of competencies and values, and providing opportunities for workers to enhance their contributions through participative processes. In addition, the capacity reservoir illustrated how the firm can derive benefits from its investment in training and organizational learning by facilitating appropriate levels of authority. It was also explained that employee competencies, commitment and involvement can lead to the release of resourcefulness, which in turn, delivers better systems review, effective decision-making and leadership at all levels.

After the implementation process, how can the firm determine whether it has been successful in bringing about behavioral change? What tools can be used to measure the health of the organizational culture? The primary means of evaluating behavioral change under productive safety management is the behavioral audit. Other measures such as incident rates, that are outcome measures, are also used in the business alignment scorecard, which will be discussed in Part 4. As explained in earlier chapters, the definition of auditing is:

> Auditing is an information gathering and evaluation activity involving three processes:
> (1) Identifying the sources of risk that the organization is exposed to as a natural system,
> (2) Evaluating the health of the organizational culture, and
> (3) Measuring the achievement of strategies in the productive safety management plan.

## Linking productive safety management to quality systems

Behavioral audits are used to evaluate the health of the organizational culture (point 2 above) with the results analyzed to improve behavior modification strategies. The overall purpose of this evaluation and feedback process is to shift the company's human resources towards optimal safety and performance and move the company along its learning curve. The management and measurement of behavioral change is critical to the ongoing performance of the firm. Failure to concurrently implement cultural development strategies that support systems change can result in the erosion of gains made from tightening up other risk management processes. Ultimately, the redesigning of machines and systems of work to 'eliminate' hazards is the preferred means of managing risk.[1] As indicated by the entropy model, however, residual risk cannot be completely eliminated nor can degradation be curtailed without continuous commitment and effort. There remains a strong onus on the firm therefore to ensure that employees are provided with the skills and attitudes to work safely within a supportive management system.

Figure 10.1 illustrates how the failure to address underlying cultural factors can sabotage the OHS management system. Even where the company invests heavily in resources to develop planned steps, unplanned organizational factors cause performance to decay over time.[2] Consequently, there may be a greater difference between what the firm hopes or plans to achieve and what is actually achieved.

There has been evidence that this problem of decay has been an issue with mine safety in Australia.

> Despite significant investment in dollars and effort, fatality statistics remain unaffected – improving and then dropping back over time, but remaining relatively stable around the trend line. It would seem that safety performance suffers a similar process of decay and for similar reasons to the decay in organizational quality performance over time. The strength of an organization's culture will determine the rate of decrease in safety performance between safety improvement initiatives.[3]

To prevent organizational decay, the total quality management (TQM) system was devised which sustains improvements by concurrently building

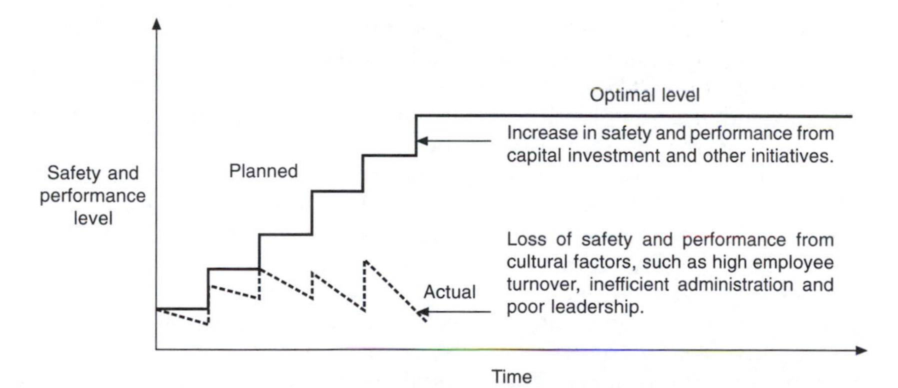


Reproduced from 'Improvement of quality and productivity through action by management', by W.E. Deming in *National Productivity Review*, 1981–1982, Winter. Copyright © 1981 Wiley Periodicals, Inc. Reprinted by permission of John Wiley & Sons, Inc.

**Figure 10.1** Results without a management system

a supportive culture. The rationale of TQM is compatible with the productive safety approach. The commonality between the two systems is that both focus on the quality of the organizational system. In addition, attention is paid concurrently to systemic procedural issues and cultural issues. Further, management commitment is critical to success. There is also a planned cyclical method of achieving progress, and auditing/ measurement methods are applied routinely. The core difference between the two systems is that productive safety is driven by safety, performance and quality simultaneously, whereas, TQM uses 'performance' as the primary objective. Figure 10.2(a) illustrates the approach taken in TQM to underpin improvements in the organizational system. An adaptation of the TQM model to the risk management process approach[3] is given in (b) and the productive safety approach is shown in (c).

Figure 10.2(a) illustrates the method of improvement using the TQM system wedge. The quality system acts as a lever that inhibits the performance decay problem. It allows the cycle of planning, doing, checking and acting to advance as a process of continuous improvement. In (b), the TQM rationale is applied to risk management. Safety performance improvement follows the four stages of risk management – hazard identification, risk assessment, implementation of controls and auditing. Safety management plans ensure that the system does not deteriorate. These are used to record and administer the risk management process. As a result, planned improvements are realized, thus planned equal actual improvements.

Figure 10.2(c) shows the productive safety wedge. This approach embraces both the principles of TQM and risk management. The productive safety management plan contains strategies for systems change, risk management and behavioral change together with relevant action plans, targets and measurement methods. The firm's standard is required to be above the minimum legal compliance level shown. Safety and performance are improved by applying the four-fold strategy in continuous cycles to manage risks and enhance system factor quality. The plan is a critical part of the management process but of equal importance are continuous improvement practices and a strong safety culture. Hence, productive safety management is an organizational 'lifestyle' and total management system integrating OHS, HRM and system factor maintenance. As systems are improved and sustained to a high quality, gains can be consolidated. This results in real and positive change.

Using the productive safety wedge, the improvement cycle is simplified and fully integrated. It contains three steps. The first is baselining – 'where are we now?'. The second step is improving – 'where do we want to be?' Finally, the change is cemented through monitoring and maintaining. Once the improved change has become consolidated, the process begins again to refine systems and behaviors towards optimal safety and performance. Along the process, the firm shifts towards the industry benchmark of best practice. The rationale is that as the firm improves its results, these achievements become the baseline for the next cycle.

This is a similar logic to an athlete's personal best (PB). An athlete always compares his performance to his highest personal achievement. By improving on the PB he moves closer towards being competitive

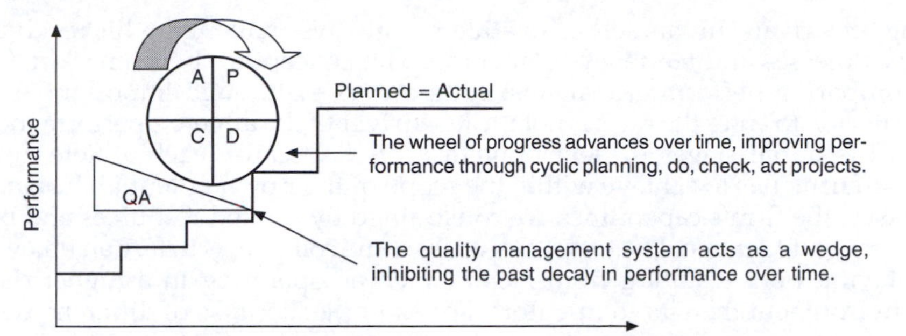


Where: P is improved performance through cyclic planning; D is doing or preparing the plan; C is checking the plan; A is acting to implement and improve projects; QA is the quality management system wedge

Reproduced from 'Improvement of quality and productivity through action by management', Deming W.E. in *National Productivity Review*, 1981–1982, Winter. Copyright © 1981 Wiley Periodicals, Inc. Reprinted by permission of John Wiley & Sons, Inc.

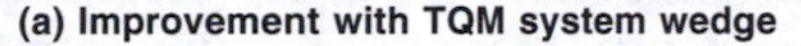

**(a) Improvement with TQM system wedge**

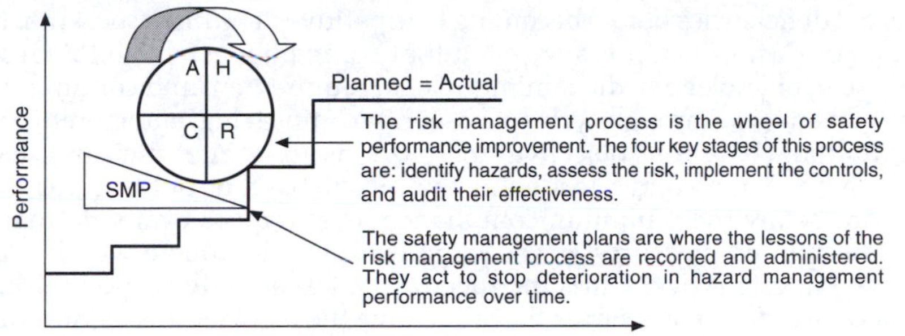


Where: H is hazard identification; R is risk assessment; C is the implementation of controls; A is auditing for effectiveness; SMP is the safety management plan wedge

Adapted from Figure 11 on page 248 in 'Improving the safety culture of the Australian mining industry', Stephan, S. in *Journal of Occupational Health and Safety – Australia and New Zealand* 2001, 17(3).

**(b) Risk management process approach**

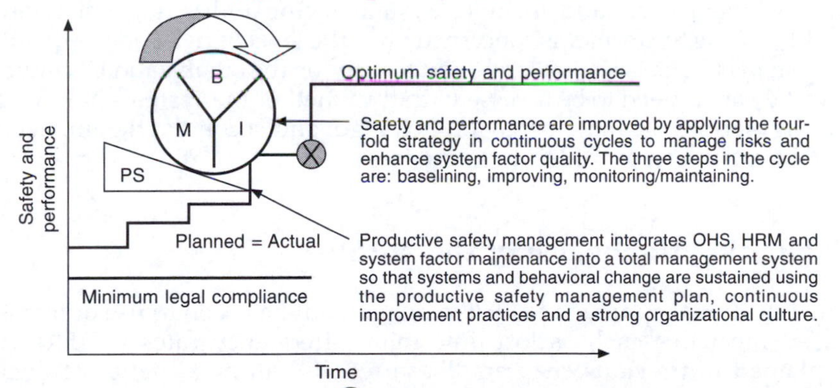


Where: B is baselining the level of current achievement; I is improving system factor quality and results using the four-fold strategy; M is monitoring/maintaining systems so that there is no regression and continuous improvement occurs; PS is the productive safety system wedge

**(c) Improvement with productive safety system wedge**

**Figure 10.2** Comparing TQM and productive safety system wedges

against rivals. In this sense, an athlete is always challenging his residual weaknesses and tendency to degrade. The concept of 'benchmarking' – comparing performance against competitors – although important as a criterion to enter the race, is not the key driver of the athlete's performance in the initial stages of development. This is because each athlete, like each firm, has to achieve within the scope of their own potential. In some cases, the firm's capabilities are constrained by limited resources and by unfavorable external factors. This causes some companies to face an unlevel playing field resulting from factors such as operating in a higher-risk environment than its competitors. For example, because of differences in geology some mining companies face greater risks of structural failures than companies mining more solid orebodies. It can be discouraging, counter-productive and unrealistic for a company that is behind industry leaders to judge itself by the standards of these leaders in the short term, in the same way as it is demotivating for an amateur runner to compare himself with an elite sportsman.

Firms also have a history that may require them firstly to overcome internal deficiencies before becoming competitive at industry best practice level. For this reason, it is suggested that companies establish PB targets that are over and above the minimum legal requirement and continuously improve on these through cycles of baselining, improving, and monitoring/maintaining. The first objective, therefore, is to define minimum legal compliance. The second is to identify the current baseline or PB. Companies that are below the minimum compliance level require dramatic change and disciplined measures, in the same way that an athlete who has put on five kilograms over Christmas needs to take corrective action through strict diet and exercise. Firms that are above the legal requirement should apply incremental consolidated change processes so that the change is sustained and creates an organizational 'lifestyle'.

Once strategies have been developed to improve risk management and organizational systems, the changes implemented have to be monitored and maintained. Auditing is part of the monitoring stage of the continuous improvement cycle and includes system factor audits and behavioral audits. While the former is concerned with the level of risk and the quality of system factors, the behavioral audit focuses on the organizational culture. Two key areas need to be assessed: firstly, whether the systems have been put in place to facilitate behavioral change; and secondly, the impact of these systems on attitudes and practices.

## Evaluating the health of the culture

The traditional approach to OHS management has been to use outcome-based measures such as lost-time injury frequency rates (LTIFR). As explained in previous chapters, these results can be a matter of 'luck' when system factors have been allowed to degrade. Although this measure is included in the business alignment scorecard, discussed in Chapter 12, it is not considered a reflection of the health of the organizational culture.

On its own, the lost-time injury frequency rate (LTIFR), and

> other measures like it, tell nothing about how OHS is managed in the enterprise and give no information which will lead to the development of improvement strategies.[4]

The LTIFR is unsatisfactory as a sole measure of safety performance for at least three reasons. Firstly, it is geared to claims and injury management processes rather than to real changes in safety performance.[5] Secondly, the results from year to year may be attributable to chance instead of changes in the condition of system factors. There is therefore no indication of the level of risk to which the firm is exposed. Thirdly, it does not give any indication about how well the most serious hazards are being managed.[5] There is a significant danger that reliance on LTI results may cause a firm to ignore the 'alarm bells' set off by near misses.

The Australia/New Zealand standard AS/NZS 4804-1997 (Occupational health and safety management systems – General guidelines on principles, systems and supporting techniques) recommends that firms incorporate positive performance indicators in their OHS management systems.[5] These evaluate the management activities implemented to prevent injury, illness and damage. OHS and management literature indicates that good performance indicators are:

- controllable or able to be influenced;
- relevant;
- assessable or measurable;
- understandable and clear;
- accepted as true indicators of performance;
- reliable, providing the same measures when assessed by different people; and
- sufficient to provide accurate information, but not too numerous.[5]

The behavioral audit uses underlying measures that provide an indication of the firm's cultural health. The wellbeing of social systems has an impact on the firm's future safety, production and quality results. In effect, it is a measure of the preparedness of the business to achieve better outcomes through the quality of its human resources and culture. A company may, for example, have 'world's best' technologies and physical environment and well-designed processes, however, the maintenance of these systems requires management practices to be in place that control residual risks and prevent degradation. The management system, which is underscored by values, beliefs and norms, sets the parameters that define appropriate organizational behavior, which in turn, determines how effectively the firm is able to sustain the quality of its physical resources. Closed communication, internal conflict over objectives and lack of management commitment can cause poor maintenance and operating practices that result in the deterioration of these physical assets. Hence, the behavioral perspective of risk management needs to be considered as part of the firm's overall risk management strategy.

In Chapters 8 and 9, a number of behavioral strategies were identified, which focused primarily on management commitment and leadership, shared ownership of productive safety as a KRA, and the development of the capacity reservoir. The first part of this audit is a scan to determine whether and to what extent management has put the systems in place to facilitate behavioral change. Specific quantitative and qualitative measures may be used such as the number of PDs containing OHS as a key outcome, the level of completion of management training, and evaluation of the quality of supervision. Such measures may be obtained from survey data, employee feedback, statistics and other tools. Table 10.1 summarizes the main areas of review in this audit.

There are six areas that may be investigated, with the first of these being management commitment and leadership. Table 10.1 identifies that scanning this behavioral issue involves determining whether the company has established the systems for change. These include the provision of management training, identification of productive safety as a KRA, development of the value statement, employment of specialist OHS personnel, formulation of an OHS management plan, and allocation of resources for risk management. Within this area, the auditor should also determine the extent to which the firm has sought to integrate its contractors into the OHS management system. Contractual constraints such as length of tenure may preclude full integration, however, there should be evidence that the firm has established criteria for safety for its contractors above the minimum standards required by legislation, as evidence of the principal firm's commitment to safety. Contractor management is assessed under this behavioral issue because the failure of the firm to integrate contractors represents a weakness in the organizational culture and the nonalignment of values within operational practice.

The auditor can use a number of tools such as the employee survey to assess the extent to which strategies have led to desirable behaviors. Surveys are particularly useful for evaluating management's commitment from employees' points of view. The second survey tool that may be used is the supervisors' questionnaire to assess their perceptions of the fit between management values and the company culture. Additionally, quantitative measures may be applied such as the level of resourcing, calculated as the percentage of the total budget, allocated to productive safety management. Budget items that may be included are resources allocated for systems change, risk management practices such as preventative maintenance programs, and behavioral change programs such as training. Expenditures that were previously unassociated with OHS management, such as equipment maintenance, may therefore be integrated into the organization's total risk management financial plan. Thus the cost items within this measure are any business costs that are derived from the productive safety KRA.

The behavioral audit also evaluates the extent of ownership of this KRA. As explained in Chapter 8, sharing proprietorship involves using PDs or other formal written tools to identify responsibilities, which are then supported by team ethic statements that establish a framework for action and operational decision-making. Some of the measures derived from these processes that determine whether behavioral change has occurred are the percentage of PDs containing OHS responsibilities and the percentage of employees who know and understand the criteria for satisfactory performance as defined by the 'achievement indicators' (refer to Fig. 8.8).

**Table 10.1** The key areas of the behavioral audit

| Behavioral issue | Implementation of systems | Measures of behavioral change and organizational learning |
|---|---|---|
| Management commitment and leadership | • Has the management team undertaken productive safety training?<br>• Has productive safety been identified as a KRA?<br>• Has a value statement been developed?<br>• Have OHS specialists been employed?<br>• Is the OHS specialist's role strategic?<br>• Has a productive safety management plan been developed?<br>• Have sufficient resources been allocated to develop and implement the management system?<br>• Has the company sought to integrate its contractors into the system? | • Employee perceptions of management commitment (employee questionnaire)<br>• Supervisor perceptions of productive safety management system (supervisors' questionnaire)<br>• % of total budget allocated to productive safety including systems change, risk management and behavioral change programs<br>• Review of contractor selection and management system |
| Shared ownership | • Has ownership of productive safety KRA been shared using position descriptions?<br>• Have employees developed team ethic statements? | • % of position descriptions containing duty of care and other OHS responsibilities<br>• % of employees who know their responsibilities (employee questionnaire)<br>• % of managers who know their own and their subordinates' responsibilities (supervisors' questionnaire) |
| Supervision and decision-making | • Have supervisors been given training in productive safety?<br>• Has the reasonableness test been applied in relation to resourcing and operational decisions?<br>• Have senior management supported these decisions?<br>• Does management respond appropriately to employee hazard reports? | • % of managers/supervisors who have completed productive safety training<br>• Employee perceptions of the quality of supervision and decision-making (employee questionnaire)<br>• Supervisor perceptions of the quality of the executive team's supervision and decision-making (supervisors' questionnaire) |
| Safety issues resolution | • Does the company have a procedure for the resolution of safety issues?<br>• Is the procedure widely communicated? | • % of employees who are aware of resolution procedure (employee questionnaire)<br>• Number of safety issues referred to senior management in the period<br>• Number of safety issues requiring external adjudication in the period |
| Safety capacity reservoir | • What training was undertaken in the past year?<br>• What training is planned for the forthcoming year? | • % of new recruits who have undertaken induction in the past year<br>• % of employees with complete |

**Table 10.1** *(Contd)*

| Behavioral issue | Implementation of systems | Measures of behavioral change and organizational learning |
|---|---|---|
| | • Does the company evaluate training quality and learning retention?<br>• Does the company use value-based recruitment practices?<br>• Does the company undertake gaps analysis for new employees?<br>• Does the company keep its employees informed of changes that affect the level of risk?<br>• How effective are these communication strategies? | job-specific competencies<br>• Post-training evaluation of learning retention<br>• Investment in training per employee per year ($)<br>• Investment in training as a % of the total budget<br>• Employee attitudes to risk and risk controls (employee questionnaire)<br>• Employee understanding of risk and risk controls (employee questionnaire) |
| Employee commitment and resourcefulness | • Does the company provide opportunities for employee participation e.g. toolbox meetings etc?<br>• Does the company have an employee suggestion program?<br>• Does the company evaluate the effectiveness of its consultative processes and participative programs?<br>• Do employees feel committed to the organization and its philosophy? | • Absenteeism rate<br>• Employee turnover for the past year<br>• Employee job satisfaction assessment (employee questionnaire)<br>• Participation rate in consultative processes<br>• Number of employee suggestions received<br>• Number of employee suggestions implemented<br>• $ value of benefits of employee suggestions after costs<br>• Number of reported incidents involving noncompliance with required practices |

In Table 10.1, the quality of supervision and decision-making is also audited as an indicator of effective behavioral change. The purpose of reviewing this issue is to determine whether training and other HRM strategies have translated into appropriate leadership styles and actions at the operational level. The scan may include an assessment of the training provided to supervisors, the extent to which the reasonableness test has been applied, and the responsiveness of managers to employee hazard reporting. Measures that may be used in this area include:

(1) The percentage of supervisors who have completed the planned training program;
(2) Upward performance appraisal assessments using employee surveys to evaluate the quality of operational decision-making;
(3) Middle-management surveys to assess the quality of strategic decision-making and supervisors' levels of job satisfaction.

From year to year, survey results can be compared to determine the extent of behavioral change, to identify areas requiring improvement, and to develop more effective strategies. As also explained in Chapter 8, the conflict resolution process and outcomes can have a significant effect on the health of the culture. The auditor should therefore determine whether the company has an appropriate procedure and the breadth to which it is known by employees. The effectiveness of the process can be measured by assessing whether or not employees are aware of their rights to raise safety issues and whether they are aware of the avenues available for resolution. The number of safety issues referred to senior management in a given period can also be tracked. It is not possible to say definitively that high numbers or low numbers of referrals are preferable. As employees become more safety conscious the number of issues may rise, or alternatively, a high result may be attributable to poor resolution at the shop floor level. Consequently, the reasons for the numbers has to be taken into consideration when interpreting the results. A further measure of the effectiveness of the safety resolution procedure is the number of matters requiring external adjudication in a given period. Intervention by regulators should not be necessary when consultative processes are centered on common goals and values.

In Chapter 9 it was explained that to develop a productive safety culture the firm has to invest in training. Conceptually, this was illustrated using the capacity reservoir. The auditor should consider whether, based on past and planned training, the investment has been and is adequate given the firm's level and types of risk. For instance, companies that have an inexperienced workforce, particularly in a high-risk environment, need to invest more extensively in training than firms with an experienced, stable workforce in a lower-risk operation. Other systems that may be audited include the use of value-based recruitment practices, training gaps analysis and communication strategies.

These training-related measures serve a dual purpose. The first is to ascertain the level of investment in human resources risk reduction and the return on that investment. The second is to determine the level of the capacity reservoir. One of the assessment criterion given in Table 10.1 is the percentage of new recruits who have undertaken induction in the past year within the first week of employment. The firm should achieve 100 per cent for this result. It is the company's responsibility to provide information about hazards and their management before exposing new recruits to the workplace. The depth of the reservoir may also be measured using the KSAs listed in the selection criteria in PDs to develop an organizational competency profile against which current employee skills can be compared. This gives the firm a measure of the level of job-specific competencies it has available and the percentage of workers with up-to-date skills, particularly those that require accreditation by legislated standards. For example, under the Occupational Safety and Health Regulations 1996 (Western Australia), boiler maintenance has to be carried out by a person approved by the Commissioner.[6] The firm must therefore ensure that personnel have the required certification to carry out this work.

Provision of training does not ensure that employees will acquire the competencies to operate safely and efficiently. For a number of reasons, such as poor delivery or relevance, training may not result in behavioral change in the workplace. The firm must therefore evaluate the effectiveness of such programs and also learning retention. Other measures that may be used to measure the reservoir include the investment in training per employee per year and the overall financial commitment to skill development as a percentage of the total budget. The level of investment in OHS training should be commensurate with the level of risk in the organization's system factors. The higher the residual risk and tendency towards degradation, the greater the need for safety training. The primary purpose of this training is to shift both competencies and attitudes towards optimal safety and performance. Employee questionnaires may be used to assess attitudinal factors specifically to determine whether workers understand the nature of risk, how their actions affect it and their willingness to modify their actions to minimize risk factors.

To evaluate the health of the organizational culture, the level of employee commitment also needs to be appraised because it has a direct impact on productivity, turnover, absenteeism, quality[7] and safety. A scan of the firm's management systems will reveal the extent to which consultation and participation occur within formal communication strategies such as toolbox meetings, suggestion schemes and team-based work organization. Two measures of employee commitment are the absenteeism and turnover rates. Absenteeism is any failure of an employee to report for or to remain at work as scheduled regardless of the reason.[8] It excludes holiday leave but includes nonattendance due to illness or accidents. The absence rate is calculated as the total labor hours lost as a percentage of the total labor hours rostered.

Turnover is an index of the number of employees who leave an enterprise in a given period as a percentage of the average workforce size during that period.[9] There is a host of factors which potentially influence an individual's turnover decision including dissatisfaction with the work, supervisory style or work-group dynamics. Unmet expectations regarding pay and training, personal factors, or other employment opportunities[10] can also be a cause. Turnover has a negative financial impact for the firm as well as implications for productivity and safety as discussed earlier in Chapter 6.

The actual costs of turnover are difficult to estimate because it varies from company to company. There will also be significant differences based on the type of job and length of time in the position. Some estimates however put the typical cost of voluntary departures between 0.5 to 2.5 times the annual salary of the position from which the incumbent leaves.[10]

The departure of employees from the firm is both a financial drain and a drain on the capacity reservoir. In addition to separation and replacement expenditure, the company incurs training costs to bring new recruits to an acceptable level of proficiency. In addition, new employees are a source of additional risk because their lack of workplace-specific KSAs is a residual hazard and they have a greater tendency to introduce entropic risk at the interface with other system factors. The expense of managing this additional risk, and the impact on safety and productivity has not been factored into the costs given above. The true cost of turnover is therefore likely to be much higher than 0.5 to 2.5 times the salary.

Measures of absenteeism and turnover are based on historic data. To anticipate future trends, the firm can use the employee survey to ascertain the level of job dissatisfaction. This level will reflect the extent to which the job and work environment is failing to fulfill the employees' economic and social needs. Again, the fit between workforce goals and values and those of the company is important. It affects the level of employee commitment to corporate objectives and the desire to remain with the firm. The survey can also be used to ascertain workers' satisfaction with the company's risk management and participative strategies.

The final measure applied to assess behavioral change is the number of reported incidents involving noncompliance with required practices. Deviations from safe work practices are an indication that formal instruction has not transferred into the workplace, or that the individuals involved may have specific traits such as high-risk tolerance, which need to be addressed through counseling or disciplinary action. This type of measure is usually referred to as behavioral sampling which:

> assesses workers' behavior on a systematic sampling basis to establish the proportion of unsafe work behaviors which might require correction, for example, by training or design improvements.[11]

The auditor's role is also to consider any underlying organizational factors that may have encouraged these deviations such as deadlines or production pressures. The company's aim should be to bring the number of noncompliances down to zero. To do so it needs to consider and rectify systemic conditions that encourage inappropriate actions and, where necessary, deal with individual risk-taking tendencies.

The behavioral audit contains both proactive and reactive measures for monitoring the culture. The latter is concerned with responding to events such as high turnover to institute practices that prevent the consequences of those events.[12] A proactive approach is to use survey data to anticipate future repercussions of unfavorable behavioral conditions and, conversely, the benefits of positive social factors. The audit therefore provides an explanation of past events as well as a direction for future prevention, management and improvement.

Overall, the auditing process evaluates the effectiveness of cultural development strategies. Central to these strategies as the two primary areas of action are training and communication. Communication and participation go hand in hand, with the main purpose being to increase the flow of ideas and information between management and workers. This is intended to make employees think and talk about safety, and work with care and vigilance, as well as improve organizational decision-making. By participating in consultative processes, risk management becomes part of the language. One of the most significant communicative changes under productive safety management should be a shift away from the understanding of risk as a singular concept. An indicator of whether this understanding has become a 'lifestyle' will be the pervasiveness of the key terms, particularly the two categories of risk, into the language. Examples of this include:

> The Geotechnical Engineer says that area of the slope is unstable. It's sent our residual risk too high. We'll have to isolate the area and let the workers know about it.

> We've got a new group of trainees starting today. We'll have to pair them up with the mechanics to make sure they stay safe and do not degrade our systems.
>
> We've got a problem with sulfur dioxide fumes on the eastside of the mine pit. Let the workers know about it. Stop drill and blast preparation for the moment. Can you give the geology department a ring and get someone out here? Let's get the OHS representative here too to assess the risks and come up with some strategies to manage the problem.

Language is a very important part of an organizational culture. The greater the use of participative processes and dissemination of information the more this language will develop. For this reason, the behavioral audit should include the measurement of the participation rate in collaborative processes – the number of workers who are exposed to the language through structured communication activities. The auditor can also make a qualitative assessment of language by attending these forums and acting as an observer.

Employee commitment is also measurable by the level of resourcefulness. As shown in Table 10.1, specific measures include the number of employee suggestions received, the number implemented, and the dollar value of benefits of these suggestions after costs. The latter is the financial return the firm gains on its investment in the reservoir and in inclusive programs. In Part 4 it will be shown that these measures should be incorporated into the business alignment scorecard with LTIFR and production output targets, to see whether the firm is achieving short-term goals and developing longer-term capacity through its human resources.

What approach should be taken if the target measures are not achieved? Where results do not meet expectations, the auditor should retrace the steps that were involved in the implementation of productive safety management so that corrective strategies can be developed. Potential weaknesses may stem from a poor fit between the system's philosophy and the common goals and values upon which the value statement was based. For instance, the statement may have been derived through 'political' process rather than honest open discussion, in which case, it may be fundamentally flawed. There may also be pockets of resistance among management personnel, particularly to the application of the reasonableness test in decision-making. Some supervisors may find it difficult to put aside personal 'glory' for the betterment of the firm as a whole, for example, if they want to be in charge of the shift with the highest level of output. Shortfalls in the results may, therefore, be attributable to eroded trust between employees and management due to operational decision-making that has not conformed to the value statement. The purpose of these investigations is to address underlying systemic problems rather than to allocate blame. It is likely that the audit process will also need to include a series of interviews with relevant personnel better to understand these issues. In this case, the following steps should be taken:

(1) Gain executive management commitment;
(2) Appoint a steering committee to guide the audit investigation;

(3) Appoint an auditing team to develop interview questions;
(4) Diagnose the corporate culture and investigate designated functional areas[12] and workplaces separately;
(5) Review the value statement and team ethic statements for suitability and alignment;
(6) Interview relevant stakeholders from each area about their perceptions of the firm's culture;
(7) Seek fundamental reasons for safety, performance, goals and values inconsistencies;
(8) Collect industry information and use it where applicable to devise solutions; and
(9) Write a report for management and the audit steering committee containing findings and recommendations.

Interviews can be used to assess a number of issues such as: management systems and accountabilities; maintenance and corrective action; discipline, rewards and communication; feedback for safe performance; and production pressure.[13] The aim of these interviews is to uncover faults in the management system such as the failure of the strategy to provide clear objectives or consequences. There may also be attitudinal barriers to taking ownership, for instance, a belief that it is the safety department's responsibility to undertake hazard inspections not the supervisors. In organizations where business unit managers have previously been given autonomy to run their departments as they see fit, there may be resistance to a management system that requires conformity and consistency in decision-making. In this regard, the interview can be used to determine the extent to which accountabilities have been identified and the level of acceptance of the need to comply with these responsibilities.

The company's maintenance practices should also come under scrutiny during the interview process, as an indicator of the firm's commitment to the concurrent pursuit of production and safety goals. In particular, evidence of how effectively management is implementing the four-fold strategy should be gathered to ascertain the depth of commitment to preventative, proactive risk management. Has the firm shifted away from 'fire-fighting' to a genuine focus on system factor quality? The approach taken to maintenance practices reflects the firm's decision-making processes. In particular, it reflects its priorities and time horizons, which are based on underlying perceptions and values defining 'acceptable' risk. Maintenance practices are therefore behavioral products stemming from beliefs and values. In a firm that has a strong safety culture, maintenance should primarily be proactive with a reducing requirement for reactive interventions. In addition, any maintenance undertaken that may have a short-term negative impact on production, such as the temporary withdrawal of equipment from operation, should be seen as a positive step towards better productivity over a medium period of time. Accordingly, supervisors who make such decisions should have the full support of management.

Audit interviews may be used to uncover other behavioral products such as disciplinary procedures, reward systems and communication strategies. The issue to be considered during the audit is how systematically and appropriately these are used. Have breaches of safe practice been dealt with fairly and consistently? Has disciplinary action been followed through with preventative measures such as additional training? Are reward systems consistent with the philosophy? These management responses are important because they can have a number of different effects on the culture – to reinforce positive behaviors, to correct undesirable behaviors, or to maintain the status quo. Research has shown that although disciplinary action is required when workers act improperly, employees do not necessarily see it as negative.[13] What is critical to workforce perceptions is the consistency and appropriateness with which these practices are applied, and this is the criterion that the auditor should use to evaluate such procedures.

> In facilities that have hazardous production processes where unsafe acts by one person can expose many others to injury, workers may complain about lack of disciplinary enforcement of safety rules.[13]

If audit interviews reveal either a pattern of deviations from safe work practices by employees and/or a lack of effective behavioral responses by management, the assessor should probe these matters for underlying antecedents. These may include poor training transfer, insufficient management support and recognition of safety performance, residual habits associated with the previous output-centered culture, or fear that punitive strategies will lead to overt conflict between management and workers.

One of the weaknesses of past OHS management systems has been an emphasis on negative performance indicators. As a consequence, criticism and pessimism have tended to become entrenched in safety cultures. The auditor should assess the types of indicators used by the firm to determine the balance of positive and negative measures and the behavioral implications of these measures. For example, do the measures encourage communication structures that recognize safe behaviors or are they primarily negative and thus inclined to lead to a climate of punitive reactions to incidents? There is a danger that managers may operate on a negative-exception basis,[13] whereby safety is given attention only when there is an injury or damage. In contrast, a primary aim of productive safety management is to promote risk management as an organizational 'lifestyle'. Positive behaviors may be encouraged through goal-setting, sound decision-making, and role modeling, balanced by clear and decisive consequences for deviations. The latter is, therefore, a safeguard or corrective mechanism not the driver of behavior.

The final area to be investigated through interviewing is whether production pressures have been allowed to override other priorities such as risk reduction and system factor maintenance. The tendency for output to dominate the culture can be attributable to a number of potential causes, particularly the failure to apply the reasonableness test to organizational decision-making. The health of the culture depends on the value statement being 'brought to life' through this test. It also depends on reporting and performance monitoring practices and how effectively these support production and safety as compatible goals. Where, for instance, the management system encourages competition between shift supervisors, values will be compromised for short-term gains and individual success, to the detriment of overall business unit achievement and the company culture.

## Summary

The behavioral audit complements the system factor audits, which were discussed in Chapters 3 to 6. The latter are used to establish the firm's baseline level of risk, for which remedies are developed using the four-fold strategy. Subsequent system factor audits determine the degree to which the firm has been successful in compressing residual risk and preventing entropic risk. It is a snapshot of the condition of system factor quality at the time of the review. The behavioral audit, on the other hand, is indicative of the firm's current culture and predictive of future safety, production and system factor quality results. The underlying beliefs, values and norms shared by the company's workforce determine the organization's potential for improvement in these areas, particularly, because of the interfacing that occurs between human resources and other system factors in the workplace. What happens at these junctures has a direct impact on the firm's ability to manage risk while generating output.

The practical implications of the entropy model and its four system factors are that the use of 'safer' technologies and the better design of workplaces and processes are significant but not comprehensive approaches to risk management. There will always need to be a concurrent strategic focus on behavior. For this reason, the behavioral audit which evaluates the tools that create such change, and the underlying human characteristics of the firm such as competencies, attitudes and social dynamics, is important. This audit assesses the firm's preparedness to shift up the organizational learning curve and to pursue continuous improvement. The review of the culture allows the firm to learn about itself and to develop plans to improve its management systems. The behavioral audit is therefore part of a strategic, long-term process of building a sustainable business centered on quality systems.

This is the final chapter of Part 3. It has focused on aligning the organizational culture with the structural systems of this approach. The behavioral strategies that have been discussed will be included with the systems change and risk management strategies in the productive safety management plan. The process of compiling and auditing the plan and developing the supporting performance management system are discussed in Chapters 11 and 12 in Part 4.

## References


1. Simpson, I. and Gardner, D. (2001) Using OHS positive performance indicators to monitor corporate OHS strategies. *Journal of Occupational Health and Safety – Australia and New Zealand*, 17 (2), 126–134.
2. Deming, W.E. (1981–82) Improvement of quality and productivity through action by management. *National Productivity Review*, Winter, 12–22.

3. Stephan, S. (2001) Improving the safety culture of the Australian mining industry. *Journal of Occupational Health and Safety – Australia and New Zealand*, 17 (3), 237–249. Quote from p. 247.
4. Shaw, A. and Blewett, V. (1995) Measuring performance in OHS: using positive performance indicators. *Journal of Occupational Health and Safety – Australia and New Zealand*, 11 (4), 353–358. Quote from p. 353.
5. National Occupational Health and Safety Commission (1994) *Positive Performance Indicators – Beyond Lost Time Injuries – Part 1 Issues*, WorkSafe Australia, Australia. Quote from p. 17.
6. *Occupational Safety and Health Regulations 1996 (Western Australia)* Western Australian Government Printers, Perth, Australia. Available on website: http://www.safetyline.wa.gov.au/sub3.htm#5
7. Cascio, W.F. (1991) *Costing Human Resources: The Financial Impact of Behavior in Organizations*, 3rd edn. PWS, Kent, Boston, USA.
8. Nankervis, A.R., Compton, R.L. and McCarthy, T.E. (1993) *Strategic Human Resource Management*. Thomas Nelson Australia, South Melbourne, Australia.
9. Kempner, T. (ed.) (1976) *A Handbook of Management A-Z*. Penguin Books Ltd, Middlesex, UK.
10. Campbell, D.J. and Campbell, K.M. (2001) Why individuals voluntarily leave: perceptions of human resource managers versus employees. *Asia Pacific Journal of Human Resources*, 39 (1), 23–41. Quote from p. 24.
11. Glendon, I. and Booth, R. (1995) Risk management for the 1990s: measuring management performance in occupational health and safety. *Journal of Occupational Health and Safety – Australia and New Zealand*, 11 (6), 559–565. Quote from p. 560.
12. Waddock, S. (2000) Corporate responsibility audits: doing well by doing good, *Sloan Management Review*, Winter. Available on website: http://www.findarticles.com/cf_dls/m4385/2_41/59522022/print.jhtml
13. Krause, T.R., Hidley, J.H. and Hodson, S.J. (1990) *The Behavior-based Safety Process – Managing Involvement for an Injury-free Culture*. Van Nostrand Reinhold, New York. Quote from p. 65.

# PART 4
# The productive safety management plan

In Part 1, the tools of productive safety management were presented. The entropy model was used in Part 2 to restructure traditional OHS practices strategically to fit the four system factors of the model. This process of restructuring resulted in the development of systems change strategies. In addition, the four system factors were considered independently to determine the sources of residual and entropic risk associated with each. In Chapter 7, a method was presented to identify, measure, and prioritize these risks so hazard management strategies could be formulated that allow managers greater degrees of confidence when allocating limited resources. From Part 2, therefore, two types of interventions were identified – systems change and risk management strategies.

In Part 3, it was shown that although it is important to improve the quality of physical capital to reduce risk by implementing better design, purchasing and maintenance systems, it is also essential to consider the impact of human behavior on system factor quality. The entropy model identifies human resources as one of four system factors that have a significant impact on the firm's level of inherent risk and susceptibility to degradation. Accordingly, risk control must also include methods to improve the quality of human resources and manage their performance. Behavioral change strategies are therefore an integral part of the risk management system.

Three types of strategies – systems change, risk management and behavioral change – comprise a total management system in which production, safety and quality are compatible goals. To facilitate implementation, each strategy must be accompanied by specific initiatives, measures, targets and action plans that identify required resources, the time frame for completion and the personnel accountable for execution and review. The focus of the following chapters is on the formulation of the productive safety management plan which includes these three strategy types.

In Chapter 11, the structure of the plan will be presented along with the basic principles of strategic management. One of these principles is to build review processes or feedback mechanisms into planning systems. In keeping with the definition of auditing given earlier, the plan itself should undergo an evaluation to assess the appropriateness of strategies, to identify any weaknesses in planning and action stages, and to pursue opportunities for improvement. This auditing process, the final type of audit in the productive safety approach, is discussed in Chapter 12.

To achieve the plan, the firm needs to have a focal point which captures its core objectives and defines how it will measure its success. This is provided by the performance management system – the achievement cycle – which will also be presented in Chapter 12. It is derived from the channel, in which there are three levels of alignment required to ensure that decision-making processes are aligned, compatible and directed towards agreed outcomes. These are external strategic alignment, internal strategic alignment and internal goal alignment. The achievement cycle summarizes the core measures of effectiveness based on these levels. Measures are selected according to the critical areas of productivity, safety, system factor quality, financial management and customer satisfaction, compliance and social responsibility, which are collectively referred to as 'alignment indicators'. The achievement cycle is the driver of continuous improvement. It provides the structure for the learning process with the 'ends' of one cycle becoming the baseline on which further improvements are developed in the next cycle.

A crucial element in this planning cycle is to define what these 'ends' will be. This requires the management team to determine at the outset those measures that most accurately define organizational achievement. It also involves ascribing a relative level of significance for each measure to obtain a balanced perspective. For an athlete, for example, measures may include improvement on her PB, conformity to the planned exercise regime, compliance with dietary requirements, the change in muscle to fat ratio, fitness level and attitudinal measures. In this case, outcome measures, such as a faster PB, would attract higher relative weighting in the evaluation of performance than compliance with dietary requirements. The selection of measures and the allocation of relative weightings are discussed in Chapter 12.

Some of the athlete's measures assess actual changes in performance, whilst others indicate how effectively interventions were implemented or alternatively, they forecast forthcoming potential. The same rationale applies to the use of the achievement cycle to assess organizational improvement. Outcome, implementation and potential measures are included. The former is used to assess realized results, such as production rates and the LTIFR. Implementation measures evaluate the effectiveness of the change process, for example, whether strategies have been acted upon on time and within budget. System factor quality measures are used to predict future potential. When system factors have low levels of inherent and entropic risk, for instance, when human resources have high competency standards and risk awareness, the firm has greater ability to improve safety and performance. Potential measures also take into consideration underlying conditions in the firm's culture. The sum of these weighted measures provides a balanced and comprehensive evaluation of the firm's level of achievement and assists in the prediction of future performance.

These measures are summarized using the achievement cycle, which contains a number of tools. The first of these is the business alignment scorecard designed to evaluate corporate results according to the three types of alignment. This scorecard is the primary driver of performance in terms of macro-issues. It provides an overall, aligned assessment of fundamental success variables by including these outcome, implementation and system factor quality measures. This multidimensional approach is particularly critical in maintaining safe work environments. In the first instance, it provides balance by including both production and safety measures within the evaluation system. These, however, do not provide a 'full picture' of corporate performance. It is possible for a firm to achieve high levels of production with low incident rates for a given period even though its systems are degraded. In effect, the business may 'survive' for a time in an environment of escalating risk. To get the 'full picture' therefore, it is also necessary to measure the quality of system factors and include these in the scorecard. These quality measures reflect the firm's susceptibility to inherent danger and degradation, its ability to maintain current performance and safety, and its potential to achieve improved results in the next cycle.

One level down from the business alignment scorecard is the business unit alignment scorecard. It is formatted to facilitate the development of measures specific to the business unit's area of specialization, whilst supporting corporate strategic initiatives. For instance, injury rates for a workshop contribute to the LTIFR for the total organization. Business unit measures therefore feed into corporate measures.

The next level down in the performance management system ensures that senior personnel maintain commitment, which as explained in Part 3, is critical to the effective implementation of the management system. Team leaders have to act as advocates and also facilitate change by demonstrating appropriate behaviors and decision-making. Primary responsibility for the implementation of structural and risk management strategies also rests with senior staff. The leader's scorecard is designed to facilitate operational goal-setting compatible with the strategic scorecards. The format provides for the setting of operational objectives specifically to increase system factor quality. Accordingly, on an annual basis, supervisors are required to develop goals in the areas of processes, technology, the physical environment and human resources to enhance the quality of these systems, to better manage risk and to pursue sustainable business practices. These workplace goals feed into and contribute to the attainment of one or more measures in the business alignment scorecard.

The final tool in the achievement cycle is the cultural health scorecard. It contains measures of the effectiveness of behavioral change strategies, such as shared ownership, management commitment and resourcefulness. This scorecard is used as a support tool, rather than a primary tool, for performance management and can be developed in conjunction with the behavioral audit.

The approach described in Chapter 12 involves linking both the business alignment scorecard and leader's scorecard to financial incentives. Specifically, above a defined level of achievement, all employees receive a bonus based on overall corporate performance and supervisors receive an additional reward geared to their individual accomplishments. Monetary benefits are not tied to achievement of the cultural health scorecard. The results are used instead to evaluate underlying variables that are predictors of future performance. This can be likened to the athlete going for a medical. Although her current results may be outstanding, the medical may reveal a systemic problem that has yet to overtly impact performance. Similarly, the cultural health scorecard evaluates whether the behavioral change strategies that have been implemented are building sustainable business performance unhindered by entrenched human factor issues.

Productive safety management provides recognition for the critical role that employees (both management and workers) play in the achievement of corporate outcomes. The success of the business is therefore linked to the success of its workforce. Gearing the achievement of scorecard measures to financial rewards reinforces this. In Chapter 12, the process of linking incentives to achievement at both corporate and individual levels will be explained. In the example that will be provided, the business alignment scorecard is applied using a 3-year cycle with bonuses apportioned to improvement in each year. In addition, it will be shown how to administer the leader's scorecard concurrently to ensure that strategic initiatives are translated into tangible changes on the shop floor. An overview of the components of the productive safety management plan which contains the achievement cycle and the scorecard tools discussed in the final part of this book, is provided in Fig. (iv).

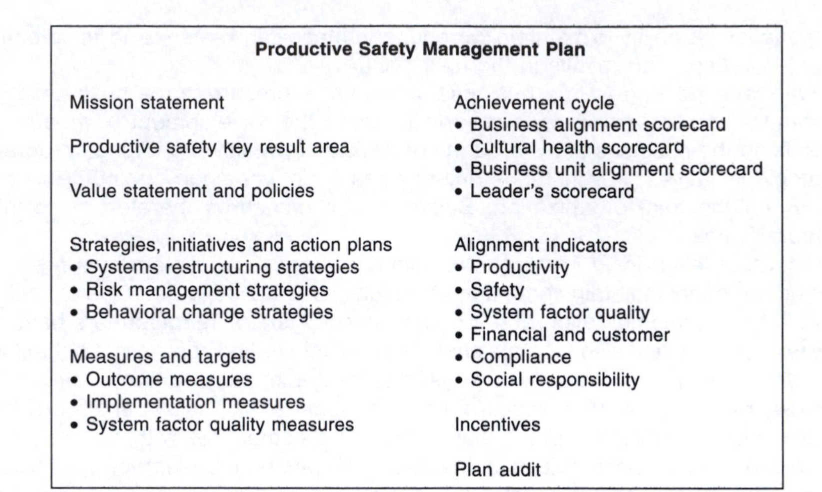


**Figure (iv)** Overview of the productive safety management plan

Chapter 11

# Formulating the productive safety management plan

There are three types of strategies included in the productive safety management plan. These are the restructuring of current OHS practices to fit the entropy model, risk management interventions based on the four-fold strategy (both of which come under systems change), and behavioral change strategies. An overview of the key management issues that are addressed within these strategic areas was provided in the Introduction in Fig. (i). It is repeated in this chapter and used as the framework for the development of the plan. In simple terms, the plan summarizes the firm's goals, the strategies to achieve these goals and the measures used to evaluate their effectiveness.

From Fig. 11.1 the main stages of the plan are:

(1) The mission statement
(2) The productive safety key result area
(3) Value statements and policies
(4) Strategies, initiatives and action plans
(5) Measures and targets
(6) The achievement cycle
(7) Alignment indicators
(8) Incentives
(9) The plan audit

## Mission and value statements

The first step in any plan is to define the firm's reason for existing – its purpose. This is captured in the mission statement.

> An enduring statement of purpose that distinguishes one organization from other similar enterprises, the mission statement is a declaration of an organization's 'reason for being'.[1]

This creed relates as much to the 'community' of the firm comprised of its stakeholders' values, beliefs and aspirations as it does to the economic perspective of organizational existence, and should, therefore, be a declaration of principles and attitudes. The components of a comprehensive

**Systems Change**

**Processes**

Legal provisions and safety standards
Process design

Risk management practices
- Job safety analysis
- Workplace inspections
- Hazard inspections
- Process audits
- Visitor safety
- Contractor management

Post-incident practices
- Incident reporting and investigation
- Evacuation procedures
- Emergency procedures
- First aid
- Workers' compensation management
- Rehabilitation

**Technology**

Planning, design and purchasing
Technological hazard management
Technology audits

Equipment maintenance
Safety and monitoring equipment
Personal protective equipment

Ergonomics
Supplier relationships

**Behavioral Change**

Productive safety culture
Human resource management systems
Shared ownership

Accountabilities and responsibilities
Management commitment and leadership
Appropriate supervision/decision-making

**Productive Safety Management Plan**

Mission statement

Productive safety key result area

Value statement and policies

Strategies, initiatives and action plans
- Systems restructuring strategies
- Risk management strategies
- Behavioral change strategies

Measures and targets
- Outcome measures
- Implementation measures
- System factor quality measures

Achievement cycle
- Business alignment scorecard
- Cultural health scorecard
- Business unit alignment scorecard
- Leader's scorecard

Alignment indicators
- Productivity
- Safety
- System factor quality
- Financial and customer
- Compliance
- Social responsibility

Incentives

Plan audit

Safety issues resolution
Capacity reservoir – building competencies and vigilance

Productive safety formula
Resourcefulness – maximizing learning potential

**Physical environment**

Environmental planning
Environmental modification for persons with disabilities

Environmental monitoring
Residual risk hazard management

Housekeeping
Physical environment audits

**Human resources**

OHS specialists
Human resource management specialists

Induction
Training
Safety promotion programs

Joint consultative committees
Communication systems
Human resources audits

**Figure 11.1** Overview of productive safety management

mission statement include customers, products/services, markets, technology, concern for survival, growth and profitability, philosophy, self-concept, concern for public image and concern for employees.[1] From the mission statement, the KRAs are derived, as explained in Chapter 8.

In the plan, productive safety is identified as a KRA and is multifunctional, applying to all business units. Other KRAs such as marketing may be specific to one or two business units. A value statement accompanies each KRA and defines how it is to be achieved. The value statement is enacted at both the strategic and operational level to ensure that organizational decision-making is aligned and appropriate leadership behavior is demonstrated.

Value statements are expressions of principles that provide direction, alignment and reinforcement of common goals and values. Policies are used to support them and cover specific areas of legal accountability, for example, the OHS policy defines employers' and employees' duties of care and the depth of the firm's commitment to OHS. While the value statement identifies the desired organizational 'personality' and belief system, policy is concerned with giving definition primarily to external strategic alignment issues. These documents provide a framework for operational practice. Policy is also used to communicate expected standards of behavior to employees and provide quality controls for organizational systems. An equal employment opportunity policy, for example, would state that the firm ensures that recruitment and selection practices use merit as the criteria for employment, promotion and access to training. An example of an occupational safety and health policy is:

> XYZ Mining Company values its employees, contractors and visitors. We are committed to their safety, health and welfare. We work continuously and diligently to promote a safe and healthy working environment. In support of this policy, management accepts responsibility for the prevention of incidents through the development of quality systems, the identification and effective management of workplace hazards and promotion of competence and vigilance among our personnel. We seek to be legally compliant and socially responsible at all times by abiding to statutory acts and regulations as a minimum standard and strive to maintain a high level of safety, which extends beyond defined legal limits. To consistently achieve this aim it is required that:
>
> - Management takes responsibility for the quality of system factors, and for ensuring that employees are provided with health and safety information, training and resources to effectively manage workplace risks.
> - All personnel accept their responsibility to work safely, encourage and assist other personnel to do likewise, and contribute to risk management by participating in operational decision-making.
> - We continually develop our productive safety system to achieve higher levels of safety, performance and quality.
> - All personnel embrace learning, competence development and safety consciousness through induction, training, OHS discussions and on-the-job problem-solving.
>
> We encourage all personnel to participate in matters affecting

safety and health, organizational performance and system factor quality and seek suggestions from them for ways to continually improve on our past achievements.

The mission statement, KRAs and value statements provide a long-term perspective of the economic and social nature of the business. They tend to remain unchanged or at most are refined from time to time. This longevity also applies to policies because they are a fit with the context and culture of the broader community in which the firm operates. These elements of the plan therefore are generally restated or referred to in consecutive plans. Strategies, initiatives and action plans, on the other hand, are develop and implemented with the objective of completing them within the lifecycle of the plan. Some strategies, for example, to reduce the level of residual risk in the maintenance workshop, may, however, apply to a number of cycles in the same way that 'to improve my PB' is on-going for the athlete. Each plan may include different interventions that pursue this strategy.

## Developing action plans strategically

Initiatives and action plans are specific to a cycle and have shorter time frames. If, for example, a company operates on a 3-year strategic time frame, the documentation will contain those initiatives that it intends to implement within this period. Subsequent plans will comprise a new set of initiatives. In this way, the firm consolidates change and develops further initiatives to pursue continuous improvement. Strategies, initiatives and action plans detail how the firm will restructure its systems, manage risk and create behavioral change. Once these have been drawn up and included in a written instrument, it should identify:

(1) What needs to be done;
(2) Why it needs to be done;
(3) What resources are required;
(4) Who will be responsible for implementation;
(5) Where it will be actioned;
(6) How it will be actioned;
(7) When it will be introduced;
(8) When it will be completed;
(9) When it will be reviewed;
(10) How it will be reviewed.

Figure 11.2 illustrates the process of developing these action plans strategically. It shows that to develop risk management strategies, the productive safety KRA has to be broken down into the four system factors so that the quality of these can be improved. Each system factor has a number of components, as shown in Fig. 11.1. The system factor given in Fig. 11.2 is 'processes'. Strategies are developed to improve each of the components within a system factor. In the example, the firm aims to improve its workplace inspection practices. The strategy is 'to better manage entropic and residual risks by improving the system of workplace

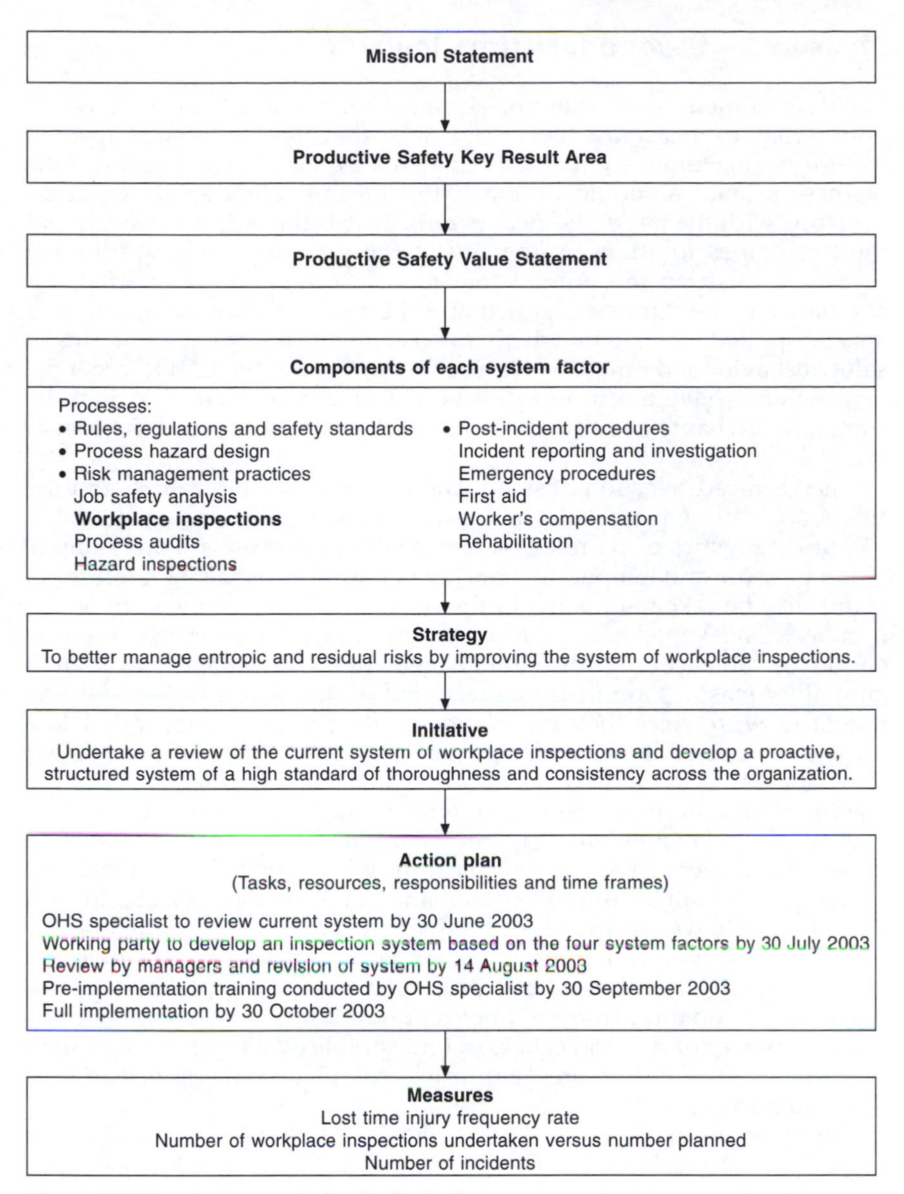


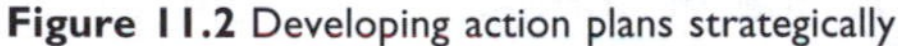

**Figure 11.2** Developing action plans strategically

inspections'. The initiative explains how the firm is going to achieve this strategy. In this case, the firm intends to 'undertake a review of the current system of workplace inspections and develop a proactive, structured system of a high standard of thoroughness and consistency across the organization'. An action plan is then drawn up to identify the resources required, time frames for completion and review, and the personnel accountable for implementation.

## Measures – beyond lost time injuries

The development of interventions is of limited value if the firm does not know what it expects to achieve. Therefore, they have to be accompanied by at least one target – either a qualitative or quantitative measure. One of these measures should be the LTIFR or equivalent safety outcome measure, with the target defined as zero. This is the optimum result that the firm hopes to attain by the end of the planning cycle. Additional measures, such as the number of workplace inspections carried out compared to the number planned and the number of incidents, should also be applied to evaluate whether real change has occurred in terms of safety behavior and outcomes, respectively. Without the LTIFR target the firm will not have a sense of direction. To maintain that direction, the company also has to identify measures which evaluate its progress towards the goal.

The balanced selection of safety and performance indicators, and the frequency with which results are measured, is the same rationale as the use of diagnostics to monitor patient health in hospitals. For example, blood pressure and temperature are taken hourly whereas the cholesterol count may be taken annually. In the firm, critical measures such as the number of near misses need to be reported at least quarterly whereas employee satisfaction with the management system may be evaluated annually. Measures are therefore recorded in the plan together with the regularity with which they are to be reported. The more frequently taken measures are used to alert management to systemic problems that need to be addressed in the short term, whereas the less frequent metrics are used for longer-term strategy development and improvement.

As explained in previous chapters, the traditional approach to assessing organizational performance, particularly the reliance on LTIFR as a measure of safety, can result in superficial evaluation of organizational conditions. A firm can achieve high levels of production and sound safety results for a period of time even though its system factor quality has deteriorated. To address this weakness, the productive safety approach applies measurement on three fronts to uncover underlying systemic deficiencies that are sources of risk and causes of undesirable events. These are called the 'what', 'how' and 'if' arcs and form a complete circle of measures as illustrated in Fig. 11.3.

The diagram in Fig. 11.3 is explained in Table 11.1. 'What' refers to what the firm needs to do. The primary objective associated with this type of measure is to improve safety and performance concurrently by developing appropriate strategies in the areas of systems restructuring, risk management and behavioral change. Strategies are accompanied by outcome measures that determine whether the interventions were appropriate and effective. Included in this measurement category are results such as LTIFR, near misses, output per employee and the turnover rate. Such measures are overt in nature – obvious and not necessarily indicative of covert system quality issues.

The measurement circle also contains 'how' or implementation measures. The objective of applying these evaluators is to determine the effectiveness of the implementation process. These diagnostics therefore evaluate how

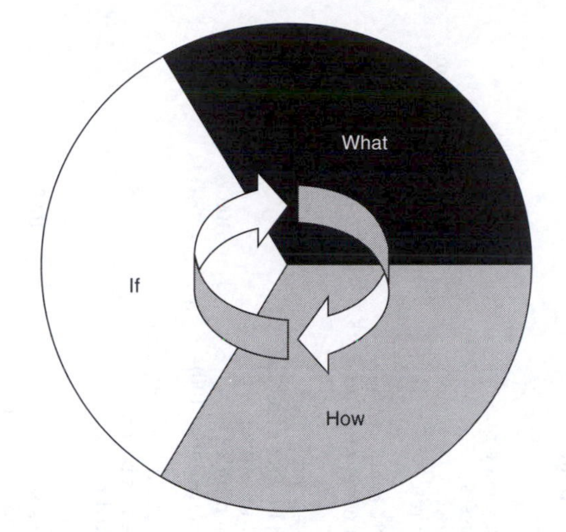


**Figure 11.3** Productive safety measurement arcs

well the change process is managed, for example, 'have initiatives been completed on time and within budget'? Other measures in this category include the extent to which employees feel that they have been kept informed of changes and their level of felt ownership, ascertained using employee surveys that measure worker satisfaction with the process.

The final measurement objective is to determine 'if' the strategies implemented have made a difference. These measures are also called 'system factor quality measures'. These are intended to uncover fundamental issues that determine the firm's efficacy in managing residual risk and preventing entropic risk, and its capacity to pursue continuous improvement. These measures also focus on evaluating the development and sustainability of the desired culture and the presence of lasting habits that lift the quality of system factors. Examples include the number of workplace inspections carried out versus the number planned, the cost of breakdown maintenance versus planned maintenance and the return on resourcefulness. These measures help managers reveal covert problems. For instance, the firm may have a satisfactory record in relation to the number of near misses (which is an outcome measure) but have a rising rate of breakdown maintenance versus planned maintenance. This suggests that the probability of near misses due to equipment failure is rising.

Measures that assess risk management in terms of the consistency and frequency of risk management practices, such as the number of workplace inspections undertaken, are positive performance indicators. The productive safety measurement arcs therefore provide a balance of positive and negative measures. Accompanying each measure is a target that defines the result the firm wishes to attain by the end of the cycle. It sets the ideal level of achievement that reflects optimum safety and performance.

As shown in Fig. 11.2, the development of risk management interventions is fairly straightforward if a strategic, aligned approach is taken. The components of each system factor are considered in turn, and strategies

**Table 11.1** Productive safety measurement arcs in action

| The measurement arc | The objective | Action required | Measure | Target |
|---|---|---|---|---|
| WHAT to do | To improve safety and performance | Develop strategies:<br>• Systems restructuring<br>• Risk management<br>• Behavioral change | OUTCOMES determine whether strategies were appropriate and effective | |
| | | | Lost time injury frequency rate | 0 |
| | | | Number of near misses | 0 |
| | | | Output per employee | 100 units/annum |
| | | | Employee turnover rate | 5% |
| HOW it is done | To implement the changes efficiently and effectively | Implement initiatives and action plans | IMPLEMENTATION evaluates how effectively changes were introduced | |
| | | | Number of initiatives completed on time and within budget as a % of the total number of initiatives | 100% |
| | | | Employee satisfaction index (Employee survey ratio out of 100) | 100 |
| IF it has made a difference | To determine whether the change is real and on-going | Evaluate practices periodically | SYSTEM FACTOR QUALITY determines whether the change has become part of organizational practice and culture | |
| | | | No. of workplace inspections carried out versus no. planned | 100% |
| | | | No. of new employees inducted as % of new employees | 100% |
| | | | Breakdown maintenance as % of total maintenance costs | 5% |
| | | | Return on resourcefulness (benefit versus cost of employee suggestions) | >1 |

devised to improve the quality of system factors. The development of restructuring strategies follows a similar process, with the result that OHS practices and supporting tools such as incident investigation forms are adapted to facilitate the concurrent pursuit of safety and production outcomes.

Behavioral change strategies are not as exact or easily defined, for example, a number of methods may be used to improve hazard communication in the workplace. The techniques chosen will depend on the culture of the firm. For example, structured formalized communications using multidiscipline, cross-departmental and toolbox meetings may be preferred over less formal OHS forums in firms that have traditional hierarchical structures and tight management systems compatible with high-risk industries. There may be a number of ways of achieving behavioral change outcomes and the firm needs to determine the HRM strategies which best fit its underlying values and beliefs. The only constraint on the selection of such interventions is that they should be assessed using the reasonableness test to ensure that they are aligned and compatible with the culture.

## Developing risk management initiatives using the four-fold strategy

In contrast, risk management strategies can be much more precise and certain based on relative risk levels and scientific method, for example, using the four-fold strategy from the entropy model or alternatively, the risk scorecard. Using the former, the resultant remedies fall into two categories – entropy prevention strategies (EPS) and residual risk management strategies (RMS), as discussed in Chapter 7. The method for quantifying these risks was also described. Specifically, it entails allocating relative measures to the risks in each system factor in a business activity. The scores may then be used to prioritize risk areas so that management has higher degrees of certainty or confidence when they develop strategies for risk reduction or control. This facilitates better use of limited resources to reduce workplace hazards. To make the process of developing risk management strategies easier the proforma provided in Fig. 11.4 may be used.

In the hypothetical example shown, on 30 July 2001 the risk management team undertook a review of the risk activity that involved climbing a 10-meter ladder up the side of a storage tank. The team identified each system factor in turn. The process is described in the proforma together with the structural characteristics of the ladder. The primary variable in the physical environment affecting the level of risk is that maintenance personnel reach an elevation of 10 meters, with exposure to the elements, when carrying out the activity. Using the risk scorers provided in Chapter 7, managers can quantify the residual and entropic risks associated with each system factor in relative terms. The risk profile shown indicates that the activity has the following risk parameters:

(1) The process has a high residual risk because of the potential energy gained by reaching a height of 10 meters.
(2) The technology has a high residual risk because of the poor design. There are no platforms or rest areas to reduce the potential energy. A fall, therefore, could involve the full 10-meter height.
(3) The physical environment exposes the worker to this full height because there are no platforms built into the infrastructure and the worker is also exposed to the elements.
(4) The relevant personnel are an average age of 34 suggesting a satisfactory level of vigilance and competence. They have an associated low level of residual risk but the physical demands of the task push this risk up to 5.
(5) The process of climbing the ladder is not complex and has been standardized using job safety analysis so that the use of a safety

**Strategy Development Format**

**Risk management team:**

| *Name* | *Position* | *Signature* |
|---|---|---|
| John Dunn | Processing Plant Supervisor | J Dunn |
| Ian Smith | Maintenance Officer | I Smith |
| Sandra Brown | OHS Officer | S Brown |

Date of review: 30 July 2001

**Risk activity** (describe system factors involved):

| | |
|---|---|
| Processes | Climbing a 10-meter ladder |
| Technology | Vertical uncaged steel ladder on side of storage tank |
| Physical environment | Elevation to 10 meter with exposure to the elements |
| Human resources | Maintenance personnel |

**Systems risk profile:**

| Processes | | Technology | | Physical environment | | Human resources | | Total risk scores |
|---|---|---|---|---|---|---|---|---|
| Residual risk: Potential energy gained with increased height | 7.5 | Residual risk: Poor design. No platforms or rest areas so no energy reduction | 8 | Residual risk: Heights involved. Worker exposed to total height of ladder and to the elements | 8 | Residual risk: High level of physical demand. Average age of workgroup is 34 years | 5 | 2400 |
| Entropic risk: Process is not complex and has been standar-dized including the use of a safety harness | 3 | Entropic risk: Aging technology poorly maintained with deteriorated nonslip tread on rungs | 8 | Entropic risk: Subject to high wind area and poor conditions | 7 | Entropic risk: Workers average age is 34. They are experienced and well trained | 4 | 672 |

**Risk assessment:**

| | | |
|---|---|---|
| Is the level of risk for this activity acceptable? | Yes | No ✓ |
| Is this activity legally compliant? If not, why not? | Yes | No ✓ |

*AS1675-1992 Fixed platform, walkways, stairways and ladders – design, construction and installation* requires landing platform every 6 meters on fixed ladders. Ladders higher than 6 meters require protective back guard.

IF NONCOMPLIANT, CEASE ACTIVITY IMMEDIATELY UNTIL RECTIFICATION HAS BEEN MADE.

**Figure 11.4** OHS Strategy development format

| **Apply four-fold strategy:** | | |
|---|---|---|
| *Entropic risks:* | *1. Eliminate entropic risk* | *2. Maintenance* |
| Processes | Process control is satisfactory. | Ensure that standardized process is adhered to and introduce equipment and conditions checks before climbs |
| Technology | Nonslip tread required on new or modified ladder. Workers to use approved scaffolding system and safety harnesses to install. | Introduce regular inspections of ladder system and develop maintenance schedule after modifications made. |
| Physical environment | Determine conditions under which climbs should not be carried out, e.g. heavy rain. | Develop guidelines for working at heights under different conditions. |
| Human resources | Supervisor to ensure maintenance worker is in good health and aware of risks before climb. | Annual medical checks to ensure workers are fit for the task Regular reinforcement of safety procedures in relation to heights. |
| *Residual risks:* | *3. Manage residual risk* | *4. Compress residual risk* |
| Processes *Unacceptable | *Process residual risk unacceptable. Cease practice immediately until height risk is reduced* | Investigate alternative methods of inspecting the tank. |
| Technology *Noncompliant | Remove lower part of ladder to prevent unauthorized access immediately. | *Residual risk unacceptable and noncompliant. Replace or modify ladder system, i.e. install cage system around ladder and at least one landing platform.* |
| Physical environment | As above | As above |
| Human resources | Include review of dangers of working at a height, i.e. danger of falls and falling objects, in induction training for relevant workers. Schedule reviews in safe work practices into toolbox meetings. | Include the physical requirements of the job in the selection criteria in the position description. |

**Figure 11.4** *(Contd)*

harness is mandatory and therefore the exposure to entropic risk is low.

(6) The ladder is poorly maintained with deteriorated nonslip tread on the rungs and thus the level of entropic risk is high.
(7) The physical environment is subject to high winds and poor conditions making the likelihood of an incident higher than in 'ideal' conditions.
(8) The experience and skill level of the workers keeps the probability of error low.

The quantification of risk in this way allows the team to better determine whether the level of risk is acceptable and also which sources of risk need most attention. In the risk assessment section, the team is prompted to investigate legal requirements. Where any system factor associated with the activity does not meet legislated standards, immediate change is required. In this case, 'AS1657-1992: Fixed platform, walkways, stairways and ladders – design, construction and installation'[2] requires landing platforms every 6 meters on fixed ladders. Ladders higher than 6 meters also have to have protective back guards. Clearly, the ladder in the example is noncompliant and requires replacement or modification.

In the following section of the OHS strategy development format, the four-fold strategy is applied. The result is a comprehensive set of short- and long-term remedies to correct the sources of entropic risk associated with the activity and to manage the residual risk, and thus to ensure legal compliance. The review of this activity is also used to identify the broader implications of working at heights and to improve the quality of supporting management systems. The scores and a check of legal requirements indicate whether or not the activity should continue. In this example, the team identified process residual risks to be unacceptable because of the height involved. In addition, the technology residual risk was identified as noncompliant. Accordingly, the process has to be ceased immediately. From the strategies developed, the first course of action is the removal of the lower part of the ladder to prevent unauthorized access. In the longer term, the replacement or modification of the ladder system is required, which entails the installation of a cage system around the ladder and at least one landing platform, as required by the Australian Standard, if the practice of inspecting the tank is to be undertaken in the future.

In this case, the application of the four-fold strategy has led to the identification of other methods to reduce or better manage risk. These included:

- The introduction of equipment and conditions checks before climbs;
- The introduction of regular inspections and maintenance of ladder systems;
- Guidelines concerning environmental conditions for tasks involving working at heights;
- Medical checks of worker fitness and health;
- Increased reinforcement of safety procedures in relation to working at heights;
- Inclusion of these procedures in induction training;
- Inclusion of the physical requirements of the job in the position description selection criteria.

The OHS strategy development format takes the assessment team through a number of analytical steps. These are:

(1) Is the activity legally compliant? If not, cease the activity immediately and take corrective action to eliminate the entropic risk and manage the residual risk.
(2) If the activity is legally compliant, is the level of risk acceptable according to the company's own standards? If not, cease the activity immediately, eliminate the entropic risk and manage the residual risk.
(3) If the level of risk is acceptable in the short term, develop strategies to reduce and better manage the risks involved by applying the four-fold strategy to improve the quality of system factors.

In the example given in Fig. 11.4, the ladder is noncompliant. This indicates that corrective action is required and the ladder should not be used again until the necessary modifications have been made. If, on the other hand, there were no legislated standards related to the installation of such technology, the company should analyze the risk scores to determine whether the level of risk is acceptable according to its own standards. In this case, the results highlight that risk management within this activity is highly dependent on the vigilance and physical capability of the maintenance personnel. Where the residual and entropic risks of technology, the physical environment and/or the process are high, the demands on the operator may exceed his capacity to manage the risk, particularly if the worker is ill, tired or distracted.

## Summary

When activities involve high-risk system factors the onus falls on management to ensure that a safe place and safe systems of work are provided in accordance with the duty of care. Where the risk is also attributable to human resources, the firm's responsibility is to provide adequate training and supervision. The risk management strategies developed should therefore address these areas of systemic weakness. When all the core business activities have been assessed in this way, priority areas can be identified so that resources are directed to maximize hazard reduction and control. The resultant interventions may be documented together with restructuring plans and behavioral change programs in the productive safety management plan. The strategies, initiatives and action plans form the 'to do' section of the plan. This is followed by the final phase of the change process which is 'to evaluate'. The achievement cycle – the performance management system – ties these strategies to targets and measures. It allows the firm to determine, at the end of the cycle, how successful it has been in implementing its strategies and creating tangible positive change. The achievement cycle also contains goal-setting tools to align the contributions of business units and leaders with corporate objectives. This performance management system is presented in the following chapter.

## References

1. David, F.R. (1995) *Strategic Management* 5th edn. Prentice Hall, New Jersey, USA. © 1995 by Prentice Hall, Inc. Reprinted by permission of Pearson Education, Inc., Upper Saddle River, NJ.
2. Standards Australia (2002) *AS1657-1992 fixed platform, walkways, stairways and ladders – design, construction and installation*, available through Standards Australia website: http://www.standards.com.au/catalogue/script/search.asp

Chapter 12

# Managing performance using the achievement cycle

## An introduction to measurement principles

Performance measurement is a critical part of any management system. It is the feedback mechanism that allows managers to determine whether goals have been achieved and whether resources have been used effectively. In the same way that an athlete should expect to enhance her performance, the firm should anticipate key performance indicators to improve from cycle to cycle. This is commonly referred to as 'continuous improvement', which involves the results achieved in one cycle becoming the baselines against which results in the next cycle are compared. To achieve a balanced perspective of organizational achievement a number of indicators need to be used. Measuring performance using a multivariable approach allows the firm to ascertain the following:

(1) The extent of improvement in specific areas;
(2) The overall level of performance;
(3) If there has been any regression overall or in specific areas;
(4) If achievements in one area have been made to the detriment of another area;
(5) Opportunities for further improvement.

Evaluation criteria results are significant in two ways. The first is their absolute value; that is the actual value for a given period of time, for example, an LTIFR of 4 in a high-risk environment may be considered to be a reasonable result. The other significance of a measure is the relative change between results in subsequent cycles. If the firm achieves an LTIFR of 3.2 in the following cycle this represents a 20% reduction which is a notable improvement on the previous cycle. When assessing measures the two issues are, therefore, whether the actual result is good in absolute terms and whether the change indicates an improvement or step in the right direction.

There are further considerations when assessing performance measure results. For example, would an LTIFR in the first cycle of 18 followed by a result of 16 in the next cycle be considered a positive improvement? Although the change is in the right direction, the baseline of 18 was very high to start with and the subsequent result of 16 remains high, therefore, the outcome would still be considered poor. To ascertain whether or not a result is a sufficient positive change it needs to be compared to a target. Consequently, the firm needs to define where it started from (the baseline), where it wants to go (the target), and how far it has come (the change in result from measurement period to measurement period). It is very similar to a hiker setting out on a journey. The hiker starts at A (the baseline) and wants to go to B (the target). Along the way he measures progress either as distance traveled from A, distance remaining to B, or as a proportion of the distance between A and B.

For some performance measures the firm will be looking for a reduction in the result such as the LTIFR, and for others an increased result, for example, the production output rate. Whether the change is positive, therefore, depends on the value of the actual result compared to the baseline and whether the change represents a shift towards the target. Table 12.1 illustrates the tracking of results by providing some examples. It identifies whether the result given for 2001–2002 is an improvement, no change or regression on the result in 2000–2001. Also shown is the frequency with which the sample measures are reported. At the business unit level for example, the LTIFR may be reported quarterly and production output monthly. The frequency of reporting depends on the information needs of the management team. The more critical the indicator is to corporate achievement, the more often it has to be measured at the operational level so that adjustments can be made to improve on the particular management issue.

How should measures be chosen? There is a saying that 'what gets measured gets done'. It is therefore absolutely essential that the firm select the right measures. In Chapter 11 it was explained that it is important to include outcome, implementation and system factor quality measures in the assessment of firm performance to ensure that underlying variables are included. The entropy model indicates that the quality of system factors has a direct effect on both performance and safety and thus needs to be incorporated into the measurement system. Concurrently, the channel highlights the importance of aligning business systems to fit the external environment, to create synergies through organizational decision-making, and to maximize the potential of human resources. The productive safety performance management system – the achievement cycle – provides a structured approach to performance measurement which ensures a balanced approach in which production, safety and system factor quality are pursued concurrently.

## The achievement cycle – a balanced approach

The achievement cycle consists of four tools. These are the business alignment scorecard, the cultural health scorecard, the business unit alignment scorecard and the leader's scorecard. The two tools used at the corporate level are the business alignment scorecard and the cultural health scorecard. The latter will be explained later in this chapter. The former is used to summarize those indicators, derived from the productive

**Table 12.1** Tracking results

| Measure | Target | Result for 2000–2001 | Result for 2001–2002 | Change in result | Type of change | Reporting frequency at business unit level | Reporting frequency at corporate level |
|---|---|---|---|---|---|---|---|
| Lost time injury frequency rate (LTIFR) | 0 | 4.2 | 3.6 | –14.3% | Improvement | Quarterly | Quarterly |
| Productivity (average units/employee/ month) | 100 | 80 | 90 | +12.5% | Improvement | Monthly | Monthly |
| No. of initiatives completed on time and within budget as % of total no. of initiatives | 100% | 75% | 75% | 0% | No change | Annually | Annually |
| Housekeeping rating | 100% | 60% | 85% | +41.7% | Improvement | Annually | Annually |
| Employee satisfaction index | 100% | 55% | 80% | +45.5% | Improvement | Annually | Annually |
| No. of employee suggestions implemented as % of suggestions received | 100% | 65% | 60% | –7.7% | Regression | Annually | Annually |
| Training completed versus training planned | 100% | 83% | 92% | +10.8% | Improvement | Annually | Annually |
| Breakdown maintenance as % of total maintenance costs | 5% | 37% | 26% | –29.7% | Improvement | Quarterly | Quarterly |
| Employee turnover rate | 5% | 8.2% | 8.8% | +7.3% | Regression | Annually | Annually |

safety KRA and other KRAs, which are critical to company performance. These are called alignment indicators because they verify that systems are aligned and balanced. In terms of the productive safety KRA, alignment is evident when measures corroborate that production, safety and system factor quality improved concurrently. Where results are mixed, for example, when production has increased and safety declined, then organizational systems are not compatible with this KRA.

As explained in previous chapters, the productive safety approach is also concerned with helping managers make optimum use of limited resources, thus requiring the inclusion of financial measures in the scorecard. Concurrently, the firm needs to assess service delivery to customers to ensure efficiency, competitiveness and product safety. Customer satisfaction is therefore included with financial management as an alignment indicator. The specific areas of measurement in the business alignment scorecard, in accordance with the entropy model and the channel, which are referred to as alignment indicators, are:

- Productivity;
- Safety;
- System factor quality;
- Financial and customer;
- Compliance (with legislation, internal standards or plans);
- Social responsibility.

This is a different approach to that proposed by Kaplan and Norton who developed the balanced scorecard. They applied four perspectives – financial, customer, internal business process, and learning and growth. The balanced scorecard involves the development of objectives, measures, targets and initiatives to address four questions that relate to these perspectives. The questions are:

(1) To succeed financially, how should we appear to our shareholders?
(2) To achieve our vision, how should we appear to our customers?
(3) To satisfy our shareholders and customers what business process must we excel at?
(4) To achieve our vision how will we sustain our ability to change and improve?[1]

The balanced scorecard uses measures to drive organizational change and success. This is a shift away from measurement as a means of controlling behavior and evaluating the past, to articulating and communicating the business strategy, and aligning individual, corporate and cross-departmental initiatives into a common goal. Multiple measures are also applied to integrate a set of cause and effect relationships. This approach, therefore, embraces both outcome measures and the underlying factors that determine these outcomes.

Kaplan and Norton's methodology has been embraced by OHS regulators for its capacity to consolidate both output and safety into a performance management system. It is an ideal system for organizations that have a low level of risk. In high-risk environments such as mining, construction, manufacturing, and oil and gas, a key variable affecting business performance is hazard management. In the balanced scorecard, safety is not considered as a separate perspective. Instead, safety may be considered in relation to process quality, the financial cost of deviations, or as a training area under the learning and growth perspective. It does not stand out as a key performance indicator or perspective. In addition, the impact of degradation of systems or the presence of residual risks, and their relationship to performance and safety are not taken into consideration. This may limit the application of this tool as a driver of improvement in OHS management in hazardous industries.

The balanced scorecard may, however, provide a rationale for the development of performance management systems for high-risk sectors. Three mechanisms are used to achieve strategic alignment in this approach – communication and education programs, goal-setting programs and reward system linkage. The business alignment scorecard presented in this book embraces these mechanisms while integrating OHS as a primary perspective upon which organizational performance is measured. It also applies the use of regular reviews because, according to Kaplan and Norton, the ultimate payoff of using the balanced scorecard occurs when organizations conduct regular strategic reviews not just operational reviews. The characteristics that differentiate the productive safety approach from the balanced scorecard are that in the former:

(1) The impact of degradation and residual risk on performance and safety are considered to specifically meet management priorities required in medium- and high-risk industries;
(2) Strategic performance management is structured according to the three levels of alignment;
(3) At the operational level the focus is on system factor quality;
(4) Specific tools required to develop a culture in which production, safety and system factor quality are compatible goals are provided;
(5) Management systems and cultural issues are addressed which explain not only what should be done but also why and how, using the strategic alignment channel and supporting tools such as the value statement and reasonableness test;
(6) The legislative framework of westernized countries such as Australia is supported by this approach.

These points will become increasingly evident as the achievement cycle is presented in this chapter. Figure 12.1 illustrates an overview of the key elements of this performance management system, which are the productive safety KRA, the three levels of alignment, alignment indicators and their respective measures. On the outer hub, the productive safety KRA (together with its value statement) acts as a driver of decision-making and facilitates alignment at the external strategic, internal strategic and internal goal levels. The alignment indicators are used to determine whether the management system is running effectively. This means that in the broader context of management, productivity, safety and system factor quality are not compromised for the sake of each other.

Certain traits are characteristic of these congruent systems. Firstly, company systems must be legally compliant with external requirements

**Figure 12.1** The alignment wheel

and internally compliant through self-regulated standards and plans. Secondly, from the financial perspective, the firm must allocate resources to OHS management to ensure that risks are managed effectively, and optimum use is made of capital resources by addressing those areas of highest risk. Other external factors are also important. For instance, to achieve external alignment, customer needs must be satisfied, which requires that the system factors used to manufacture goods and services must be maintained to ensure that products are of the required quality. Alignment is also enhanced when the firm derives benefits from acting with social responsibility, for example, by investing in research and development, by providing employees with assistance programs, and by supporting community ventures, thereby attracting the loyalty and support of workers, customers and the community.

To determine whether these alignment indicators are being achieved to the desired level, they have to be measured. A mix of outcome, implementation and system factor quality measures is used to ensure that short-term results are balanced against long-term objectives. The measurement system is at the center of this alignment wheel and therefore, the achievement cycle provides the focal point for the implementation of productive safety management.

## Business alignment scorecard

As suggested by the channel, alignment is the key factor affecting the performance of the firm and accordingly, the business alignment scorecard is structured to focus on this issue. Measures are developed within three

| Business Alignment Scorecard (year) | | | | | | |
|---|---|---|---|---|---|---|
| Level of alignment | Target | Baseline | Actual | Weighting (%) | Weighted actual this year (%) | Progress required this year (%) |
| External Strategic Alignment | | | | | | |
| | | | | | | |
| | | | | | | |
| | | | | | | |
| | | | | | | |
| *Total weighted score – External strategic alignment* | | | | | | |
| Internal Strategic Alignment | | | | | | |
| | | | | | | |
| | | | | | | |
| | | | | | | |
| | | | | | | |
| | | | | | | |
| | | | | | | |
| | | | | | | |
| | | | | | | |
| *Total weighted score – Internal strategic alignment* | | | | | | |
| Internal Goal Alignment | | | | | | |
| | | | | | | |
| | | | | | | |
| | | | | | | |
| | | | | | | |
| | | | | | | |
| | | | | | | |
| | | | | | | |
| *Total weighted score – Internal goal alignment* | | | | | | |
| TOTAL ACHIEVEMENT SCORE | | | | | | |

Target — Result to be achieved by end of achievement cycle
Baseline: — Level of achievement at the beginning of the cycle
Actual: — Result for the current year
Weighting: — Relative weight of the measure for the achievement cycle
Weighted actual: — Calculation of the progress made towards the target this year multiplied by the weighting
Progress required: — Progress required this year to achieve the target at the end of the achievement cycle

**Figure 12.2** The business alignment scorecard

perspectives – external strategic alignment, internal strategic alignment and internal goal alignment, as shown in Fig. 12.2. The first step in developing such a system is to determine the length of the cycle to which measures will apply. In the examples provided in this chapter a 3-year cycle is used. At the beginning of the cycle, the measures, targets, baselines and the relative weights applied to each measure are recorded. Within this cycle, progress is evaluated on an annual basis. As an example, a business alignment scorecard 2001–2004 would be the blueprint for achievement over 3 years and the scorecards for 2001–2002, 2002–2003 and 2003–2004 used to document progress each year.

As explained in Chapter 11, not every measure of corporate achievement has the same weighting. Some measures are more critical to the success of the firm than others. This weighting is defined at the beginning of the cycle and recorded in the scorecard as a percentage such that the weighting of all measures sums to 100 per cent. It is recommended that the total weight given to external strategic alignment is 20, for internal strategic alignment 60 and for internal goal alignment 20, although each firm should determine these according to its specific circumstances. Internal strategic alignment attracts the largest proportion of the weighting because as shown later, it contains core business measures. At the end of the each year the weighted actual is calculated for each measure as follows:

$$\text{Weighted actual} = \left\{\frac{\text{Actual result} - \text{Baseline}}{\text{Target} - \text{Baseline}}\right\} \times \text{Weighting}$$

The weighted actual is a calculation of the progress made towards the target for a given year multiplied by the weighting. The scorecard also contains a number expressing the progress required this year. If, for example, the business operates the achievement cycle over 3 years, it would expect at least one-third of the required improvement to be attained each year in order to be on track to reach the target. Firms that require substantive improvement because they have difficulty achieving legal compliance and desired output results, may adjust these progress proportions, for example, to 50, 30 and 20 per cent in successive years to raise performance to the required standards. These calculations are explained further using the hypothetical example provided in Fig. 12.3, which applies a 33.3, 66.7 and 100 per cent progress requirement over a 3-year period.

As explained earlier, to achieve alignment and balance, the business alignment scorecard must contain a mix of alignment indicators and there must be at least one of each type of indicator used in the scorecard. In the example given, the measures that reflect alignment with the external environment are included in the external strategic alignment section. In this case, the chosen measures affirm the company's concern for customer satisfaction, shareholder value, market development, regulatory compliance and community involvement, which are measures that assess the fit with external requirements. In addition, these evaluators quantify the firm's ability to take advantage of external opportunities and manage external threats, for example, the competitive opportunities and threats posed by the market place.

The external strategic alignment measures that this example company has used communicate its priorities – those areas it considers most critical to achieving compatibility with the external environment comprising external stakeholders, sources of constraint and sources of opportunity. The sample firm has set a target of 100 per cent customer satisfaction. (Targets set the ideal scenario or at the very least a stretch target. In reality it may not be totally achievable, however, as explained later in the chapter, the firm also sets a cutoff point for bonuses/incentives which is a proportion of the overall level of achievement, for example, 80 per cent. In this way, the firm pursues significant improvements, which become

| Business Alignment Scorecard 2001–2002 | | | | | | |
|---|---|---|---|---|---|---|
| Level of alignment | Target | Baseline | Actual | Weighting (%) | Weighted actual 2001/2002 (%) | Progress required 2001/2002 (%) |
| External Strategic Alignment | | | | | | |
| Customer satisfaction index | 100% | 85% | 91% | 6 | 2.40 | 2.00 |
| Price/Earnings ratio | 15 | 9.1 | 11.2 | 6 | 2.14 | 2.00 |
| Market share | 50% | 32% | 36% | 4 | 0.89 | 1.33 |
| No. of OHS, environmental and IR issues requiring regulatory intervention | 0 | 7 | 5 | 2 | 0.57 | 0.67 |
| Community sponsorship (% of annual profit) | 5% | 3.1% | 3.5% | 2 | 0.42 | 0.67 |
| *Total weighted score – External strategic alignment* | | | | **20** | 6.42 | 6.67 |
| Internal Strategic Alignment | | | | | | |
| Production (units of output per employee per month) | 100 | 87 | 91 | 10 | 3.08 | 3.33 |
| Re-work and defect rate per 100 units of output | 0 | 4 | 2.7 | 6 | 1.95 | 2.00 |
| Profitability (return on company investment) | 15% | 8% | 10.2% | 8 | 2.51 | 2.67 |
| Actual/budget expenditure ratio | 1 | 1.03 | 1.01 | 4 | 2.67 | 1.33 |
| Lost time injury frequency rate | 0 | 8.5 | 6.2 | 7 | 1.89 | 2.33 |
| No. of near misses | 0 | 50 | 30 | 7 | 2.80 | 2.33 |
| No. of initiatives completed on time and within budget as % of total no. of initiatives | 100% | 86% | 94% | 4 | 2.29 | 1.33 |
| System factor quality audit index | 100% | 66 | 78 | 8 | 2.82 | 2.67 |
| Breakdown maintenance as % of total maintenance costs (infrastructure and equipment) | 5% | 45% | 28% | 6 | 2.55 | 2.00 |
| *Total weighted score – Internal strategic alignment* | | | | **60** | 22.56 | 20 |
| Internal Goal Alignment | | | | | | |
| Employee turnover rate | 5% | 11.4% | 9.3% | 4 | 1.31 | 1.33 |
| Absenteeism | 3% | 9.5% | 6.2% | 3 | 1.52 | 1.00 |
| Employee satisfaction index | 100% | 62% | 74% | 4 | 1.26 | 1.33 |
| No. of employee suggestions implemented as % of suggestions received | 100% | N/a | 63% | 3 | 1.89 | 1.00 |
| Training completed versus training planned | 100% | 88% | 96% | 2 | 1.33 | 0.67 |
| Return on resourcefulness Benefit of employee suggestions ratio/cost of training) | >1 | N/a | 0.6 | 4 | 2.40 | 1.33 |
| *Total weighted score – internal goal alignment* | | | | **20** | 9.71 | 6.67 |
| TOTAL ACHIEVEMENT SCORE | | | | **100** | **38.69** | **33.33** |

Target: Result to be achieved by end of 2004
Baseline: Result as at 1 July 2001
Actual: Result for 2001–2002 financial year
Weighting: Relative weight of the measure for the 2001–2004 achievement cycle
Weighted actual: Calculation of the progress made in 2001–2002 towards the target multiplied by the weighting
Progress required: Progress required in 2001–2002 to achieve the target at the end of the 2001–2004 cycle

**Figure 12.3** The business alignment scorecard for company A

the baseline for the next cycle.) As shown in the figure, the baseline level of customer satisfaction at the beginning of the cycle, in July 2001, was 85 per cent. At the end of the first year, 2001–2002, the firm achieved a rating of 91 per cent obtained from its customer feedback system. The weighting given to this measure was 6 per cent of the overall performance result. The weighted actual is calculated as:

$$\text{Weighted actual} = \left\{\frac{\text{Actual result} - \text{Baseline}}{\text{Target} - \text{Baseline}}\right\} \times \text{Weighting}$$

$$= \left\{\frac{91 - 85}{100 - 85}\right\} \times 6$$

$$= 2.40$$

Given that the weighting for this particular measure is 6 per cent for the full cycle, the progress required in the first year of a 3-year period is 2.00 per cent, as shown in the final column. The total weighted score for external strategic alignment is 20 per cent. The weighted actual achieved in 2001–2002 is 6.42 (the sum of weighted actuals for the measures in the area), which is slightly short of the progress required of 6.67.

Internal strategic alignment measures attract the highest weighting with 60 per cent of the total allocation, because without efficiency and effectiveness at the core of the business, it will not remain competitive. In this section, internal strategic measures are included such as productivity, profitability, budgetary and management control, system factor quality and OHS outcomes. Inclusion of at least one of each type of alignment indicator ensures that production, safety and system factor quality are not compromised for each other. The weightings are also used to provide balance and symmetry. If, for example, production output was given a very high weighting and the LTIFR and number of near misses given a low weighting, the scorecard would not fit the productive safety value statement. The total of the weightings given to production, safety and system factor quality measures have to be reasonably proportional to reflect the intent of the value statement.

What do the results in this section of the scorecard indicate? When the weighted actuals are compared with the progress required it is apparent that the firm has made significant inroads into reducing the cost of breakdown maintenance as a proportion of total maintenance costs, has improved system factor quality significantly, and has tightened up on budgetary and initiative implementation controls. In some other areas it needs to improve, for example, employee turnover and market share.

The final section of the scorecard addresses internal goal alignment issues. The firm has allocated 20 per cent of the total weighting to this section. These measures center on maintaining the quality of human resources available to the firm and maximizing the return on labor investment. The indicators given measure the human resources system factor for leakage (turnover), availability (absenteeism) and capacity (training and suggestions). Some of the measures are new as indicated by the lack of a baseline, shown by N/a (not applicable). In this case the baseline, for the sake of the calculation, is assumed to be zero. The total weighted score indicates that the firm is achieving sound results in this area. In the 2001–2002 financial year, the firm has attained a total achievement score of 38.69, shown in the bottom line of the scorecard, which is above the required progress level of 33.33. It is therefore on track in its pursuit of the targets.

The business alignment scorecard shown in Fig. 12.3 contains a balanced mix of alignment indicators with at least one type of each included in the evaluation tool. Figure 12.4 illustrates how, in the example given, the firm has developed this mix to meet its specific needs. The measures associated within each alignment indicator have been coded. Included in the 'productivity indicator' are four measures – output production, the re-work and defect rate per 100 units of output, the turnover rate and the absenteeism rate. The last two measures affect the availability of human resources and therefore, the capacity of the firm to produce more given fixed capital resources. In this case, the productivity measures are a total of 23 per cent (10 + 6 + 4 + 3 obtained from Fig. 12.3) of the total weighting of 100 per cent.

Two safety measures have been included, which are the LTIFR and number of near misses, with a total weight of 14 per cent (7 + 7). In the example given, the system factor quality measures used are the audit results, maintenance of technology and the physical environment, and a measure of human resources quality that is the 'number of employee suggestions implementable as a percentage of the suggestions received'. Process measures may also be used in this alignment area, such as the number of processes for which JSA has been undertaken as a proportion of the number of known standard processes ordinarily carried out.

The scorecard takes into consideration the primary reason for the existence of the private sector firm, which is profit-oriented and dependent on having customers for the output produced. The 'financial and customer' indicator covers those measures related to the viability of the firm. In the public sector there is currently a push for increasing levels of accountability and return to the community to the extent that the 'financial and customer' indicator is also highly relevant. Figure 12.4 shows that the firm has allocated these types of measures in each of the alignment areas. At the external level, customer satisfaction, price/earnings ratio and market share measures have been applied to evaluate the extent of alignment with customers, shareholders and the market place. At the internal strategic level, there is concern for profitability and budget control as financial measures. In the internal goal perspective, the firm has selected the measure 'return on resourcefulness' as an assessment of the benefits derived from its investment in employee training. In addition, it has treated employees as internal customers by gauging their satisfaction using an employee survey.

The final alignment indicator is 'social responsibility'. This indicator increases the fit between the firm and the external environment and reflects the firm's role as a responsible, corporate citizen. In this example, the company uses its community sponsorship program as a measure of social responsibility. Other measures that may be used in this category include research and development (R & D), work experience opportunities and sponsorship programs. R & D fits this indicator if it is taken from the

| Business Alignment Scorecard 2001–2002 | | | | | | | |
|---|---|---|---|---|---|---|---|
| Level of alignment | Target | Baseline | Actual | Weighting (%) | Weighted actual 2001/2002 (%) | Progress required 2001/2002 (%) | Indi-cators |
| External Strategic Alignment | | | | | | | |
| Customer satisfaction index | | | | | | | F |
| Price/Earnings ratio | | | | | | | F |
| Market share | | | | | | | F |
| No. of OHS, environmental and IR issues requiring regulatory intervention | | | | | | | C |
| Community sponsorship (% of annual profit) | | | | | | | SR |
| *Total weighted score – external strategic alignment* | | | | | | | |
| Internal Strategic Alignment | | | | | | | |
| Productivity (units of output per employee per month) | | | | | | | P |
| Re-work and defect rate per 100 units of output | | | | | | | P |
| Profitability (return on company investment) | | | | | | | F |
| Actual/budget expenditure ratio | | | | | | | F |
| Lost time injury frequency rate | | | | | | | S |
| No. of near misses | | | | | | | S |
| No. of initiatives completed on time and within budget as % of total no. of initiatives | | | | | | | C |
| System factor quality audit index | | | | | | | SFQ |
| Breakdown maintenance as % of total maintenance costs (infrastructure and equipment) | | | | | | | SFQ |
| *Total weighted score – internal strategic alignment* | | | | | | | |
| Internal Goal Alignment | | | | | | | |
| Employee turnover rate | | | | | | | P |
| Absenteeism | | | | | | | P |
| Employee satisfaction index | | | | | | | F |
| No. of employee suggestions implemented as % of suggestions received | | | | | | | SFQ |
| Training completed versus training planned | | | | | | | C |
| Return on resourcefulness (Benefit of employee suggestions/cost of training ratio) | | | | | | | F |
| *Total weighted score – internal goal alignment* | | | | | | | |
| TOTAL ACHIEVEMENT SCORE | | | | | | | |

P Productivity
S Safety
SFQ System factor quality
F Financial and customer
C Compliance (to legislation, internal standards or plans)
SR Social responsibility

**Figure 12.4** Types of measures needed for alignment and balance

perspective that it may be used to reduce the negative impact of the firm's operations on the society. If it is used solely as a strategy for product development it is less likely to fit this indicator, unless the research leads to goods/services which are safer, more environmentally friendly or which reduce the wastage of the firm's resources.

The business alignment scorecard not only ensures that key areas are addressed by including the alignment indicators, it also correlates these measures to a mix of 'what', 'how' and 'if' questions. Each measure therefore falls into one of the measurement arcs. Reiterating, these are:

(1) What to do, measured using outcomes that determine whether strategies were appropriate and effective;
(2) How it is done, evaluated using implementation measures that assess how effectively changes were introduced;
(3) If it has made a difference, measured by system factor quality measures that determine whether the change has become part of organizational practice and the culture.

The measures therefore address the three levels of change. These are: knowing what needs to be done, knowing how to go about it and doing it in an appropriate manner, and making a difference by forming lasting constructive habits and developing high quality systems. Each step is as important as the next in creating sustained improvement at both corporate and operational levels. To ensure that each of these areas of change is covered, the management team should check the scorecard for this mix. In the example given in Fig. 12.3, 'outcomes' include the LTIFR and production, 'implementation measures' include the number of initiatives completed on time and within budget and employee satisfaction index, and 'system factor quality measures' include the audit result and breakdown maintenance as a percentage of total maintenance costs.

The business alignment scorecard allows the management team to evaluate the results at the end of each year in the cycle to reveal any weaknesses in the organizational system. In the case given in Fig. 12.3, for instance, management can see that the firm is behind the required progress level in expanding its market share. This provides an opportunity to revisit its strategy and to ascertain whether there were any incorrect assumptions made at the planning stage. It may also be appropriate to scan the external environment for unforeseen threats and new opportunities that may allow a revised approach and increase the likelihood of success in the forthcoming year. In addition to providing a measurement methodology, the scorecard can be used as an information and strategy development tool.

## Cultural health scorecard

The second scorecard used at the corporate level is the cultural health scorecard. As will be explained later, the business alignment scorecard is linked to the incentive/bonus system. The cultural health scorecard stands separately from this system and is used primarily as an information source to evaluate the health of the culture and the effectiveness of the management team in implementing cultural development strategies.

Why would it be necessary to measure the culture? In the same way that system factor quality is an underlying condition affecting the firm's susceptibility to risk, the culture is a fundamental state affecting its ability to pursue production, safety and system factor quality concurrently. It also determines the dynamics of human factors such as management behavior, communication and the extent of authority versus autonomy, that in turn affects the firm's ability to maximize employee contributions. Essentially, an unhealthy culture is an obstacle to achievement and a healthy culture is a catalyst, and therefore, it is important for the firm to diagnose these conditions.

The format of the cultural health scorecard is provided in Fig. 12.5. It may be evaluated on a yearly basis and contains targets, baselines, actuals,

| Cultural Health Scorecard (Year) | | | | | | |
|---|---|---|---|---|---|---|
| Culture measures | Target | Baseline | Actual | Weighting (%) | Weighted actual this year (%) | Progress required this year (%) |
| Shared ownership | | | | | | |
| | | | | | | |
| | | | | | | |
| | | | | | | |
| Management commitment and leadership | | | | | | |
| | | | | | | |
| | | | | | | |
| | | | | | | |
| Employee commitment | | | | | | |
| | | | | | | |
| | | | | | | |
| | | | | | | |
| Supervision and decision-making | | | | | | |
| | | | | | | |
| | | | | | | |
| | | | | | | |
| Safety issues resolution | | | | | | |
| | | | | | | |
| | | | | | | |
| | | | | | | |
| Capacity reservoir | | | | | | |
| | | | | | | |
| | | | | | | |
| | | | | | | |
| Resourcefulness | | | | | | |
| | | | | | | |
| | | | | | | |
| | | | | | | |
| TOTAL CULTURAL HEALTH SCORE | | | | | | |

Target: Result to be achieved by end of achievement cycle
Baseline: Level of achievement at the beginning of the cycle
Actual: Result for the current year
Weighting: Relative weight of the measure for the achievement cycle
Weighted actual: Calculation of the progress made towards the target this year multiplied by the weighting
Progress required: Progress required this year to achieve the target at the end of the achievement cycle

**Figure 12.5** Cultural health scorecard

weightings, weighted actuals and progress scores using the same rationale as the business alignment scorecard. It is structured on the key elements of behavioral change that were discussed in Part 3. They are shared ownership, management commitment and leadership, employee commitment, supervision and decision-making, safety issues resolution, the capacity reservoir, and resourcefulness.

A worked example of the cultural health scorecard is provided in Fig. 12.6. The 'shared ownership' category includes any measure that determines the extent to which employee accountabilities have been formalized, for example, using PDs, or any measure that evaluates the extent of participation in information provision and decision-making processes driven by the productive safety KRA. Management commitment and leadership can be measured by assessing the levels of turnover and absenteeism in this group. Turnover is particularly important because for the firm an experienced manager represents a substantial investment. Not only is the cost of recruiting and selecting senior personnel more expensive than lower level employees, it is not unusual for firms to invest heavily in management training and development. When a manager leaves the company this represents a significant drain on organizational competencies and 'local knowledge' that is difficult to replace.

The extent of management's commitment and the level of leadership effectiveness in terms of bringing about change can also, to a certain degree, be evaluated by their rigor in implementing initiatives on time and within budget. This in turn influences employees' perceptions of management's commitment to change and thus has secondary effects on the culture.

Employee commitment should also be evaluated using turnover and absenteeism figures. According to the entropy model, when the level of competence in the human resources system factor declines, which is the case when skilled employees leave the firm or when the required skills are not available due to absenteeism, the firm's exposure to risk rises. This affects both the level of safety and performance. In a healthy culture, employees are motivated to stay with the company and to attend work. Absenteeism, by definition, being any failure of an employee to report for or to remain at work for any reason,[2] includes absences due to sickness and accidents. An investigation into the causes of absenteeism can provide management with additional insight into health issues beyond the obvious consequences that result from OHS incidents. If, for example, a significant proportion of workers from a particular area are absent due to headaches, the cause may be linked to airborne toxins in the workplace. Absenteeism may therefore be an indicator of either the physiological or psychological demands of work.

The next area of measurement on the cultural health scorecard is supervision and decision-making. This is assessed because the quality of supervisors' managerial skills has a significant impact on the culture. The employee satisfaction index obtained through an employee survey is the primary evaluator of the quality of supervision as perceived by the workforce. In effect, this qualitative assessment aims to determine the extent to which the value statement is being applied at the operational level. The statement is specifically designed to modify decision processes

| Cultural Health Scorecard 2001–2002 | | | | | | |
|---|---|---|---|---|---|---|
| Culture measures | Target | Baseline | Actual | Weighting (%) | Weighted actual this year (%) | Progress required this year (%) |
| **Shared ownership** | | | | | | |
| No. of position descriptions fitting productive safety as % of total no. of pds | 100% | N/a | 65% | 4 | 2.60 | 1.33 |
| No. of hazardous area employees involved in OHS forums as % of average workforce population in these areas (participation rate) | 100% | 35% | 58% | 5 | 1.77 | 1.67 |
| **Management commitment and leadership** | | | | | | |
| Management turnover | 5% | 8% | 7.3% | 6 | 1.40 | 2.00 |
| Management absenteeism | 3% | 7.2% | 5.9% | 6 | 1.86 | 2.00 |
| No. of initiatives completed on time and within budget as % of total no. of initiatives | 100% | 85% | 94% | 5 | 3.00 | 1.67 |
| **Employee commitment** | | | | | | |
| Employee turnover | 5% | 11.4% | 9.3% | 6 | 1.97 | 2.00 |
| Employee absenteeism | 3% | 9.5% | 6.2% | 6 | 3.05 | 2.00 |
| **Supervision and decision-making** | | | | | | |
| Employee satisfaction index | 100% | 62% | 74% | 5 | 1.58 | 1.67 |
| Supervisory training completed versus training planned | 100% | N/a | 85% | 5 | 4.25 | 1.67 |
| **Safety issues resolution** | | | | | | |
| No. of safety issues requiring regulatory intervention | 0 | 12 | 8 | 4 | 1.33 | 1.33 |
| Days lost due to industrial disputes (all issues) | 0 | 18 | 10 | 5 | 2.22 | 1.67 |
| **Capacity reservoir** | | | | | | |
| % of new employees completed induction within 1 week of starting | 100% | 76% | 96% | 7 | 5.83 | 2.33 |
| Post-induction evaluation rating (learning retention) | 100% | N/a | 74% | 4 | 2.96 | 1.33 |
| No. of employees with certification as % of those requiring certification by law | 100% | 92% | 96% | 7 | 3.50 | 2.33 |
| OHS training completed versus training planned | 100% | 88% | 95% | 5 | 2.92 | 1.67 |
| Competencies available versus competencies required | 100% | N/a | 76% | 5 | 3.80 | 1.67 |
| **Resourcefulness** | | | | | | |
| Average number of suggestions received per employee | +1 | N/a | 0.25 | 4 | 1.00 | 1.33 |
| No. of employee suggestions implemented as % of suggestions received | 100% | N/a | 63% | 4 | 2.52 | 1.33 |
| Return on resourcefulness (benefit of employee suggestions/cost of training ratio) | >1 | N/a | 0.6 | 5 | 3.00 | 1.67 |
| TOTAL CULTURAL HEALTH SCORE | | | | **100** | **50.56** | **33.33** |

Target: Result to be achieved by end of 2004
Baseline: Result as at 1 July 2001
Actual: Result for 2001–2002 financial year
Weighting: Relative weight of the measure for the 2001–2004 achievement cycle
Weighted actual: Calculation of the progress made in 2001–2002 towards the target multiplied by the weighting
Progress required: Progress required in 2001–2002 to achieve the target at the end of the 2001–2004 cycle

**Figure 12.6** The cultural health scorecard for company A

to ensure appropriacy at all hierarchical levels. Employees are more likely to perceive outcomes as fair and equitable, thus creating a more trusting and positive organizational climate, when decisions fit the value statement and the reasonableness test.

Within this category, measures may also be included to evaluate the effectiveness with which the firm has modified and developed its managerial competencies. A suitable measure is 'supervisory training completed versus training planned'. As explained in Chapter 8, the onus is on the firm to provide supervisors with the instruction required for successful decision-making and leadership. An alternative measure is to identify the gap between the management team's current KSAs and those required for sound supervision, and measure whether the gap is being addressed through training. In other words, the level of competencies held as a proportion of the skills required can be measured. These measures can be used to assess the current level of satisfaction with supervisory behaviors and determine whether the satisfaction gap is being closed through management training.

A further factor related to the quality of management is the effectiveness of conflict resolution processes. In a healthy climate, these procedures are clear, and understood by employees. Differences of opinion are perceived as opportunities to find better solutions to workplace issues. The value statement encourages disputing parties to put aside their personal differences and seek mutually beneficial outcomes. With the appropriate processes in place and with constructive attitudes shown by management and employees, there should seldom be a need for intervention by external regulators. The measure, 'number of safety issues requiring regulatory intervention', may be used as an indicator of the effectiveness of these processes and the underlying group dynamics required to reach agreement. An additional measure that may be applied is 'days lost due to industrial disputes'. This measure evaluates the firm's ability to resolve conflict in all areas internally, not just safety issues, and therefore reflects the degree of internal goal alignment. It is a further barometer of the quality of relations between management and workers. In a healthy climate, the company should be able to negotiate constructive outcomes for both parties, without overt conflict arising, using common goals and values to drive the intended change.

As suggested by the entropy model, firms need to ensure that they have the competencies employed to perform work safely and efficiently. The capacity reservoir is important to cultural health because it holds these collective competencies and attitudes. As explained in Chapter 9, the first step in building human resources capacity is recruitment and selection, which involves the identification of the candidate with the KSAs that are the best fit with the firm's requirements from both the economic and social perspective. The next step is assimilation into the culture during induction. In all workplaces, new employees are required, under duty of care legislation, to be given information about the types of hazards present and how these are to be managed. In high-risk environments induction is extremely important as it helps to develop the new recruits' workplace-specific competencies and vigilance.

There are a number of measures that may be used to evaluate the reservoir. In medium- and high-risk workplaces, measures of the application rate and effectiveness of induction should be included to ensure that the firm fulfills its legal obligations and that new employees receive adequate pre-operational training. These include 'the percentage of new employees that have completed induction within one week of commencement' and 'post-induction evaluation rating'. The latter can involve a short written test to assess learning retention. Employee feedback should also be sought to ascertain the transferability of the learning into the workplace – the extent to which the induction has assisted the new recruit to be productive and safe in his new position.

The capacity reservoir illustrates that a productive safety culture requires investment in training to enhance the quality of the human resources. The 'depth' of the reservoir reflects the business' commitment to building such a culture as well as its current skills levels, both of which can be ascertained using measurement criteria. Sample evaluators are included in the cultural health scorecard in Fig. 12.6, for instance, the 'number of employees with certification as a percentage of those requiring certification by law'. The target for this should be 100 per cent – the mandatory level for legal compliance. Other measures include 'OHS training completed versus the training planned', which is an implementation measure, and the 'competencies held by the current workforce versus the competencies required', as identified in the firm's PDs. The level of commitment to training may also be gauged in dollar terms such as training investment as a percentage of the total budget or cost of training per employee per annum.

The final area used to evaluate the health of the culture is resourcefulness. Prerequisites for resourcefulness are competent, well-trained employees, a climate that encourages innovation and participation, and management that balance assertiveness/responsiveness/collaboration according to the level of risk to achieve optimal decision-making outcomes. The more evidence there is of resourcefulness, the healthier the culture. Several indicators may be used, for example, in Fig. 12.6 the assessors proposed are the 'average number of suggestions received per employee', the 'number of suggestions implemented as a percentage of those received', and the 'return on resourcefulness' expressed as the benefit of suggestions versus cost of training ratio.

How can the results in the cultural health scorecard be interpreted? In Fig. 12.6 the 'total cultural health score' indicates that the firm is performing well in building a productive safety culture with a score of 50.56 per cent being achieved when the required progress is only 33.33 per cent for the year. There are some areas that need to be improved including the rate of management turnover and absenteeism. The 'employee satisfaction index' and 'average number of suggestions received per employee' were also below the progress required. At the end of the first year in the 3-year cycle, these results suggest that some morale issues need further investigation at both operational and management levels. This is not to say that the results are conclusively negative, for it may be the case that those managers who were resistant to the new culture have left and been replaced by new staff with competencies and values that fit the company's current vision. The underlying reasons for less than the desired level of performance should be investigated and strategies revised to improve these outcomes in forthcoming years.

Concurrently, it is also worth exploring the reasons for the company's successes. In this hypothetical example, the firm has done extremely well in attaining a high implementation rate for the employee suggestions received. What can this success be attributed to? It may be found that through a series of OHS forums management were able to effectively propose a number of risk control challenges that led to a significant number of high quality suggestions being put forward. From this experience, the company may decide to hold such participative forums to solve problems in other areas of the business. At a strategic level, the process of shifting along the organizational learning curve involves learning through experience – through the analysis of both triumphs and tribulations.

The cultural health scorecard is therefore intended to be a learning tool. It is a gauge of the firm's capacity to address underlying social dynamics that affect its ability to maximize the potential of its human resources and therefore, the potential of the firm as a whole. By building a positive culture, the firm can enhance existing strengths, address systemic weaknesses, for example in terms of competency levels, create internal opportunities through employee suggestions and other participative systems, and reduce threats such as industrial disharmony. The condition of the culture is a measure of its readiness and willingness to change. This scorecard allows management to ascertain whether or not the climate is conducive to the pursuit of the targets in the business alignment scorecard. If, for example, the firm has a very high level of management turnover, it is likely to face difficulties in achieving some of these strategic objectives because of the time required for new managers to reach proficiency. In particular, its implementation measures may be threatened. Similarly, in a firm that has had a history of antagonistic industrial relations, it is likely to take longer for employees to gain sufficient trust to become actively involved in participative processes. The cultural health scorecard helps management to understand these underlying cultural variables. Some of the main issues are:

(1) Are all members of the executive team committed to productive safety management and the targets in the business alignment scorecard?
(2) Do all employees, including management, understand their current roles and responsibilities?
(3) Is there a significant problem with the level of management turnover and/or absenteeism that could derail or hinder the process?
(4) Does the management team have sufficient time management and financial skills to effectively implement the change?
(5) Does this team understand the need to build employee commitment and are they willing to take the lead?
(6) Has sufficient commitment been made for supervisory and competency training?
(7) Has the company a history of conflict, intervention and/or noncompliance that could hinder the building of trust and obstruct participative processes?

(8) Does the company know what competencies it requires to be efficient and safe?
(9) Does the company know what competencies its workforce possesses?
(10) Does the company have support available from OHS and HRM professionals?

If the answer to any one of these questions is no then the firm is not in full readiness to achieve its targets, which means that areas of weakness need to be addressed concurrently during the implementation phase. Only in circumstances where there is an acute deficiency in the level of commitment from the top should the rollout be halted to address this underlying prerequisite for change. Where the lack of preparedness is significant it may be appropriate to revise progress requirements, particularly for the early years. For example, the proportional progress within a 3-year cycle may be changed from 33.3, 66.7 and 100 per cent to 25, 50 and 100 per cent to allow for a more gradual transformation process and for the correction of cultural factors. If, for example, the company does not know what competencies its current workforce possesses, part of the first year strategy should be to undertake a gaps analysis to identify any competency shortfall and to develop a training plan from this. Further, if the firm has a history of risk-taking driven by production pressures that has eroded trust between supervisors and workers, there needs to be a strong emphasis on applying the value statement in operational decision-making and clear internal procedures for the handling of OHS grievances.

## Business unit alignment scorecard

The business alignment scorecard and cultural health scorecard presented thus far operate at the corporate level. The achievement cycle also includes a scorecard that may be applied at the business unit level. The format is provided in Fig. 12.7. To maintain strategic alignment, the measures in the business unit's scorecard support those applied at the corporate level. Business units can develop the external strategic alignment section of the scorecard by identifying key stakeholders in the external environment, then developing indicators that assess the fit between the business unit's activities and this environment.

As was the case with the business alignment scorecard, core business activities are included in the internal strategic alignment section. Specific measures that relate to the alignment of the business unit's management with its employees fall within the internal goal alignment section. In this way, the department can develop a specific set of priority areas to suit its needs, provided that these areas are also compatible with corporate outcomes and encourage departmental interdependencies. A high-risk production team, for example, may develop a strategy to increase participative risk control for hazards that do not require immediate correction. Participation rates may therefore be selected as a measure of unit effectiveness. In a technical environment consisting of multiple professional disciplines, communication systems and information sharing may be an area addressed under internal goal alignment.

| Business Unit Alignment Scorecard (Year) | | | | | | |
|---|---|---|---|---|---|---|
| Level of alignment | Target | Baseline | Actual | Weighting (%) | Weighted actual this year (%) | Progress required this year (%) |
| External Strategic Alignment | | | | | | |
| | | | | | | |
| | | | | | | |
| | | | | | | |
| *Total weighted score – external strategic alignment* | | | | | | |
| Internal strategic alignment | | | | | | |
| | | | | | | |
| | | | | | | |
| | | | | | | |
| | | | | | | |
| | | | | | | |
| | | | | | | |
| | | | | | | |
| *Total weighted score – internal strategic alignment* | | | | | | |
| Internal goal alignment | | | | | | |
| | | | | | | |
| | | | | | | |
| | | | | | | |
| | | | | | | |
| *Total weighted score – internal goal alignment* | | | | | | |
| TOTAL ACHIEVEMENT SCORE | | | | | | |

Target: Result to be achieved by end of achievement cycle
Baseline: Level of achievement at the beginning of the cycle
Actual: Result for the current year
Weighting: Relative weight of the measure for the achievement cycle
Weighted actual: Calculation of the progress made towards the target this year multiplied by the weighting
Progress required: Progress required this year to achieve the target at the end of the achievement cycle

*Measures to be included:*

- Productivity
- Safety
- System factor quality
- Financial and customer
- Compliance (to legislation, internal standards or plans)
- Social responsibility

**Figure 12.7** The business unit alignment scorecard

On the business unit alignment scorecard, measures are again developed using the six alignment indicators. Recapping, these are productivity, safety, system factor quality, financial and customer, compliance, and social responsibility. For an administration environment, for example, a social responsibility measure may be to reduce the wastage of resources used by the office or to provide work experience opportunities to local students.

### Leader's scorecard

The final scorecard in the achievement cycle is the leader's scorecard. It is used to develop goals at the individual level. The leader's scorecard is directed towards operational improvement and derived from the entropy model. The model indicated that to improve performance and safety results concurrently, it is necessary to maintain system factors at a high standard of quality. The role of the leader is to establish and maintain operational systems to ensure that the system factors for which he is responsible are utilized in a manner consistent with the productive safety KRA and value statement. For example, where the role involves financial authority, it is the leader's responsibility to ensure that both performance and safety issues are taken into consideration when acquiring physical capital, in accordance with the company's purchasing criteria. If the supervisor is directly accountable for the management of the supplies inventory, then he should not allow goods to be stored in a manner which is unsafe or which hinders the efficient intake, storage or removal of such goods. The rationale of the leader's scorecard is to ensure that at the operational level, team leaders develop methods that maximize the safe and efficient usage of company resources. These shop floor improvements should assist the company to achieve its objectives within the business unit and corporate-level scorecards.

The format of the leader's scorecard is provided in Fig. 12.8. It shows that the supervisor is required to set at least one and preferably two goals in each of the system factor areas each year. In other words, as a minimum, one goal is set for each of the following: processes, technology, the physical environment and human resources. The purpose of these goals is to improve the quality of system factors under the supervisor's scope of authority. To ensure that these goals are aligned, they are derived from one or more alignment measures on either the business alignment scorecard or the business unit alignment scorecard. As stated in the upper portion of the leader's scorecard, the goal must improve results in these measures by shifting the system factor towards optimal performance and safety. Once the goal has been set, an operational measure is recorded that will be used to determine whether the goal has been attained. The operational measure states what will be done, to what quality/quantity and by when. The final part of the goal-setting process is to develop an action plan that defines the steps involved in achieving the goal; specifically, who will do what, by when, to reach the goal.

The leader's scorecard shown in Fig. 12.8 has a structure that involves an initial goal-setting meeting, followed by reviews at quarterly intervals. There is no hard and fast rule that this appraisal tool should have a quarterly structure, so each firm should develop a system to suit its own needs. The higher the company's level of risk, however, the more frequent the reviews should be to ensure that changes are being implemented effectively. In the example given, the goal-setting meeting and first quarterly review formats are provided.

There is a number of benefits of having more frequent rather than less frequent reviews. The first is that it provides greater opportunity for the supervisor's manager to ensure that the supervisor is on track to achieve his goals. This is particularly important because the goals feed into the business alignment scorecard and therefore have an impact on corporate success. The second reason is that it provides both parties with the opportunity to address any obstacles that may hinder the attainment of goals, such as competency shortfalls, resourcing problems or procedural

| Leader's Scorecard (year) | | |
|---|---|---|
| System factor | Alignment Measure | Identify one or more alignment measures in the business or business unit alignment scorecard. |
| | Goal | Develop a goal which will improve results in the alignment measure/s by shifting the system factor towards optimal performance and safety. |
| | Operational Measure | Develop an operational measure for this goal which states what will be done, to what quality/quantity, and by when. |
| | Action Plan | Identify who will do what by when to make this goal attainable. |
| Processes | Alignment measure/s:<br>Goal 1:<br>Operational measure:<br>Action plan: | |
| | Alignment measure/s:<br>Goal 2:<br>Operational measure:<br>Action plan: | |
| Technology | Alignment measure/s:<br>Goal 1:<br>Operational measure:<br>Action plan: | |
| | Alignment measure/s:<br>Goal 2:<br>Operational measure:<br>Action plan: | |
| Physical environment | Alignment measure/s:<br>Goal 1:<br>Operational measure:<br>Action plan: | |
| | Alignment measure/s:<br>Goal 2:<br>Operational measure:<br>Action plan: | |
| Human resources | Alignment measure/s:<br>Goal 1:<br>Operational measure:<br>Action plan: | |
| | Alignment measure/s:<br>Goal 2:<br>Operational measure:<br>Action plan: | |

**Figure 12.8** Leader's scorecard

| Quarterly Review | |
|---|---|
| Employee:<br>Review date:<br>Review conducted by: | |
| **System factor** | **Comments: progress and adjustments** |
| Processes | Goal 1:<br>Goal 2: |
| Technology | Goal 1:<br>Goal 2: |
| Physical environment | Goal 1:<br>Goal 2: |
| Human resources | Goal 1:<br>Goal 2: |

| System factor | Achievement Level | | | | | | |
|---|---|---|---|---|---|---|---|
| | Efficiency | | Effectiveness | | Leadership | | Score (average for each system factor) |
| | Goal 1 | Goal 2 | Goal 1 | Goal 2 | Goal 1 | Goal 2 | |
| Processes | | | | | | | |
| Technology | | | | | | | |
| Physical environment | | | | | | | |
| Human resources | | | | | | | |
| TOTAL ACHIEVEMENT SCORE FOR THIS REVIEW (OUT OF 40) | | | | | | | |

**Figure 12.8** *(Contd)*

obstacles. These hindrances should be addressed in a timely manner, for example, skill inadequacies can be rectified through appropriate training and support. When establishing performance management systems such as this, firms should ensure that employees are provided with the avenues to seek assistance when organizational factors constrain their achievement. In addition, once the required support or resources have been provided, the leader should have adequate time to complete the goal before the final review.

In Fig. 12.8, in the final part of the leader's scorecard, the level of achievement is calculated. At each review scores are determined using the criteria 'efficiency', 'effectiveness' and 'leadership' for each goal set. Again, the firm should establish its own evaluation criteria to fit its requirements and its culture. In this sample format, the scores are averaged for each system factor so that a total achievement score is obtained for the review. The sum of these scores, weighted by the relative importance of the review (the last review being more important than the first), is the total achievement score for the year. This will be explained further using the worked example provided in Fig. 12.9.

The hypothetical example given in Fig. 12.9 is for a supervisor in charge of a mobile plant workshop at a mining operation. This middle manager works for company A that developed the business alignment scorecard given in Fig. 12.3, so therefore, his goals are derived from company A's corporate objectives. The example shows that in the initial goal-setting meeting at the beginning of the year, he set two goals to improve processes, one goal for technology, one goal for the physical environment, and two goals to enhance the return on human resources. (The minimum requirement is to set at least one goal for each system factor.)

How were these goals developed? The first step in the goal-setting process is to refer to the business alignment scorecard and identify those measures which are relevant to the supervisor's position. The maintenance supervisor and his manager have determined that the LTIFR, number of near misses and system factor quality audit index are measures, that in part, are affected by the management of processes in the workshop. These alignment measures are then used to develop an operational goal. The basic question that is asked is, 'what can be done in the workshop to help the company achieve sound results in these alignment measures?'

The goal that has been selected is to 'improve safety by increasing the application of JSA techniques in nonstandardized workshop tasks'. The operational measure accompanying this goal is the number of injuries resulting in lost time in the workshop for the year with the target set at less than 5.0 for the 2001–2002 financial year. (By the end of the cycle in 2004, the result is to be zero.) In addition, the supervisor has added the measure 'days lost due to lost time injuries' with the target for the year set at less than 5 days. This makes it clear that while the target for the number of LTIs has not been reduced to zero yet, the firm is concerned with reducing the severity of such injuries. An action plan is developed outlining who will do what, by when, to make the goal attainable. In many cases, the action plan will involve liaising with specialist personnel and managers to provide the resources and support needed.

The parties have also elected to address productivity and re-work within the processes system factor. The second goal is to 'reduce re-work on vehicles by 40 per cent in 2001–2002'. The operational measure accompanying this goal is the 'cost of re-work on vehicles maintained by the workshop', which contributes directly to the reduction of re-work at the organizational level. In the business alignment scorecard in Fig. 12.3, company A identified 'breakdown maintenance as a percentage of total maintenance costs (infrastructure and equipment)' as a key alignment measure, which suggests that historically, this cost must have been disproportionately high. When plant breaks down it not only represents a loss of equipment availability which therefore has a negative impact on production output, but it also represents a rise in risk, particularly if the breakdown leads to loss of control by the operator. In response to this organizational priority, the maintenance supervisor has set a goal of

| Leader's Scorecard (year) | |
|---|---|
| System factor | **Alignment measure:** Identify one or more alignment measures in the business or business unit alignment scorecard. |
| | **Goal:** Develop a goal which will improve results in the alignment measure/s by shifting the system factor towards optimal performance and safety. |
| | **Operational measure:** Develop an operational measure for this goal which states what will be done, to what quality/quantity, and by when. |
| | **Action plan:** Identify who will do what by when to make this goal attainable. |
| Processes | Alignment measure/s: LTIFR, No. of near misses, system factor quality audit index<br>**Goal 1: Improve safety by increasing the application of JSA techniques in nonstandardized workshop tasks**<br>Operational measure: No. of workshop lost time injuries (reduce to less than 5.0 for 2001–2002)<br>Days lost due to LTIs (reduced to less than 5 days)<br>Action plan: Report and communicate LTIs and near misses to managers/personnel. (1/7/01 onwards)<br>Liaise with OHS Specialist to arrange JSA training for workshop personnel (by 31/8/01)<br>OHS specialist to conduct JSA training (by 31/10/01)<br>Monitor the application of JSA techniques in the workshop (1/11/01 onwards)<br>Reinforce and review JSA application during monthly toolbox meetings (1/11/01 onwards) |
| | Alignment measure/s: Productivity, re-work and defect rate per 100 units of output<br>**Goal 2: Reduce re-work on vehicles by 40% in 2001–2002**<br>Operational measure: Cost of re-work on vehicles maintained by workshop<br>Action plan: Investigate causes of re-work in 2000–2001 and identify processes requiring improvement (by 1/8/01)<br>Liaise with OHS Specialist and maintenance personnel to establish safe standardized quality procedures in problem areas (by 31/12/01)<br>Implement and monitor changes (1/1/02 onwards) |
| Technology | Alignment measure/s: Breakdown maintenance as % of total maintenance costs<br>**Goal 1: Reduce breakdown maintenance as % of total maintenance costs on vehicles to 25% in 2001–2002**<br>Operational measure: Breakdown maintenance as % of total maintenance costs<br>Action plan: Investigate causes of breakdowns in 2000–2001 and identify problem areas (by 31/8/01)<br>Compile a service schedule based on manufacturer's instructions for all vehicles (by 30/9/01)<br>Develop a 3 year Maintenance Plan based on manufacturer's instructions, problem areas and equipment age/wear and tear (by 31/12/01)<br>Implement plan and monitor progress (1/1/02 onwards) |
| | Alignment measure/s:<br>Goal 2:<br>Operational measure:<br>Action plan: |
| Physical environment | Alignment measure/s: System factor quality audit index<br>**Goal 1: To improve the safety/efficiency of the physical environment using good housekeeping practices and structural improvement strategies**<br>Operational measure: Physical environment audit index (workshop result)<br>Action plan: Analyze results of last physical environment audit (by 31/7/01)<br>Investigate causes of workshop incidents in 2000–2001 to identify enviro. problems (by 31/7/01)<br>Liaise with workshop personnel to include housekeeping practices in team value statements and implement (by 31/8/01)<br>Develop a 3-year structural improvement plan for the workshop (by 30/12/01) |
| | Alignment measure/s:<br>Goal 2:<br>Operational measure:<br>Action plan: |
| Human resources | Alignment measure/s: No. of employee suggestions implemented as % of suggestions received. Return on resourcefulness (benefit of employee suggestions/cost of training ratio)<br>**Goal 1: To increase the quality, quantity and return on employee suggestions in the workshop**<br>Operational measure: $ value of workshop employee suggestions in 2001–2002<br>Action plan: Liaise with the OHS and HR depts. to introduce the suggestions scheme to the workshop (by 30/9/01)<br>Make presentation to workshop personnel to reinforce company priority areas in the Business Alignment Scorecard (by 30/9/01)<br>Discuss safety and performance issues in toolbox meetings/informal situations. Call for suggestions.<br>Encourage problem solving and contributions (1/10/01 onwards) |
| | Alignment measure/s: Employee satisfaction index<br>**Goal 2: To improve the quality of team leadership**<br>Operational measure: Employee feedback (self-evaluation)<br>Action plan: Attend supervisory training course (by 30/9/01)<br>Develop a personal plan to refresh/reinforce training knowledge and skills (by 30/9/01)<br>Immediate supervisor to provide guidance and feedback (on going) |

**Figure 12.9** Example of leader's scorecard (goal setting meeting and first quarterly review)

| First Quarterly Review | |
|---|---|
| Employee:<br>Review date:<br>Review conducted by: | John Brown, Maintenance Supervisor<br>1 October 2001<br>Joe Green, Maintenance Manager |
| **System factor** | **Comments: progress and adjustments** |
| Processes | Goal 1: LTIFR reporting systems completed. JSA training scheduled for 15/10/01. JSA software purchased and installed. OHS specialist to provide supervisors with training before 30/10/01.<br><br>Goal 2: Investigation completed. Quality procedures to be developed after JSA training i.e. between 1/11/01 and 31/12/01. Procedures to be written up and assured in 2002. |
| Technology | Goal 1: Manufacturers contacted to obtain up-to-date service instructions. Breakdown problem areas report currently in draft. Final report to be presented to management by 25/10/01.<br><br>Goal 2: N/A |
| Physical environment | Goal 1: Audit analysis incomplete. Team value statements updated. Initial preparation of improvement plan started.<br><br>Goal 2: N/A |
| Human resources | Goal 1: Suggestion scheme introduced and presentations completed. Toolbox meeting agenda revised to include employees' areas of concern, brainstorming and potential suggestions. Employees to be provided with feedback regarding progress on implementing suggestions during toolbox meetings (from 1/11/01)<br><br>Goal 2: Supervisory training course attended. John intends to review Business/ Business Unit Alignment Scorecards weekly and to implement appropriate decision-making using the reasonableness test daily. |

| System factor | Achievement level | | | | | | |
|---|---|---|---|---|---|---|---|
| | Efficiency | | Effectiveness | | Leadership | | Score (average for each system factor) |
| | Goal 1 | Goal 2 | Goal 1 | Goal 2 | Goal 1 | Goal 2 | |
| Processes | 8 | 8 | 8 | 7 | 7 | 7 | 7.5 |
| Technology | 6 | – | 7 | – | 7 | – | 6.7 |
| Physical environment | 4 | – | 6 | – | 6 | – | 5.3 |
| Human resources | 8 | 8 | 8 | 7 | 8 | 7 | 7.7 |
| TOTAL ACHIEVEMENT SCORE FOR THIS REVIEW (OUT OF 40) | | | | | | | 27.2 |

**Figure 12.9** *(Contd)*

reducing breakdown maintenance as a percentage of the total vehicle maintenance costs to 25 per cent in 2001–2002. To achieve this goal, all of the shift supervisors would need to contribute and therefore, some of the goals in the leaders' scorecards would be common to a number of personnel. The manager should advise the supervisor when goals are shared to emphasize the need for co-operation, and where all such goals are common, it may be appropriate for goal-setting to be a collective process involving the manager and all relevant supervisors.

In the example given, the action plan identifies a key outcome – to develop and implement a 3-year maintenance plan. This goal effectively involves the adoption of a proactive approach to the management of the technology system factor with the emphasis on preventing entropic risk.

As explained earlier, the business alignment scorecard contains six alignment indicators one of which is 'system factor quality'. As a result the scorecard must include at least one measure directly related to quality to track the firm's susceptibility to degradation. A metric that may be used is the audit index. At the operational level, the maintenance supervisor can contribute to this corporate objective by addressing risks in the workshop's physical environment. The goal given in the example identifies two ways of achieving this. The first is to implement sound housekeeping practices, and the second, is to facilitate these practices by developing a structural improvement program, which may include installing shelves and storage spaces for the safe keeping of tools and equipment. The operational measure accompanying this goal is the physical environment audit index result for the workshop. The action plan involves referring to the results of the last audit to identify any problem areas and undertaking an investigation into the causes of workplace incidents in the past year to identify any environmental issues that may have had an additive effect. From this information a 3-year structural improvement plan will be formulated. Concurrently, the supervisor plans to hold a series of forums to involve workshop employees in the development of housekeeping practices to fit their team ethic statement. The goal-setting in this operational scorecard is therefore used not only to direct resources and efforts to priority areas, but also to stimulate participative processes and share the responsibility for achieving the objective.

The final system factor addressed in the leader's scorecard is human resources. In the example, the supervisor and his manager have extracted 'number of employee suggestions implemented as a percentage of suggestions received' and 'return on resourcefulness' from the business alignment scorecard. The supervisor is able to contribute to these measures by increasing the quality, quantity and return on employee suggestions in the workshop. The operational measure applied is the dollar value of workshop employee suggestions in 2001–2002. The action plan identifies the initial involvement of OHS and HRM professionals in the development of a suggestions scheme and the presentation of company priorities to employees. After this, the supervisor's responsibility is to discuss safety and performance issues with his team members and encourage their input wherever possible, channeling their ideas through the formal suggestion program.

As part of its risk management and human resources quality development strategies, the company also wishes to improve the level of employee satisfaction. Accordingly, it has included the 'employee satisfaction index' in the internal goal alignment section of the business alignment scorecard. The supervisor can have a positive impact on this alignment measure by enhancing his skills as a team leader. The action plan outlined for this goal identifies the training and support that the organization will provide for the supervisor to achieve this goal. The supervisor is encouraged to measure his own performance by seeking feedback from subordinates, undertaking self-evaluation and responding to the feedback and guidance provided by his manager. Therefore a supervisor is expected to take some of the initiative for his own development and evaluation.

As explained earlier, many of the goals identified in this example are common to a number of employees at an equivalent level in the organizational hierarchy. In this case, most if not all the shift supervisors will have these goals in their scorecards. Shared goals promote teamwork, discourage unhealthy internal competition, and reduce the workload involved in the performance management system. The key message conveyed, in this example, through the sharing of goals among shift supervisors, is that any safety or performance problems that arise in a given shift should, where practicable, be corrected immediately and not left to the next shift to solve. This approach encourages leaders and their teams to take ownership of the issues within their respective workplaces and to work collaboratively within the business unit. The performance management system therefore not only drives outcomes and results but is used to model appropriate leadership behaviors. Concurrently, as explained later in this chapter, the supervisor's scorecard is linked to incentives so that these results and behaviors are rewarded with financial bonuses that recognize individual achievement. In addition, team leaders have access to further monetary rewards geared to the attainment of objectives within the business alignment scorecard, which are payable to all employees proportional to the level of collective accomplishment.

Once goal-setting has been undertaken, regular reviews are scheduled throughout the year to monitor progress and to identify and remedy any obstacles to goal attainment. The next section of the leader's scorecard contains the first quarterly review. It is important for the manager and supervisor to agree on the progress comments that are recorded, particularly if timeframe and output targets have not been achieved. For example, if one item in the action plan is to develop a maintenance plan, the employee cannot be held accountable for progress if the manager has failed to complete the document review process on time. Where necessary, adjustments should be made to action plans to reflect these minor changes to the time frame or to clarify responsibilities. Employees must only be held accountable therefore, for those parts of the work that are within their scope of control. As the parties become more proficient in the goal-setting process, the action plans that are developed should become more accurate and readily implemented. The achievement cycle is intended to facilitate a learning process and through its alignment with other management systems assist the firm to progress along its learning curve.

As shown in the example, at the end of each review the leader's achievements are given a score out of ten for each goal using the criteria – 'efficiency', 'effectiveness' and 'leadership'. Efficiency indicates whether the action plan is being implemented according to the schedule developed in the goal-setting meeting. Effectiveness refers to the extent to which the goal and its implementation are having a positive impact and are creating the desired change at the operational level. Leadership measures whether the supervisor is applying the value statement and enlisting the support of subordinates during the change process. From these results, an average score out of ten is calculated for each system factor. The summation of these gives a total achievement score for the review out of forty. (As will be explained later, the result for each review is multiplied by the weighting given to each review, then added to obtain a total achievement score for the year. This score is converted to a percentage and used to determine the level of incentive/bonus payable for individual achievement.)

Figure 12.10 illustrates the structure of the achievement cycle and how measurement methods are linked to ensure alignment and balance. In a cycle, for example, a 3-year turnaround, all three goal-driven scorecards (which exclude the cultural health scorecard) are used to facilitate improvement. The business alignment scorecard is applied on an organizational scale to the full cycle and measures are developed according

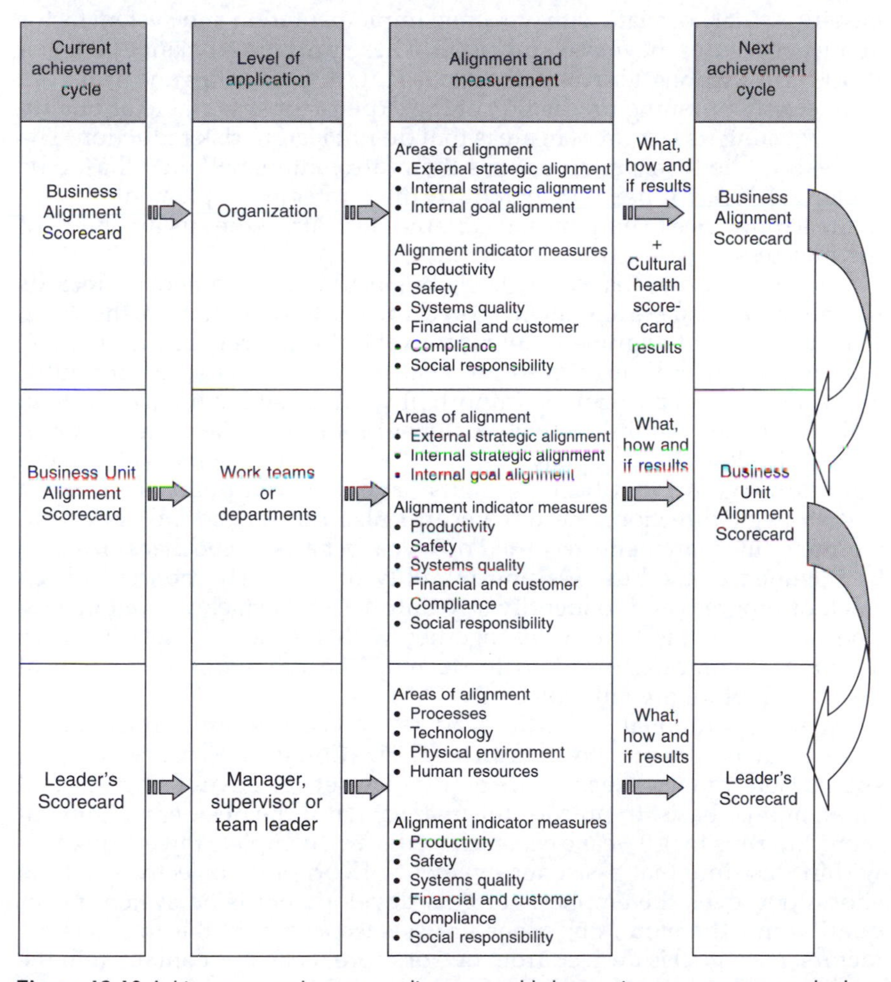


**Figure 12.10** Achievement cycle ensures alignment and balance using measurement methods

to the three levels of alignment – external strategic, internal strategic and internal goal alignment, as shown in the third column of the figure. The alignment indicator measures used in this corporate level scorecard are productivity, safety, system factor quality, financial and customer, compliance, and social responsibility. At least one measure in each of these areas is included so that the firm pursues a balanced approach such that weightings are not skewed towards production measures without compensating weighting given to safety and system factor quality.

The business alignment scorecard runs for the full length of the cycle. Progress is measured on a yearly basis and recorded in the annual version of this scorecard. Some measures are evaluated frequently and reported in business reports periodically, however, the annual result is recorded in the scorecard as a measure that reflects achievement for the entire year. The results of the first cycle are used, together with the cultural health scorecard results to develop the organizational scorecard for the next cycle. The results are also appraised to ensure that 'what', 'how' and 'if' measures, that correlate with outcome, implementation and system factor quality measures, have been included. These provide a balance between short-term and long-term objectives, particularly preventing the firm from erroneously pursuing production at the expense of system factor quality or by wasting its resources in areas that do not lead to risk reduction. The analysis of the firm's interventions therefore centers on 'what' has to be done, 'how' has it been done and 'if' the strategies implemented have made a difference to bottom-line variables that affect the sustainability of the business.

The figure shows that strategically business unit alignment scorecards fit under the business alignment scorecard. The two formats are the same, with the areas of alignment falling into the three levels in the channel. The alignment indicators used are also the same to ensure continuity, and this becomes particularly important when business units operate in isolation either administratively or geographically. The consistency of approach across the company and the linkage to the business alignment scorecard ensures that business units have a shared point of reference and strategic direction. These scorecards also run for the full cycle with summary measurements recorded on a yearly basis. Some measures may be included in the business unit monthly or quarterly reports to keep track of progress and to identify any unforeseen obstacles or anomalies. The results of this first cycle, together with the next cycle's business alignment scorecard, are used to develop the following generation of business unit alignment scorecards.

At the operational level, the leaders' scorecards are applied to set goals that contribute to overall organizational success. Managers, supervisors and team leaders are required to set objectives using this tool on an annual basis. In practice this means that if the business alignment scorecard runs for a 3-year cycle, each leader will complete three scorecards within this time that assist achievement of corporate objectives. At the shop floor level, the emphasis of goal development is on system factor quality and therefore, objectives are set according to the four system factors. Each goal is derived from one or more measures contained in the business or business unit alignment scorecard to ensure that individual goals have a direct impact on the firm's overall level of performance. Consequently, operational measures also fall within one of the categories of the alignment indicators. Once annual results have been achieved the changes that led to these results become operational practice. For example, if a strategy implemented in 2001–2002 involves pre-start equipment inspections this becomes a standard operating practice. The target for this strategy may be 100 per cent compliance to the pre-start check procedure. This target, once achieved, becomes the baseline for future measurement. Accordingly, less than 100 per cent compliance becomes unacceptable. In addition, the baseline may be used to develop goals in the next cycle, thereby facilitating continuous improvement of factor and management system quality. For instance, a follow-on goal for the next cycle may be to instruct all new recruits in pre-start check requirements within the first week of commencing employment.

## Using incentives to reward achievement

As mentioned earlier, the business alignment scorecard and the leader's scorecard results are linked to incentives. The achievement cycle not only serves to document the firm's achievement in KRAs but also provides a source of motivation for the workforce. Tying the business alignment scorecard to incentives helps to enlist employee commitment. It also shares the corporate gains that result from attaining strategic objectives.

How are these incentives determined? Using cost-benefit analysis, management should predetermine the benefits expected from planned interventions, for example, a reduction in the LTIFR should reduce the costs associated with lost productivity, workers' compensation and rehabilitation. Historical data may be used to estimate these costs. Benefits derived from improved results in the alignment indicator measures in the business alignment scorecard, less the cost of implementation, equals the financial gain of applying these interventions. Ordinarily, this financial return should be proportional to the level of achievement, for example, the greater the reduction in the LTIFR, the greater the reduction in OHS costs and the more workers should be rewarded with financial incentives. After the cost-benefit analysis, management should determine a cutoff point or gauge line that reflects the minimum level of organizational achievement that will attract a corporate bonus. Above this gauge line, the higher the level of achievement the greater the financial return to the workforce.

How is the gauge line determined? Realistically, it is extremely difficult for a firm to achieve the target in all measures in the business alignment scorecard to a 100% success rate in the achievement cycle. For instance, the company is unlikely to reach zero near misses because of residual risks. The firm therefore has to determine a level of improvement that it considers sufficiently positive to warrant the payment of bonuses. This cutoff depends on the benefits correlated to various levels of achievement and how these impact the 'bottom-line'.

## Corporate incentives

An example of an annual incentives gauge is shown in Fig. 12.11, based on the business alignment scorecard given in Fig. 12.3. (The method presented is intended to be a suggestion and it is recommended that each firm develops a system that meets its specific needs.) The example company has graphed the target level of progress for each of the 3 years in the cycle. These are 33.3, 66.7 and 100 per cent for the years 2001–2002, 2002–2003 and 2003–2004, respectively. The progress required is gradual so that gains can be consolidated, rather than trying to achieve sharp and substantial improvements that may be difficult to sustain. In the example, the gauge line for the payment of employee bonuses has been set at 80 per cent of the target for each year, as shown by the solid line.

In 2001–2002, for example, 80 per cent of 33.3 equals 26.7 per cent, therefore, 26.7 per cent of progress towards the target is expected in the first year. In 2001–2002, if the weighted actual in the business alignment scorecard is greater than 26.7 per cent, then employees receive a bonus for the year of $470 as shown by the right-hand axis of the graph. A result below 26.7 per cent would result in no bonuses being paid. In 2002–2003, a weighted actual equal to 53.3 per cent results in a bonus of $730 per employee and in 2003–2004, 80 per cent leads to a $1000 bonus.

In Fig. 12.11 the minimum bonus has been set at $470 per employee. If the company achieves 26.7 per cent then each employee receives this amount. What happens if the company attains a greater level of achievement than the minimum gauge? Should employees be rewarded and encouraged for returning greater than expected benefits to the

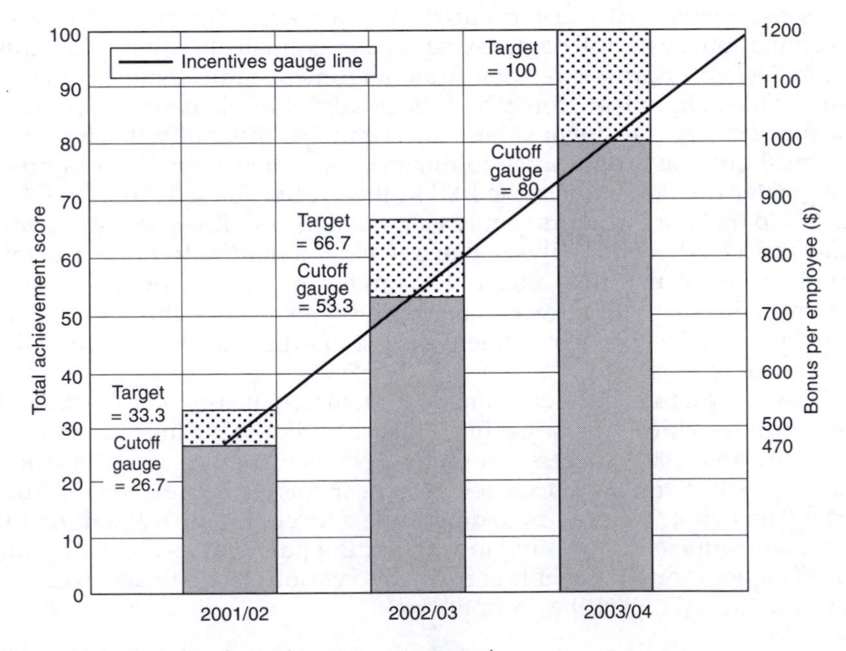


**Figure 12.11** Example of annual incentives gauge graph

company? The productive safety approach rewards the workforce according to the level of benefit it provides to the company through its efforts. The greater the effort and results, the higher the financial incentive available. Figure 12.12 illustrates how it may be applied.

In this hypothetical example, the levels of achievement for each year of the 3-year cycle are marked circle. A circle shows that the level of achievement in 2001–2002 was well above the gauge level of 26.7 per cent with the result being 40 per cent. Looking at the right-hand side of the graph this correlates with a bonus level of $600 per employee. In the second year, 2002–2003, the level of achievement was also above the gauge line (set at 53.3 per cent) with the result being 60 per cent, as shown by the square. The bonus amount for this year equaled $800 per employee. In 2003–2004, the business alignment scorecard result was below the gauge for that year, marked by the triangle. The result was 69 per cent when the company was aiming for 80 per cent. As this still represents an improvement on the previous year, the amount of incentive that applies is the minimum amount of $470 per employee. The result of undertaking the achievement cycle in this example, is a 69 per cent improvement overall on the baseline results at the beginning of the cycle. Employees receive bonuses to the total of $1870 ($600 + $800 + $470) each over 3 years.

What happens to bonus payments if during the cycle the level of achievement regresses? Figure 12.13 provides an example. In the first year, the company attained a very high success rate with a result of 57 per cent shown by the circle. This first year result was above the weighted

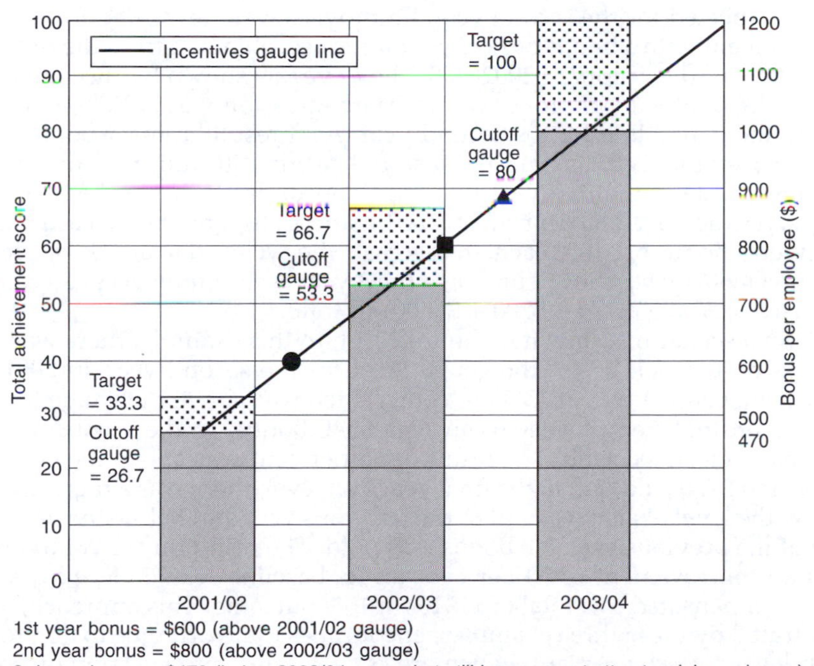


**Figure 12.12** Annual incentives gauge graph with yearly achievement levels for company A

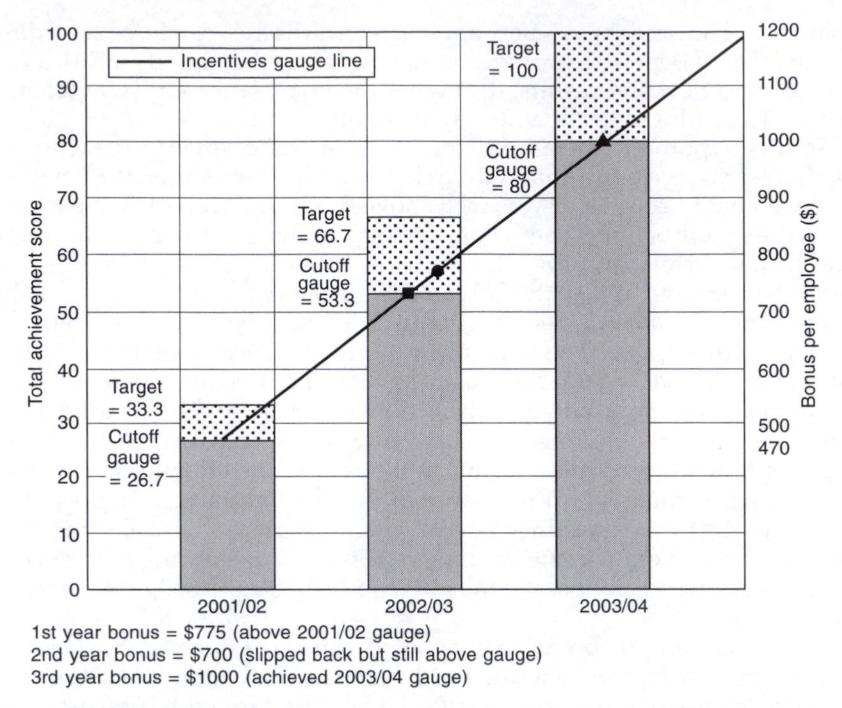


**Figure 12.13** Annual incentives gauge graph with yearly achievement levels for company B

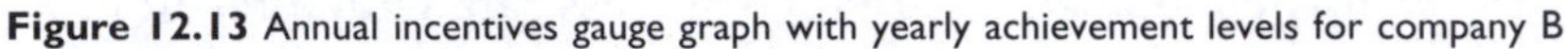

actual expected for the second year. Employees were rewarded for their achievement with a bonus of $775 per employee, obtained from the right-hand side of the graph. In 2002–2003, however, as shown by the square, they were unable to improve on this. The regression was still, however, above the gauge line for the second year. As a result, a lesser bonus of $700 per employee was paid because the result still indicated that the changes made in the first year had been consolidated. In the final year, employees achieved 80 per cent for the business alignment scorecard and received a bonus of $1000 each. In this case, the cycle led to an 80 per cent improvement on baseline figures over the 3-year cycle. Employees received bonuses of $2475 ($775 + $700 + $1000) in total.

A final scenario is shown in Figure 12.14. In this example, there was a regression that fell below the gauge level for the second year. In 2001–2002 achievement was at 33.3 per cent, which was above the gauge line, and accordingly, employees received a $540 bonus. In the second year progress continued well. The result of 55 per cent was above the cutoff point and $770 paid out. In the final year, however, there was a regression below the level of achievement of the previous year and below the gauge line of the previous year. No bonus was paid. The result for the company was an improvement of 50 per cent on its baseline results. Employees were compensated to a total of $1310 for this outcome. This approach, as illustrated by these three examples, encourages the workforce to aim for high levels of achievement, which are consolidated and sustained from year to year.

Where a firm adopts this method of linking incentives to corporate

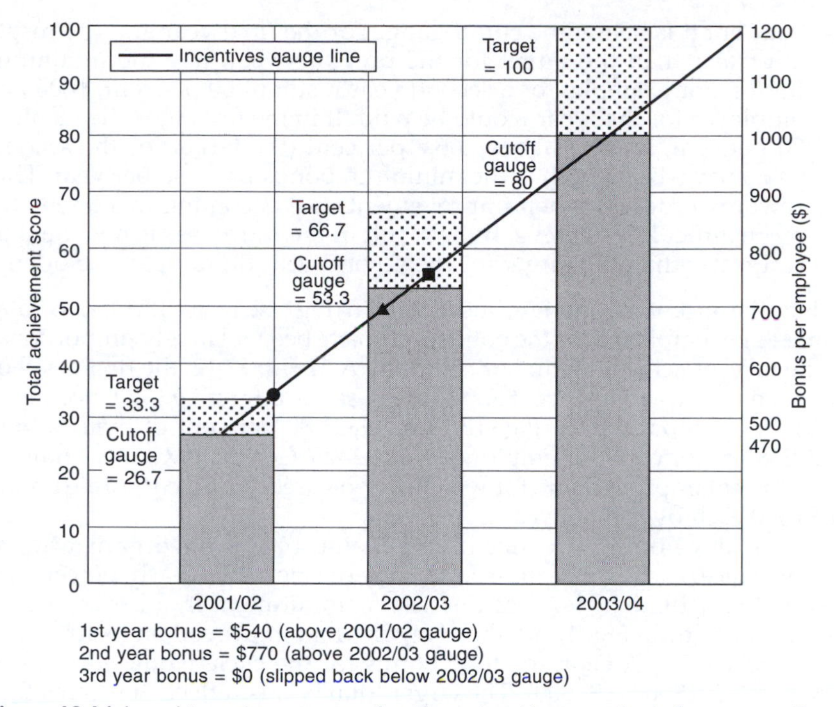


**Figure 12.14** Annual incentives gauge graph with yearly achievement levels for company C

results it should develop specific payment criteria. These should be communicated to employees so that they know the rules of the 'game'. Applying these to the 3-year cycle described above, these rules are:

(1) Divide the 100 per cent achievement of targets by the number of years of the cycle to obtain the target level of progress per year. For a 3-year cycle these are 33.3, 66.7 and 100 per cent. Alternatively, gear the level of achievement for each year according to the firm's readiness or to the severity of external pressures. If pressures are high then greater achievement is required in the early years of the cycle.
(2) Determine the gauge line upon which incentives will be based and the corresponding levels of incentive. (In the example given the gauge line was set at 80 per cent.)
(3) Below the first year cutoff gauge (in this case 26.7 per cent) do not pay any incentive.
(4) For levels of achievement above the yearly cutoff gauge, pay the corresponding level of bonus.
(5) If there is regression in the subsequent year, pay the corresponding incentive provided that it is still above the cutoff gauge for that year. Treat it as a consolidation and seek explanations.
(6) If the regression is below the previous year's cutoff gauge for any year do not pay the incentive. For example, if in the third year the achievement level is below the second year's cutoff gauge do not pay the incentive.
(7) In the final year of the cycle, if there is an improvement on the previous

year but it is below the cutoff gauge for the final year and the result is greater than the target for the previous year, pay the minimum bonus. For example, for a second year result of 60 per cent, $800 per employee for that year would be paid. If in the final year the result is 70 per cent, which is above 66.7 per cent (the target for the second year) pay $470, which is the minimum bonus payable per year. This rewards progress when improvements are becoming incrementally more difficult to achieve. If the result in the third year is not equal to or greater than the target for the second year do not pay the bonus.

For the incentives gauge line given in Fig. 12.12 to 12.14, the total bonuses per employee for the entire cycle have been relatively proportional to the level of achievement. For company A in Fig. 12.12, the final level of improvement was 69 per cent above the baseline attracting total incentives of $1870. Company B, in Fig. 12.13 reached 80 per cent of their targets and the total payout per employee was $2475. Company C only attained a 50 per cent improvement, for which workers received a corporate bonus of $1310 each over the 3 years.

When developing an incentive scheme based on organizational performance, each firm will need to consider possible yearly percentage results to calculate its level of financial commitment. If, for example, in the second year in Fig. 12.13, the company had improved on its first year and reached 70 per cent, the total bonus for the cycle would have been $2675 ($775 + $900 + $1000). The larger total bonus reflects the gains that the company makes from having increasingly high levels of success in consecutive years of the cycle. As also shown in these figures, a cutoff has been determined to define the maximum incentive payable ($1200 per annum for 100 per cent achievement yearly in this case). The firm can therefore state its maximum financial commitment for the attainment of goals on the business alignment scorecard. In the examples given this would be $3600 per employee for the 3-year cycle under the ideal, but unlikely, conditions that all targets are achieved in each year to 100 per cent standard.

The business alignment scorecard is designed to ensure that should one or more of the critical measures such as profitability not improve against the baseline, then it is more difficult for the company to achieve the required total achievement score on which incentives are based. In the sample scorecard in Fig. 12.3, if the company fails to improve its profitability by the end of the third year, then it loses 8 per cent of the overall weighting. If the profitability goes down compared to the baseline, the weighted actual will be a negative result, thus having an even greater impact on the final total. The higher the gauge line is set, the more difficult it is for the workforce to receive a corporate bonus if critical, highly weighted measures are not achieved. If, for example, the gauge is set at 80 per cent then the firm can only afford to be a total of 20 per cent below its combined targets before the incentives cut out, whereas a gauge of 70 per cent allows an additional 10 per cent leeway.

The scorecard ensures that the firm has the capacity to pay the incentives by the inclusion of financial alignment indicators, with sufficient combined weighting to influence the total achievement score for the year. This is important as it links incentives to the firm's capacity to pay. These matters, however, depend on the firm's strategy. For instance, if the firm is in a trough in the business cycle because of external market conditions, it may, in the current business scorecard place less emphasis on profitability and more on building sustainability. In such a case, it may choose to reduce the weighting on financial measures and reward employees for systems improvements in readiness for the upturn in the business cycle. Accordingly, the scorecard should be formulated to reflect the business' strategy and balance short-term returns against long-term viability.

There are two levels at which achievements are linked to bonuses under productive safety management. As discussed thus far, the business alignment scorecard is used as a centralized performance evaluation tool, to focus on those measures that have the most significant effect on the firm's success. The mix of alignment indicator measures provides balance and congruence. When achievements are above a specified yearly cutoff gauge, all employees receive an equivalent bonus as a reward for their collective contribution to the company's success.

### Individual incentives

The second tool linked to incentives is the leader's scorecard. Additional rewards are provided at supervisory and management levels to reinforce that commitment and leadership are prerequisites for effective change implementation. Productive safety management, or any other management system, cannot be introduced effectively without the upfront and sustained support of senior staff. The leader's scorecard helps managers maintain their focus on achieving the targets in the business alignment scorecard, and provides a structured approach to operational goal-setting that is aligned with corporate objectives. A similar rationale of gearing the level of incentive to the level of achievement is applied to the leader's scorecard, as was done in the business alignment scorecard. The difference is that the goals are implemented within a year rather than over the full cycle. In addition, progress is measured quarterly or according to the chosen review structure.

In the annual incentives gauge example given in Figs 12.11 to 12.14, the final level of achievement expected for the business was 80 per cent. Progress towards the targets was incremental to ensure that sufficient time was allowed for good habits to be formed and for results to be sustained over the longer term. On the leader's scorecard, the individual is expected to achieve the required outcomes to completion within one year. When setting the final cutoff gauge for the leader's scorecard, a number of issues need to be considered. The first is the relative weight given to each of the review scores in terms of its contribution to the total cycle result. Should the supervisor fail to attain progress by the first review and therefore receive poor results, it can become very difficult to make up the difference in the following reviews if all the evaluations are given the same level of importance. The supervisor is therefore placed under undue stress to attain a much higher level of performance in subsequent reviews to make up for the initial poor result. In some cases, 'failure' may be attributable to issues outside the individual's control such as lack of resources, support or training. For this reason, the reviews have to be weighted so that the final review carries the greatest weight in the calculation of the total achievement score for the year. Fig. 12.15 illustrates how this weighting may be determined proportionally to the level of progress expected. The example assumes a system that involves reviews every quarter. Companies may vary the number of reviews with their own needs and adjust the proportions accordingly.

In the figure, the graph is divided along the bottom line into quarters that show the intervals between reviews. The first meeting at the beginning of the year involves goal-setting. No scoring is undertaken at this stage. Every quarter thereafter a review is undertaken to assess the progress that is being made against the points outlined in the action plan for each goal. The upright of the graph indicates that all goals are expected to be 100 per cent completed by the final review date. The weighting of each

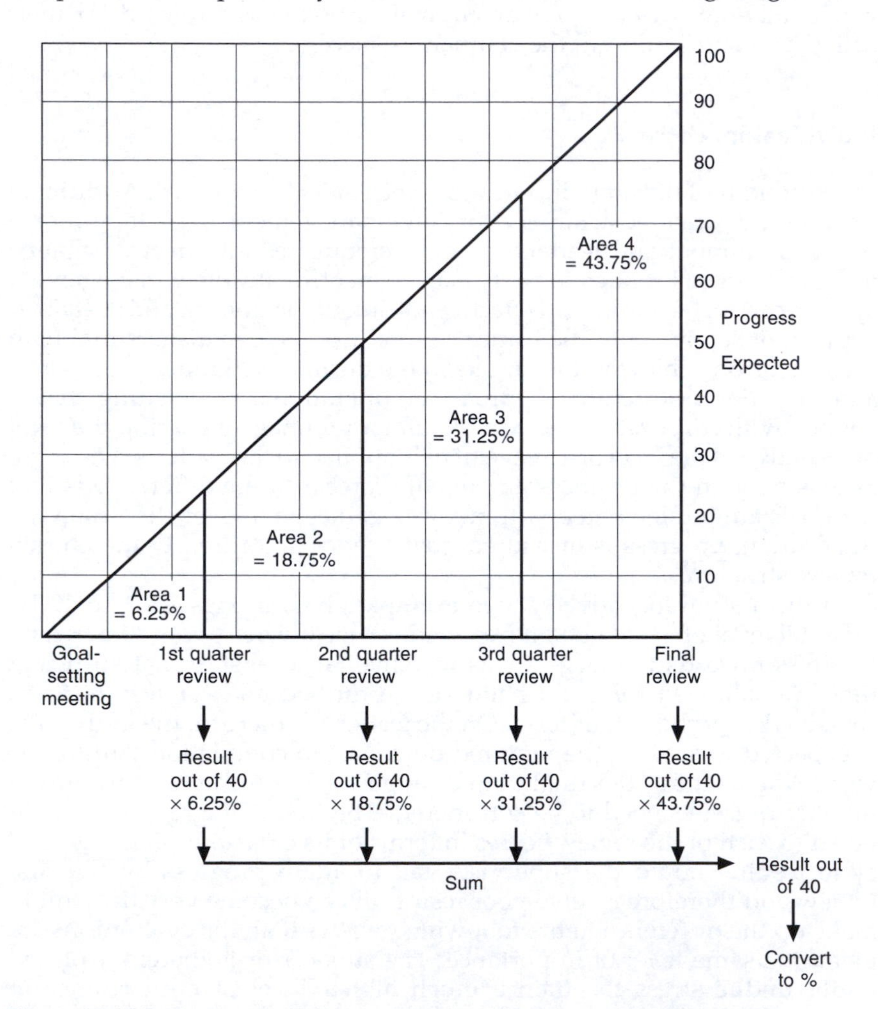

**Figure 12.15** Determination of the proportional weighting for each quarterly review

quarterly review is calculated using the area of each triangle. Area 1 is 6.25 per cent of the largest triangle so therefore, the first quarterly review contributes 6.25 per cent to the final annual score. Area 2, which excludes area 1, equals 18.75 per cent of the largest triangle. The weighting given to the second review is thus 18.75 per cent. The third and the final reviews are calculated the same way with the weightings being 31.25 and 43.75 per cent, respectively.

By using increasing weightings, the leader's scorecard allows for the learning involved in developing new goals. It encourages experimentation and the pursuit of innovation in the early stages of implementation. This is likely to allow higher levels of achievement in the final quarters than adhering rigidly to a plan specified at the beginning of the year; in other words, it permits revision and improvement of the action plans. In addition, it provides scope for remedial action to be taken to correct hindrances to goal attainment such as competency shortfalls, reliance on the input of other personnel, and unclear expectations.

As a general guide, the manager should expect to see progress in each of these four stages for those goals that require a full year cycle. In the first interval after the goal-setting meeting, research and planning should be undertaken. Second, comes the development of systems and the initial implementation steps, between the first and second quarter reviews. From the second to the third review these systems should be executed further, and in the final quarter the changes should be consolidation and monitoring. In some cases goals may not require the full year to be implemented, so progress should be assessed according to the leader's ability to sustain the change and to generate results.

As each review is completed, a result out of 40 is obtained for the leader's performance, in the sample format given in Fig. 12.9. This score is then multiplied by the relevant review weighting, for example, a score of 36 for the first review is multiplied by 6.25 per cent to obtain a result of 2.25. As shown at the foot of Fig. 12.15 the results for each review are summated to obtain a total out of 40 for the year, which is converted to a percentage. Incentives are geared to this final percentage. An example showing how to obtain a final score is given in Fig. 12.16(a).

In the example above, the manager has achieved a final score of 81.1 per cent for the implementation of goals in the leader's scorecard for the year. How can this achievement be linked to the individual bonus? An example of how it may be applied is given in Fig. 12.16(b). The firm in this case has decided that the cutoff percentage is 70 per cent. This means scores of 70 per cent or more attract a bonus. It should be noted that very high cutoffs of 80 per cent or higher discourage experimentation in the early stages of goal-setting and implementation. It could lead to 'playing catch-up' later in the year and to demoralizing the employee. The team leader may feel that because of initial poor results, it is too difficult to attain compensating high scores in the latter part of the cycle. Under circumstances where the employee initially struggles with the goals, it is absolutely essential that the manager provides the support, leadership and encouragement to generate motivation and to show trust in the employee's abilities. In addition, the scoring of employees' achievements is based directly on whether and to what extent goals have been attained.

| Review meeting | Score out of 40 | Weighting for the review (%) | Weighted score (score out of 40 × review weighting) |
|---|---|---|---|
| 1st quarter | 36 | 6.25 | 2.25 |
| 2nd quarter | 32 | 18.75 | 6.00 |
| 3rd quarter | 34 | 31.25 | 10.63 |
| Final review | 31 | 43.75 | 13.56 |
| Total out of 40 | | | 32.44 |
| Percentage | | | 81.1% |

**(a) Calculation of the total achievement score**

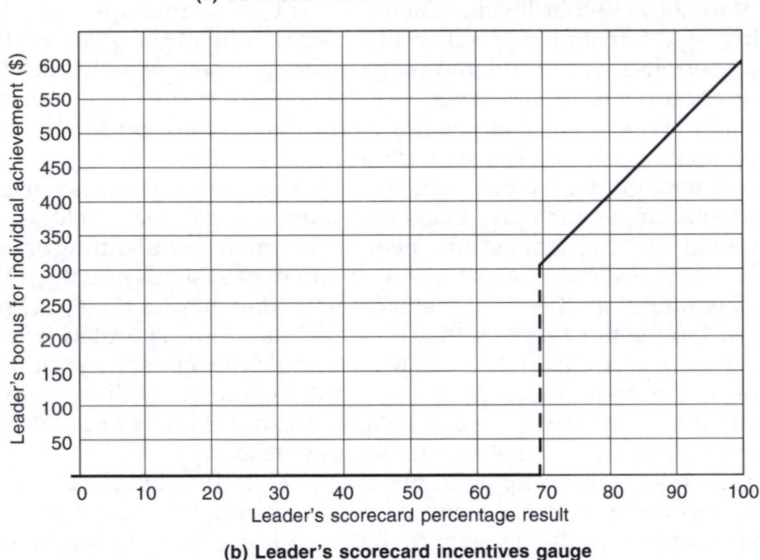


**(b) Leader's scorecard incentives gauge**

**Figure 12.16** Determination of leaders' incentives

It is not designed to compare one employee against another with the result that there are some 'winners' and some 'losers'. Ideally, all leaders should achieve their goals. In this way, both the supervisors and the organization benefit, resulting in win–win outcomes.

According to the approach taken in Fig. 12.16(b), there is no individual incentive payable below the cutoff score of 70 per cent. After this there is a steep incentive gauge line. This means that the higher the level of achievement the greater the reward. Supervisors who receive a score of 80 per cent, for example, are paid $400 and those who attain 90 per cent receive $500. The level of financial reward should be commensurate with the benefits derived by the company from improved results. When 'selling' the incentive system to the organization's supervisory staff it is important to highlight the interdependence of the financial rewards in the business alignment scorecard and the leader's scorecard. Individual incentives do not stand alone as each supervisor's contribution adds to the attainment of goals in the corporate-level scorecard. These employees have the opportunity to be rewarded twice for their efforts – firstly, on the basis of their individual achievement, and secondly, on the basis of their input into overall organizational success. The individual bonus scheme is used to stimulate and maintain management commitment, which is the most critical factor affecting the successful implementation of change in an organization.

The incentive system assists the firm to take a strategic approach to goal-setting and provides a structure for gain-sharing. It reinforces common goals and values from both the economic and social perspectives of the firm. In relation to the latter, the reward system helps to focus on win–win outcomes for equity holders and employees. The linkage between the performance management and incentive systems is a fundamental component of the management approach that builds and sustains alignment. At the external strategic level, the company should become a more attractive employer making it easier to acquire the people with the required competencies from the labor market. Internal strategic alignment is enhanced when the firm is better able to retain quality human resources and to maximize the return on labor investment by optimizing employee participation in organizational decision-making. At an operational level, the incentive system reinforces interdependencies between teams and discourages internal competition. Internal goal alignment is achieved and sustained when management–workforce relations are based on mutual benefit.

## Closing the loop – auditing the plan

The payment of these incentives is one of the final steps in the achievement cycle. The only remaining process to close the feedback loop is to undertake a holistic review of the firm's strategies and results to determine their effectiveness using the final auditing phase. From earlier chapters:

Auditing is an information gathering and evaluation activity involving three processes:

(1) Identifying the sources of risk that the organization is exposed to as a natural system.
(2) Evaluating the health of the organizational culture.
(3) Measuring the achievement of strategies in the productive safety management plan.

What areas of the management plan should be audited? The plan needs to be reviewed in its totality not only for the effectiveness of specific strategies but also for alignment. This includes, for example, confirming that systems, tools, strategies and the choice/balance of measures reinforce the compatibility of production, safety and system factor quality. Other matters to be considered are the actual levels of achievement in any of the alignment indicator areas. At the end of the cycle, management should compare the cost of incentives paid to the benefits derived from the improvements that have been made. This is the true 'bottom-line' impact of undertaking the plan and tests the accuracy of assumptions made when developing the incentive system. The results of this auditing process can then be used to refine the plan development process for the next cycle, so that future strategies and the methods of implementation result in continuous improvement. The cycle is therefore not only concerned with generating results but also with organizational learning and development.

This final auditing process involves analyzing each element of the productive safety management plan to assess whether fundamentals have changed according to the alignment indicator measures. Other broader issues include whether there is greater alignment between the firm and the external environment in terms of legal compliance and social responsibility, better use of resources and improved decision-making in strategic matters, and more productive management–employee relationships as a result of changes to internal systems. Some of the specific issues to be audited include:

(1) Key result area
  - Is productive safety identified as a KRA?
  - Has a value statement been developed to support the productive safety KRA?
  - How well is the value statement known, understood and applied by management?

(2) Strategies, initiatives and action plans
  (a) Systems change
    (i) Structural change
      - Do OHS practices fit the four system factors?
      - Have the necessary tools, e.g. incident investigation questionnaire, been developed?
      - Are the systems and tools applied consistently throughout the firm?
      - Are the systems and tools readily available to relevant employees?
      - Are records kept effectively?

    (ii) Risk management strategies
      - Have audits been conducted and what were the results?
      - Have priorities been determined and strategies developed to minimize/manage risks?
      - Are employees involved in the development of these strategies for their workplaces?
      - Have the risk management strategies been effective? What were the measured results for production, safety and system factor quality? Were these sufficient improvements on the baselines?
      - How can these results in the business alignment scorecard be improved?
      - Is the company now consistently legally compliant?

  (b) Behavioral strategies
    - Has management commitment been sustained throughout the cycle?
    - Are employees' responsibilities clearly understood now?
    - Are participative and information dissemination processes effective?
    - Have the strategies for building the capacity reservoir been

effective? Have sufficient controls been implemented to select, retain and optimize human resources? Have higher levels of competency and vigilance resulted?
- Has the quality of supervision and decision-making improved?
- Have there been any conflicts that have required regulatory intervention?
- How effectively were conflicts managed and what impact has this had on the culture?
- Does the firm apply the productive safety formula?
- What has been the level of return on resourcefulness and has it exceeded the costs of training and development plus the implementation costs of the suggestion scheme?

(3) Measures and results
- Were measures appropriate and comprehensive?
- Were appropriate time frames applied for the regular monitoring of progress?
- Were the targets achieved and what has been the effect on overall firm performance?
- What percentage of leaders, who undertook a leader's scorecard, achieved the required level of performance to receive an individual bonus?
- For those leaders who did not achieve the bonus were sufficient training, support and resources provided by the company?
- What was the per employee cost of corporate incentives over the cycle?
- What impact has this had on motivation, morale, turnover and absenteeism?
- Did the level of incentive fairly reflect the achievements of the company?
- Did reporting systems work effectively and were employees kept informed of the results?
- Were sufficient resources allocated to achieve the strategies in the plan?
- Were any resources wasted?
- How could strategies, measures and the process be improved?

(4) 'Big picture' analysis
- Is the company operating as a legally compliant and socially responsible enterprise?
- Has it improved in other underlying fundamentals such as resource management?
- Does management feel that they are now more proactive and less reactive?
- Has the relationship between management and employees improved?
- Have system factors been raised towards optimal performance, safety and quality?
- Where is the company now compared to at the beginning of the cycle?
- How does the company see its future?

## Concluding comments

The plan and supporting achievement cycle are the centerpieces of productive safety management. The plan is a product of the company's vision, aspirations and its espoused values. It is the pathway designed to allow the firm to pursue production and safety as compatible goals. In this firm, all employees understand the importance of maintaining system factor quality to optimize performance and safety and to minimize systemic risks. This is also a business in which managers apply analytical methods to prioritize risks and to allocate limited resources to best mitigate hazards, thus optimizing the use of financial and physical capital for risk management.

The plan is the culmination of this holistic approach embracing multiple disciplines. It contains both systems change and behavioral change strategies. The focus on human factors, together with organizational and systemic weaknesses, is important. This is because risk management is a social construction and not simply a mechanical process. The plan must, therefore, reflect the structural design of the OHS management system and also the human and social dynamics that affect the firm's level of risk. In addition, these dynamics are fundamental to the change process.

A holistic approach must therefore account for the human factor without limiting it to this perspective. Risk management is incomplete if accident prevention focuses on behaviors alone. Productive safety management is, therefore, a substantial shift away from the perspective of 'human error' as a primary cause of incidents and inefficiencies. The human factor is instead one of four system factors for which management has a responsibility balanced with employees' obligations to fulfill their duty of care and to perform the job to the best of their ability. Safety and performance are thus shared responsibilities.

In this context, management is not simply the cycle of planning, organizing and controlling. In addition to these, the role of management is to optimize financial, physical and human capital while aligning the firm externally and internally. The strategic alignment channel illustrates this. The implication is that future managers need to take greater responsibility for the selection, maintenance and optimization of capital resources, including human capital, to control the risks that affect company safety and performance.

Some of the greatest opportunities for improvement in management systems exist in relation to the employment of this human capital. In the business sector there are likely to be a number of 'reality checks' associated with these opportunities. The first is that productivity problems cannot be remedied simply by greater mechanization. Regardless of the degree of automation, without human competencies, a firm is unable to sustain itself. The future is likely to see greater emphasis on the socio-economic perspective of the firm rather than purely an economic perspective.

This financial view of the firm has been particularly pervasive since the 1980s. It has been accompanied by the rise of accounting to define not only the firm's financial parameters but also social constructs. The result has been that much greater value is now placed on monetary and physical capital than human capital. In fact, human capital is accounted for as a cost, whereas, financial and physical capital items are treated as assets. The irony is that while a machine depreciates with time, reflecting its declining contribution to performance and safety, an experienced employee has the potential to provide rising returns to the firm in terms of effectiveness and risk management. As competitive forces become more aggressive, rivalry between firms will no longer be restricted to consumer markets. There will also be greater competition among companies in the labor market for individuals with high levels of competency, safety consciousness and motivation. An outcome of this must be the adoption of accounting for human asset value. This means that the realistic worth of an employee consists of the investment made as a result of selection, maintenance and optimization. The asset value is thus the cost of recruitment and selection, accumulated salaries/wages and on-costs (reflecting the value of on-the-job experience) and training and development.

In this context, the organizational restructure will no longer be centered on 'do we need this position?' This often results in the loss of investment in people and a drain on organizational competencies when employees are retrenched. In future the response will be, 'can we afford to lose these skills and do we want to write-off our investment in this employee?' The tasks carried out within a position may become redundant but the KSAs held by the incumbent will not necessarily lose their relevance within the company. There will also be greater consideration given to the transferability of an employee's competencies within the firm. This will allow businesses to retain multifunctional capabilities such as safety consciousness and resourcefulness.

The final reality check in the western world will be changing social values centered on the restoration of balance between work and family lives. Accompanying this will be employees' desires to obtain greater intrinsic rewards from work. This includes being part of a positive organizational culture in which relationships are based on mutual benefit and trust. On a national and global level, the cost of economic growth on the environment will increase the external forces requiring firms to be consistently legally compliant and socially responsible. This is likely to be accompanied by escalating financial and social penalties for deviations from expected standards of firm behavior. In the current economic setting, a large proportion of the social costs of production, particularly the costs associated with occupational accidents, are borne by the community through the health system and other government services. Tighter legislative controls will be used to ensure that firms bear more of this cost and take greater responsibility for the by-products and consequences of their activities.

As a result of these social changes, the human factor in risk management is unlikely to remain skewed towards error, disposability and cost. It will instead focus on selection, maintenance and optimization of the human capital that possesses the competencies and attitudes needed to build sustainable businesses that are legally compliant and socially responsible. It is for this reason that in this book much of the content has been dedicated to addressing these human factors in risk management within an integrated socio-economic perspective of the firm. This perspective is underscored by values-driven systems and an organizational culture that balances the needs of key stakeholders.

Productive safety management contains the models and tools that provide structure to this needed change process. The entropy model allows a strategic systems-wide approach to risk management and explains the relationship between system factor quality and both organizational performance and safety. The model provides the four-fold strategy to address risks in the short and long term. It facilitates a shift away from reactive risk management to proactive solutions.

At a fundamental level, the model highlights a basic and confronting reality missing from many traditional theories of accident causation. This basic truth is that employees do not want or choose to be injured at work. Accidents are, therefore, very rarely attributable solely to worker error. Instead, there are numerous contributing factors including the demands that residual risks and degradation place on employees, that in turn, lead to workers' coping capacities being exceeded. This has been evident in the incident cases described in this book, not least of which, was the case in which Ray had both wrists shattered when the dozer ripper fell towards him. All these cases show evidence of systemic weaknesses within the organizational system that are not entirely within the control of the individual worker.

The entropy model therefore provides a systems-based explanation of the risks faced by businesses, and in particular, highlights the need for strategies to manage and compress residual risks. Hence, it has a high level of application to hazardous industries such as oil and gas, mining and construction, which concurrently are the sectors that historically have faced the greatest difficulty in balancing production and safety objectives. Accordingly, the challenge in writing this book has not only been to develop a new perspective of accident causation and the nature of risk but also to establish systems that support this new perspective on risk management. The management system described provides a balanced, strategic approach for the simultaneous pursuit of production, safety and quality, embracing those social values associated with sustainable enterprises. These, as proposed by the channel, are characterized by legal compliance as the minimum standard of firm behavior and future business development that embraces socially responsible outcomes.

The channel provides a strategic, integrated approach to OHS management. It explains the external context in which the firm operates and how factors in this environment provide a framework for company behavior. In particular, it sets the boundaries defining minimum legal compliance and social responsibility. It also explains the three levels of alignment required of fully integrated systems. To date, the singular concept of alignment has made it difficult for managers to address the divergent needs of company stakeholders and to create compatible strategic systems. Alignment is the key to developing sustainable businesses in which financial, physical and human capitals are optimized.

To achieve this alignment the company must have supportive HRM systems. Productive safety management provides the required tools and aligns them to the entropy model and the channel. These are designed to build a positive organizational culture and quality human resources capable of working safely and efficiently. These management systems are metaphorically speaking, the engine of productive safety management, which must be kept fuelled and fired by commitment and leadership. It is imperative that managers demonstrate this commitment using the value statement and the reasonableness test when implementing the approach. These ensure that management decisions and actions are appropriate and aligned with core goals and values. Fundamentally, therefore, the firm must evaluate itself not only from an economic but also from a social and cultural perspective to reflect the importance of both when developing a sustainable business.

The economic perspective of the firm has dominated company management over recent decades. The current financial framework has strongly emphasized the 'primary' costs of production within the firm's accounting systems as the main sources of business expenditure. Concurrently, part of the 'secondary' costs of production, such as workplace accidents, environmental damage and product defects, has been borne by the community. The shift in social values towards greater concern for employee health, safety and the environment, towards increased gain-sharing and recognition for worker effort, and higher expectations of work's social value, mean that this economic perspective is being eroded in favor of the socio-economic view. Accordingly, firms through the legislative framework are being increasingly held accountable for these 'secondary' costs.

This emerging view of the firm centers on basic human values. In particular, these include the sanctity of human life, quality of life, shared sense of purpose, and above all else, the right of the employee to come home safe to family and loved ones at the end of the working day. The path is therefore open for organizational leaders, on whose shoulders accountability ultimately rests, to embrace these values and to build sustainable businesses that are safe, productive and socially responsible.

## References


1. Kaplan, R.S. and Norton, D.P. (1996) *Translating Strategy into Action – The Balanced Scorecard*. Harvard Business School Press, Boston, Massachusetts, USA.
2. Nankervis, A.R., Compton, R.L. and McCarthy, T.E. (1993) *Strategic Human Resource Management*. Thomas Nelson Australia, South Melbourne, Australia.

# Glossary of terms

**Acceptable risk** The level of risk resulting when all practicable measures have been taken to mitigate it, given the severity of any potential harm that may be involved. The state of knowledge and the means of preventing harm in terms of resource availability, which is considered acceptable by management, employees and regulators.

**Achievement cycle** The productive safety performance management system which comprises four tools:

- Business alignment scorecard
- Business unit alignment scorecard
- Cultural health scorecard
- Leader's scorecard

and is used to measure organizational performance in terms of productivity, safety, system factor quality and other key performance indicators.

**Achievement indicators** The performance criteria used in the leader's scorecard to set individual goals for managers and supervisors that focus their efforts on improving the quality of system factors. The achievement indicators are:

- Processes
- Technology
- Physical environment
- Human resources

**Alignment** Alignment involves the compatibility of three levels of organizational decision-making that is determined by:

(1) the fit between the firm and the external environment;
(2) the balance of resources allocated to strategic management systems;
(3) operational systems, practices and behaviors that reinforce the compatibility of organizational goals and values with employees' goals and values.

**Alignment indicators** The performance criteria used in the business and business unit alignment scorecards to ensure that a balanced approach to corporate goal-setting is achieved that allows production, safety and system factor quality to be pursued concurrently, and that considers both short-term performance and long-term sustainability. The alignment indicators are:

Productivity
Safety
System factor quality
Financial and customer
Compliance
Social responsibility

Auditing — An information gathering and evaluation activity involving three processes:

(1) Identifying the sources of risk that the organization is exposed to as a natural system,
(2) Evaluating the health of the organizational culture, and
(3) Measuring the achievement of strategies in the productive safety management plan.

Baseline level of residual risk — The average of the residual scores for each system factor for all activities undertaken in a given workplace. An example of the formula is:

Baseline level of process residual risk

$$= \frac{\Sigma(\text{Activity } 1P_{RR} \text{ Score} + \text{Activity } 2P_{RR} \text{ Score} + \ldots \text{Activity } N\,P_{RR} \text{ Score})}{N}$$

where $P_{RR}$ is the process residual risk score

Business activity — An activity results when system factors combine, specifically, when a process is undertaken by human resources using technology in a given physical environment.

Business alignment scorecard — The performance management tool used at the corporate level to ensure that the business is aligned with the external and internal environments. It applies a balance of measures to ensure that production, safety and system factor quality are pursued concurrently, and that short-term achievement and long-term sustainability are both taken into consideration.

Business unit alignment scorecard — The performance management tool used at the business unit level that is strategically linked to the business alignment scorecard, and ensures that the business unit is aligned with the external and internal environments. It applies a balance of measures to ensure that production, safety and system factor quality are pursued concurrently, and that short-term achievement and long-term sustainability are both taken into consideration.

Capacity reservoir — The total competency pool of the firm's human resources system factor comprising the collective

| | |
|---|---|
| | knowledge, skills, abilities and attitudes of its employees. |
| Catalyst | A system factor condition that caused a rapid rise in entropic risk or the sudden translation of residual risk into imminent danger. |
| Competency | The knowledge, skills, abilities and attitudes expected of an employee in the workplace that embodies the ability to transfer and apply such capabilities to new situations and environments. |
| Cultural health scorecard | The performance management tool used at the corporate level to measure the health of the company culture and which is used to evaluate the firm's readiness for change. |
| Entropic risk | The risk associated with the degradation of a firm's system factors. |
| Entropy | The degradation of a firm's system factors that results from the tendency of natural systems to become disorganized or shift towards a state of chaos. |
| Entropy model | A model of accident causation that illustrates how firms are exposed to two types of risk – entropic risk and residual risk. These risks are characteristic of organizational system factors and have a negative impact on safety and performance. |
| Entropy prevention strategies (EPS) | Strategies that involve the use of corrective action and maintenance practices to eliminate entropic risk as far as practicable. |
| External environment | The environment outside the firm in which the firm operates comprising those forces that define the scope of strategic and behavioral choices. |
| External strategic alignment | The alignment of the firm's business strategy with the external environment, shown by the strategic alignment channel, which leads to legal compliance and socially responsible organizational behavior. |
| Four-fold strategy | A comprehensive strategic approach to address entropic and residual risks that involves:<br>(1) taking immediate corrective action to eliminate entropic risk;<br>(2) establishing maintenance strategies to prevent future entropic risk;<br>(3) managing residual risk in the short term; and<br>(4) compressing residual risk in the longer term. |
| Hazard | In relation to a person, means anything that may result in injury or harm to the health of the person. In relation to property or physical conditions, means anything that may result in damage. |
| Human resources | The firm's employees including managers, supervisors, and workers. Concurrently, human resources are the human capital employed by the firm. |
| Implementation | Measures which determine whether strategies and |

| | |
|---|---|
| measures | interventions have been implemented effectively, for example, number of initiatives implemented on time and within budget as a percentage of planned initiatives. |
| Inherent risk | The risk in all organizational, human and natural systems which cannot be completely eliminated. Refer to residual risk. |
| Integration | Integration is the overlapping of various management disciplines under an 'umbrella' management system that has a shared set of decision-making and performance criteria. |
| Internal environment | The firm's internal operations comprising strategic management systems that determine the balance of capital resources, and operational systems, practices and behaviors that determine whether the organization's goals and values are aligned with employees' goals and values. |
| Internal goal alignment | Internal goal alignment results when operational systems, practices and behaviors are developed based on the alignment of the firm's goals and values with the goals and values of its workforce, as shown by the strategic alignment channel. |
| Internal strategic alignment | Internal strategic alignment results when strategic management systems align human, physical and financial capital. This is characterized by an appropriate mix and level of resource allocation for the firm to achieve its strategic objectives and compliance with external requirements. |
| Job safety analysis | The process of breaking a work activity into steps, analyzing the hazards and the associated risks at each step, and developing safe work procedures to control these risks. |
| Leader's scorecard | The performance management tool used at the operational level to facilitate goal-setting by managers and supervisors, to focus their efforts on improving the quality of system factors and balancing production, safety and quality objectives. |
| Legal compliance | Firm behavior that meets or exceeds the minimum requirements set out in the legislation that applies to the firm's operations. |
| Outcome measures | Measures which evaluate key areas of organizational performance which are 'overt' in nature, for example, lost time injury frequency rate, product defect rate and output per employee. |
| Physical environment | The locational and fixed-structural factors within and surrounding the firm's operational site. An alternative definition is the firm's physical capital resources that are of a fixed nature. |
| Position | A formal document outlining the tasks to be under- |

description — taken by a particular position or role within the organization.

Practicable — Reasonably practicable having regard, where the context permits to:

(a) the severity of any potential injury or harm to health that may be involved, and the degree of risk of it occurring;

(b) the state of knowledge about:

  (i) the injury or harm to health referred to in paragraph (a);

  (ii) the risk of that injury or harm to health occurring; and

  (iii) means of removing or mitigating the risk or mitigating the potential injury or harm to health; and

(c) the availability, suitability, and cost of the means referred to in (b) (iii).

Adapted from Occupational Safety and Health Act 1984 (Western Australia)

Primary risk parameter — The key variable that determines the level of risk in a system factor.

Principle of separation of system factors — The process of breaking business activities into system factors and analyzing the relative risk levels in each system factor independently of the other system factors involved in the activity. For instance, analyzing process risks separately from technological, physical environment and human resource risks.

Processes — The work practices undertaken by the firm.

Productive safety formula — A formula that states:

σ System factor safety and performance = σ resourcefulness – σ risk, where σ means change in.

It shows that if resourcefulness is increased and risk decreased concurrently, then system factor safety and performance improves.

Productive safety management plan — The plan containing the strategies for:

OHS management system restructuring

Risk management

Behavioral management

which provides the firm with a strategic direction for the concurrent pursuit of production, safety and system factor quality.

Residual risk — The inherent danger in all organizational, human and natural systems which cannot be compressed in the short term. The risk that remains when entropic risk is eliminated.

Residual risk management strategies (RMS) — Strategies that involve the implementation of longer-term interventions, such as investment in employee training, purchasing of safer technologies, and better physical environment and process design, to reduce the level of inherent danger in organizational systems.

| | |
|---|---|
| Resourcefulness | The overflow of the capacity reservoir that stores the organization's learning potential and collective competencies of the workforce. Resourcefulness results in higher levels of safety consciousness and employee participation, which translate into better problem-solving and risk management. |
| Risk | The probability of injury or harm occurring that result from either inherent danger (residual risk) or the tendency of systems to degrade (entropic risk) or a combination of these. |
| Risk modifier | A risk modifier is a parameter or characteristic of a system factor that either increases or decreases the level of risk. Negative risk modifiers increase the level of risk, whereas positive risk modifiers lower this level. |
| Scorer | A risk ranking tool used to determine the relative level of risk for a given system factor. There are eight scorers in total in the productive safety quantitative risk assessment method – an entropic risk scorer and a residual risk scorer for each of the four system factors. |
| Social responsibility | Firm behavior that exceeds the minimum requirements set out in the legislation that applies to the firm's operations, accompanied by self-regulation, including internal standards, quality controls, monitoring and continuous improvement systems. |
| Strategic alignment channel | A model illustrating that there are three levels of alignment – external strategic, internal strategic and internal goal alignments – that a firm needs to achieve for the firm to fit the external environment and have internal systems that create synergies. |
| System factors | The factors of an organization that combine into business activities. These are processes, technology, the physical environment and human resources. |
| System factor quality measures | Measures which evaluate key areas of organizational performance which are 'covert' in nature and indicative of underlying systemic conditions, for example, audit results and cost of breakdown maintenance versus planned maintenance. |
| Technology | The firm's physical capital of a nonfixed nature, including plant, equipment, tools and chemicals that are utilized by the firm or are stored at the operation's site. |
| Unsafe act | An unsafe act is a behavior consciously taken by an employee in defiance of the safety rules, procedures, training and expected norms of behavior, and independently of external pressures stemming from the management system. This act must occur after the firm has provided sufficient training and supervision for the employee to have a reasonable |

| | |
|---|---|
| | understanding of the risks involved and potential consequences, for it to be considered unsafe. |
| Vigilance | Maintenance of a state of watchfulness and alertness by an employee resulting from an awareness of workplace risks and their potential consequences. |

# Index